NASA SP-87

Proceedings of the
APOLLO UNIFIED S-BAND TECHNICAL CONFERENCE

Held at Goddard Space Flight Center,
July 14-15, 1965

Prepared by Goddard Space Flight Center

Scientific and Technical Information Division   1965
NATIONAL AERONAUTICS AND SPACE ADMINISTRATION
Washington, D.C.

PROGRAM COMMITTEE

Chairman	K. E. Peltzer
Vice Chairman	G. E. Abid
Program Coordinator	R. Burns
Secretary	J. R. Soares
Public Affairs	A. Shehab

FOREWORD

The history of the Manned Space Flight Network reveals that each successive mission is considered more complex than the previous one. As a result, the tracking and communication problems become more complicated, requiring more sophisticated equipment.

The Apollo missions differ considerably from past manned missions in that there is a requirement, for the first time, to send astronauts to the moon and return them safely to earth. For this reason, the position of the spacecraft must be known at all times and continuous communications must be maintained between the earth and the spacecraft during most of the mission phases. This requirement has dictated incorporating the Unified S-Band System into the Manned Space Flight Network. This system will provide the primary tracking and communications data between earth and the spacecraft in the later Apollo missions.

The proceedings contained herein are the conference records of papers presented at the Technical Conference on the Apollo Unified S-Band System, which was held at the Goddard Space Flight Center on July 14 and 15, 1965. This conference brought together about 500 participants from the various NASA centers and Apollo contractors. These proceedings will constitute a first handbook pertaining to the Apollo Manned Space Flight Network.

Each person concerned with the Apollo Manned Space Flight Network, either from an engineering or operational viewpoint, will find that this document contains a reasonably comprehensive description of the primary equipment used at the Apollo ground stations.

Kenneth E. Peltzer
Manned Flight Support Office (T&DS)
Goddard Space Flight Center

PARTICIPANTS

SESSION I

INTRODUCTORY SESSION

E. W. Wasielewski, Associate Director
   Goddard Space Flight Center
   Greenbelt, Maryland

W. P. Varson
   Manned Flight Support Office
   Goddard Space Flight Center
   Greenbelt, Maryland

W. D. Kahn
   Systems Analysis Office
   Goddard Space Flight Center
   Greenbelt, Maryland

SESSION II

ANTENNA SYSTEM

L. E. Hightower
   Engineering Support Office
   Goddard Space Flight Center
   Greenbelt, Maryland

N. A. Raumann
   Antenna Systems Branch
   Goddard Space Flight Center
   Greenbelt, Maryland

J. Flowers, Jr.
   Network Engineering Branch
   Goddard Space Flight Center
   Greenbelt, Maryland

SESSION III

TRANSMITTER/RECEIVER SYSTEM

J. B. Martin
   Network Engineering Branch
   Goddard Space Flight Center
   Greenbelt, Maryland

R. Bunce
   Jet Propulsion Laboratory
   Pasadena, California

J. H. Jacobi
   Manned Flight Support Office
   Goddard Space Flight Center
   Greenbelt, Maryland

SESSION III (Cont.)

G. Hondros
   Manned Flight Support Office
   Goddard Space Flight Center
   Greenbelt, Maryland

T. E. McGunigal
   R. F. Systems Branch
   Goddard Space Flight Center
   Greenbelt, Maryland

SESSION IV

DIGITAL SYSTEMS

P. Lindley
   Jet Propulsion Laboratory
   Pasadena, California

W. M. Hocking
   Network Engineering Branch
   Goddard Space Flight Center
   Greenbelt, Maryland

R. L. Granata
   Network Engineering Branch
   Goddard Space Flight Center
   Greenbelt, Maryland

SESSION V

IMPACT OF USB SYSTEM ON ORBIT DETERMINATION

J. H. Donegan
   Data Operations Branch
   Goddard Space Flight Center
   Greenbelt, Maryland

J. Barsky
   Data Operations Branch
   Goddard Space Flight Center
   Greenbelt, Maryland

SESSION VI

NETWORK SYSTEMS

C. O. Roberts
   Manned Flight Engineering Branch
   Goddard Space Flight Center
   Greenbelt, Maryland

SESSION VI (Cont.)

W. A. Dental
  Manned Flight Engineering Branch
  Goddard Space Flight Center
  Greenbelt, Maryland

W. E. Willis
  Manned Flight Engineering Branch
  Goddard Space Flight Center
  Greenbelt, Maryland

C. B. Knox
  Manned Flight Engineering Branch
  Goddard Space Flight Center
  Greenbelt, Maryland

G. N. Georgeadis
  Manned Flight Engineering Branch
  Goddard Space Flight Center
  Greenbelt, Maryland

SESSION VII

SPACECRAFT USB SYSTEM

B. H. Hood
  Systems Analysis Branch
  Manned Spacecraft Center
  Houston, Texas

W. E. Kuykendall
  Systems Analysis Branch
  Manned Spacecraft Center
  Houston, Texas

A. Travis
  System Engineering and Test Branch
  Manned Spacecraft Center
  Houston, Texas

B. Reed
  Marshall Space Flight Center
  Huntsville, Alabama

SESSION VIII

S-BAND IMPACT ON OPERATIONS, NASCOM, APOLLO SHIPS, AND APOLLO AIRCRAFT

R. H. Newman, Jr.
  Manned Flight Support Office
  Goddard Space Flight Center
  Greenbelt, Maryland

W. B. Dickinson
  Communications Engineering Branch
  Goddard Space Flight Center
  Greenbelt, Maryland

M. D. Greene
  Office of Instrumentation Ships
  Goddard Space Flight Center
  Greenbelt, Maryland

L. C. Shelton
  Manned Flight Operations Branch
  Goddard Space Flight Center
  Greenbelt, Maryland

O. M. Covington, Deputy Assistant Director
  Goddard Space Flight Center
  Greenbelt, Maryland

CONTENTS

	Page
Program Committee	ii
Foreword	iii

SESSION I: INTRODUCTORY SESSION

    Introduction

        E. W. Wasielewski 1

    Functional Description of Unified S-Band System and Integration into the Manned Space Flight Network

        W. P. Varson 3

    Tracking Studies for Project Apollo

        W. D. Kahn 13

SESSION II: ANTENNA SYSTEMS

    USB Antenna Structures

        L. E. Hightower 21

    USB Servo System

        N. Raumann 29

    Antenna Feeds and Acquisition Antennas

        J. Flowers 39

SESSION III: TRANSMITTER/RECEIVER SYSTEM

    Parametric Amplifier, and Noise Figure and Test Signal Network

        J. B. Martin 47

    Receiver-Exciter Subsystem

        R. Bunce 59

    Verification Receiver, SCO Oscillator and Up-Data Modems

        J. H. Jacobi 75

    Signal Data Demodulator

        G. Hondros 83

CONTENTS (Cont.)

	Page
The Unified S-Band Power Amplifier	
T. E. McGunigal	91

SESSION IV: DIGITAL SYSTEMS

JPL Ranging System	
P. Lindley	99
Doppler Counter, Antenna Programmer, and Tracking Data Processor	
W. M. Hocking	109
Apollo Precision Frequency Source and Time Standard	
R. L. Granata	125

SESSION V: IMPACT OF USB SYSTEM ON ORBIT DETERMINATION

Apollo Mission Profile	
J. J. Donegan	135
Computer Test Program to Qualify USB System	
J. Barsky	145

SESSION VI: NETWORK SYSTEMS

Network Systems	
C. O. Roberts	151
Apollo Network PCM Decommutation Systems	
W. A. Dentel	165
Apollo Network Remote Site Computer Systems	
E. Willis	181
Apollo Digital Command System	
C. B. Knox	191
Apollo Remote Site Display System	
G. N. Georgeadis	205

CONTENTS (Cont.)

SESSION VII: SPACECRAFT USB SYSTEM

Page

Command and Service Module Unified S-Band System

    B. Hood 223

Lunar Excursion Module Unified S-Band System

    W. Kuykendall 233

Unified S-Band RF System Compatibility Test Program

    A. Travis 243

Command and Communication System

    B. Reed 247

SESSION VIII: S-BAND IMPACT ON OPERATIONS, NASCOM, APOLLO SHIPS AND APOLLO AIRCRAFT

Typical Acquisition Procedure

    R. H. Newman 261

Impact of Apollo Unified S-Band System on NASA Communication Network

    W. Dickinson 269

Role of Apollo Ships

    M. D. Greene 275

Apollo/Range Instrumented Aircraft

    L. C. Shelton 283

STATUS OF THE PROGRAM

    O. M. Covington 293

APPENDIX A: GLOSSARY 297

APPENDIX B: LIST OF ATTENDEES 299

INTRODUCTION

Mr. Eugene W. Wasielewski, the Associate Director of the Goddard Space Flight Center, opened the Technical Conference on the Unified S-Band System by extending a welcome to Goddard employees, contractors, members of DOD, and three other NASA centers namely: Manned Space Flight Center, Marshall Space Flight Center, and the Jet Propulsion Laboratory. He gave a brief description of Goddard's missions and the role it is playing in the Apollo Unified S-Band System.

FUNCTIONAL DESCRIPTION OF UNIFIED S-BAND SYSTEM AND INTEGRATION INTO THE MANNED SPACE FLIGHT NETWORK

by
W. P. Varson
Goddard Space Flight Center

ABSTRACT

The lunar phases of the Apollo missions require techniques and equipment exceeding the capability of those previously used in the Manned Space Flight Network. This improvement in network capability is necessary to provide reliable tracking and communications of the Apollo spacecraft at lunar distances. To fulfill this requirement, the unified S-band (USB) system has been introduced into the network. The USB system used with 85-foot antennas will provide the only means of tracking and communications at lunar distances. The USB system with 30-foot antennas will be used to fill the gaps in the coverage provided by the three 85-foot antennas. The USB system with the 30-foot antennas will also be used to provide data during the earth-orbital and post-injection phases of the missions.

In order to insure reliability, the USB system utilizes existing, proven techniques and hardware. These items of equipment developed and used by the Jet Propulsion Laboratory and the Scientific Satellite Network have been adapted to the USB system. The more significant of this equipment is the range and range rate equipment supplied by the Jet Propulsion Laboratory to the program and the antenna systems which are nearly identical to those used in the Scientific Satellite program.

INTRODUCTION

The Apollo program is significantly more complex than either the Mercury or Gemini programs and has consequently presented a corresponding increase in the complexity of the support required from the Manned Space Flight Network (MSFN). This has affected the quantity of data that must be handled, the geographic areas that must be covered, and the technical capability of equipment. For the first time, the network is required for provide reliable tracking and communications to lunar distance. This has required the incorporation of the unified S-band (USB) system into the network. The existing network instrumentation is capable of supporting the earth-orbital phases of the mission and, in fact, will be the sole support for the initial Apollo flights. Since the USB system will be the only means of tracking and communicating with the spacecraft during the lunar phases of the mission, it is mandatory that it be installed, checked out, and proven operational during the early Apollo missions.

FUNCTIONAL DESCRIPTION OF USB SYSTEM

The USB system utilizes a single carrier frequency in each direction to provide tracking as well as communications with the spacecraft. This is depicted in Figure 1, where all of the functions are accomplished with a single system. The interface with the network equipment is the same whether the data comes from the USB system or the Gemini equipment.

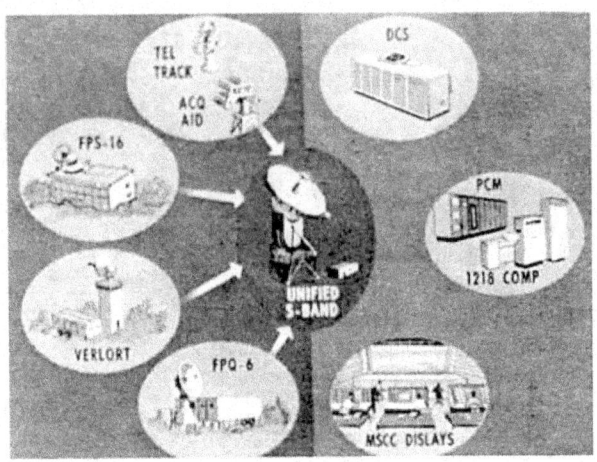

Figure 1—Apollo network evolution.

Perhaps the first thing that should be discussed is why the unified systems approach was adopted rather than extending the range of the existing network equipment. It was adopted primarily because it was considered to offer a superior technical solution with a minimum of new development. To expand the range of the existing type of network equipment would have required development of high-powered radar beacons, the use of coherent radar techniques and a major expansion of the range capability of the VHF and UHF equipment. Systems capable of operating to lunar distance which employ the unified systems techniques were already in operation. In addition to requiring considerably less development and expense, the unified systems approach also reduced the equipment required aboard the spacecraft.

One of the major decisions was the selection of the basic techniques to be used in the unified systems approach. It is desirable to use the best equipment available in support of the Manned Space Flight Missions; however, it is also desirable to use proven techniques and equipment to minimize development and to afford the highest probability of success. There have been several approaches to the unified systems concept, but perhaps the most thoroughly developed is that used by the Jet Propulsion Laboratory. This system has been employed successfully in support of lunar and planetary programs and, with minor modifications, was applicable to the Apollo tracking and communications requirements. Therefore, it was a logical choice for the USB system.

The design of the USB system is based on a coherent doppler and the pseudo-random range system which has been developed by JPL. The S-band system utilizes the same techniques as the existing systems, with the major changes being the inclusion of the voice and data channels.

A single carrier frequency is utilized in each direction for the transmission of all tracking and communications data between the spacecraft and ground. The voice and up-date data are modulated onto subcarriers and then combined with the ranging data (Figure 2). This composite information is used to phase-modulate the transmitted carrier frequency. The received and transmitted carrier frequencies are coherently related. This allows measurements of the carrier doppler frequency by the ground station for determination of the radial velocity of the spacecraft.

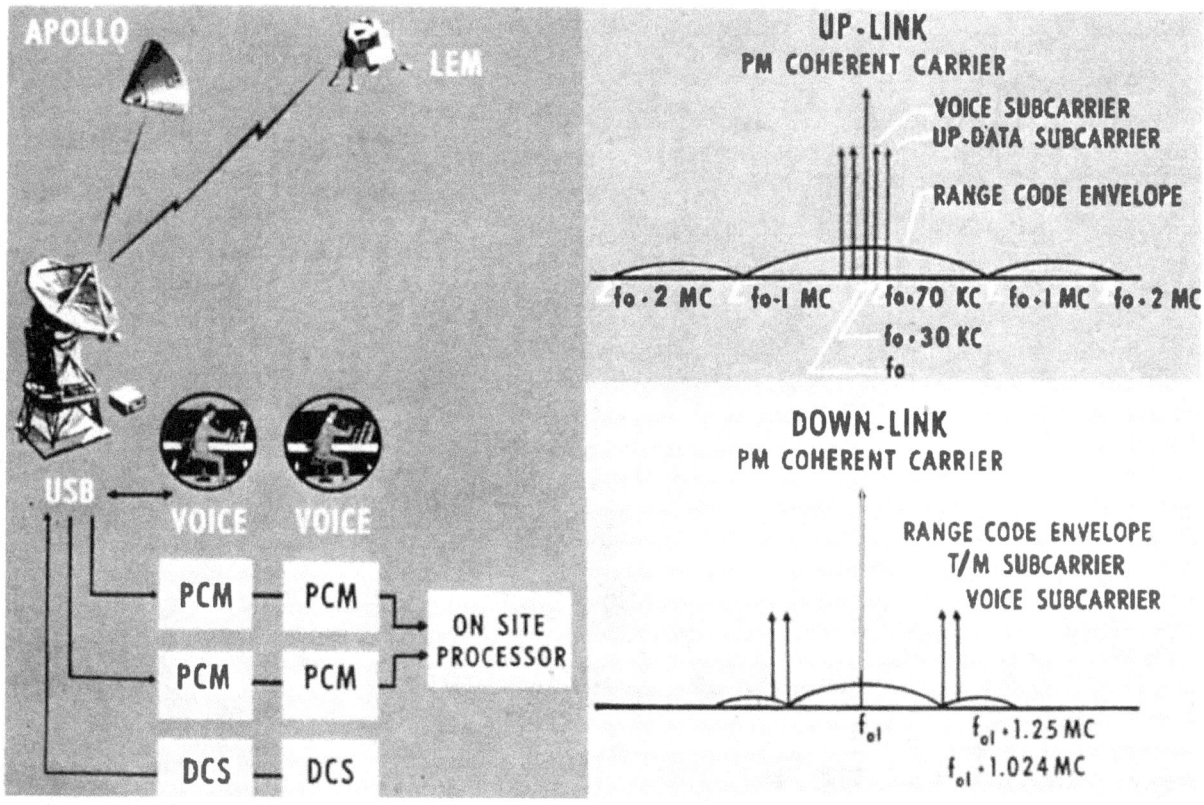

Figure 2—USB modulation technique.

In the transponder the subcarriers are extracted from the RF carrier and detected to produce the voice and command information. The binary ranging signals, modulated directly onto the carrier, are detected by the wide-band phase detector and translated to a video signal.

The voice and telemetry data to be transmitted from the spacecraft are modulated onto subcarriers, combined with the video ranging signals, and used to phase-modulate the down-link carrier frequency. The transponder transmitter can also be frequency-modulated for the transmission of television information or recorded data instead of ranging signals.

The basic USB system has the ability to provide tracking and communications data for two spacecraft simultaneously, provided they are within the beamwidth of the single antenna. The primary mode of tracking and communications is through the use of the PM mode of operation. Two sets of frequencies separated by approximately 5 megacycles are used for this purpose (Figure 3). In addition to the primary mode of communications, the USB system has the capability of receiving data on two other frequencies. These are used primarily for the transmission of FM data from the spacecraft.

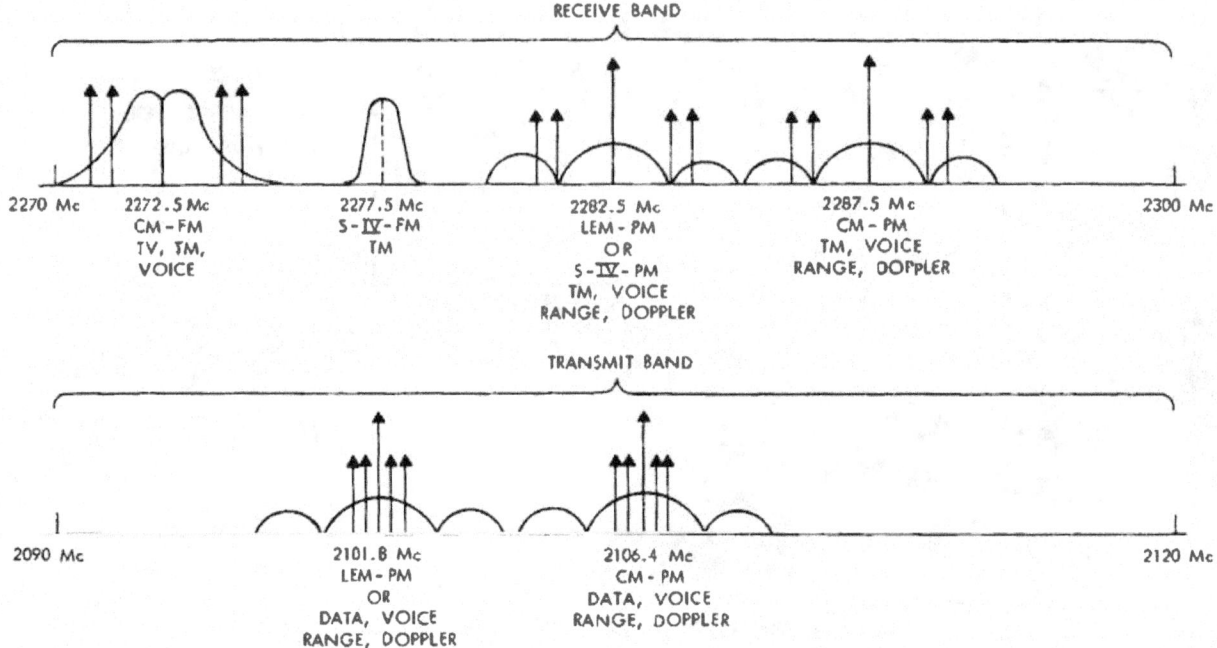

Figure 3—USB system frequency spectrum.

INTEGRATION OF USB SYSTEM INTO MSFN

Since JPL had developed some of the equipment to be used in the unified systems, it was decided to have them continue to provide these elements of the system. This includes the receiver, the transmitter/exciter, the ranging system, and test transponders. This equipment will be supplied to the USB system contractor for integration into the system.

A typical USB site is shown in Figure 4, which identified the equipment to be supplied by the contractor, equipment supplied by JPL, and the network equipment. The majority of equipment shown in Figure 4 is the network equipment, which points out the significance of the system engineering and interface job facing the USB system contractor. The network equipment for the most part is the same equipment which is utilized for the Gemini program. The USB system has been designed so that the data inputs and outputs into the network equipment are identical to those of the Gemini equipment. This approach was selected to allow the USB system to be integrated into the MSFN without disrupting the normal network operations or requiring equipment changes or modifications.

The tracking and communications with the spacecraft during the lunar missions will be provided by three primary deep-space facilities, employing 85-foot antennas, spaced at approximately equal intervals of longitude around the earth to provide the continuous coverage of the lunar missions (Figure 5). Three of the deep space instrumentation facilities (DSIF) located at approximately the same locations will be equipped to serve as backup to the primary stations. Each of these facilities, both the primary and backup stations, will be equipped to track and provide communications with both the Lunar Excursion Module (LEM) and the Command Module simultaneously.

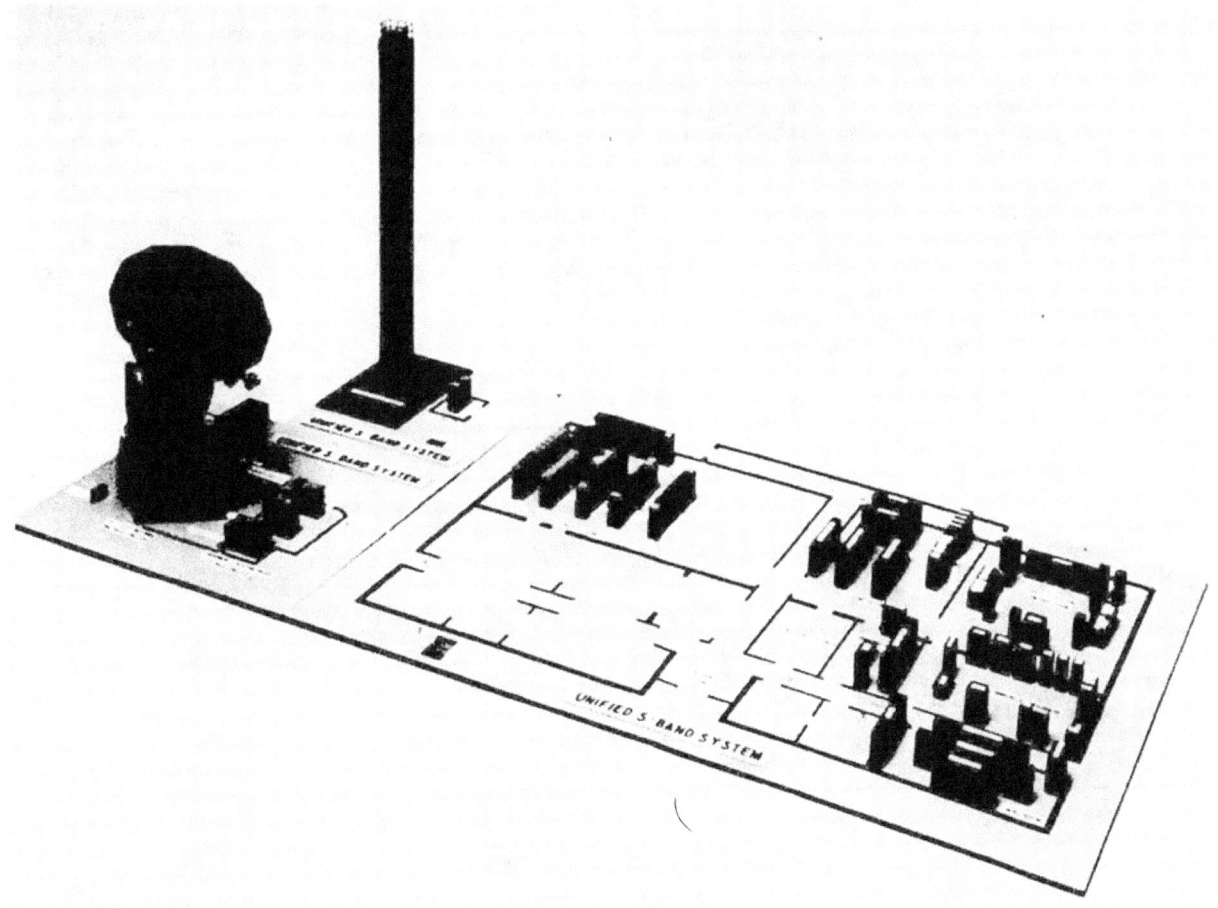

Figure 4—Typical USB site.

In addition to the stations with the 85-foot antennas, a number of other stations employing 30-foot antennas are also required in the network. These systems are needed for launch coverage in-flight checkout of the spacecraft, to fill gaps in the coverage of the three lunar stations, and to provide instrumentation coverage for testing the spacecraft in earth orbit.

Four land stations (Cape Kennedy, Grand Bahama, Antigua, and Bermuda) and one instrumentation ship are required to provide continuous USB coverage from launch through insertion. Seven land stations (Canary Islands, Guaymas, Texas, Ascension Island, Carnarvon, Guam, and Hawaii) and two additional instrumentation ships are required to complete the USB system coverage requirements. In addition to these stations (Figure 6) the Apollo networks will also include two reentry ships and eight instrumented aircraft.

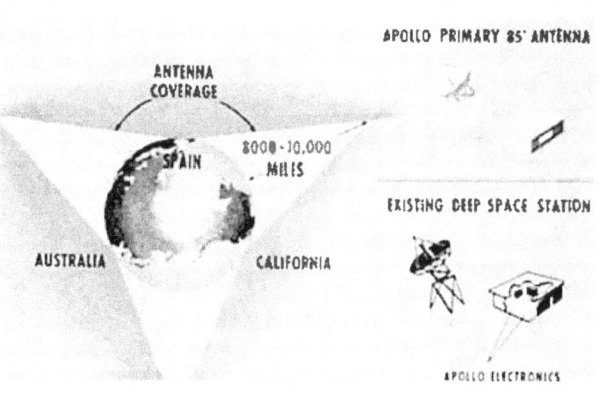

Figure 5—85-foot antenna stations for Apollo.

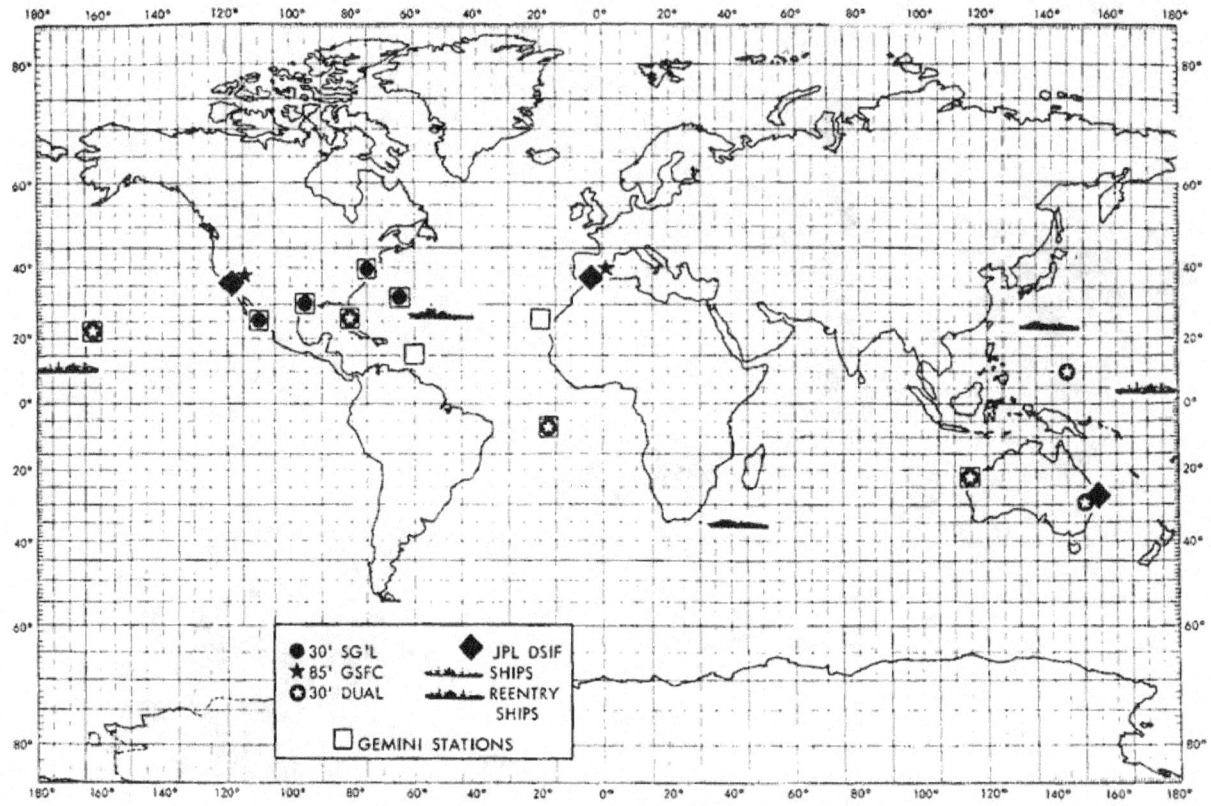

Figure 6—Apollo USB network.

The 30-foot antenna system was selected as a result of a study of the gain and tracking accuracy requirements and comparative costs of several systems. The 30-foot systems must provide data in earth orbit as well as during the realignment of the spacecraft during the lunar phases of the missions. During the lunar missions, these systems should be capable of tracking the spacecraft to a range of approximately 15,000 nautical miles, using the spacecraft omnidirectional antenna. This represents the most stringent requirement and comes about because it is desirable to complete the spacecraft transposition prior to deployment of the directional antennas. These systems will also be capable of providing tracking data at lunar distance.

The 30-foot antenna has a beamwidth of approximately one degree at 2300 megacycles. This requires that acquisition information be provided in order that the system can acquire the target. This information could be provided by a separate acquisition system operating at a lower frequency. However, the USB system may be used when 2300 megacycles is the only signal radiated from the spacecraft. This requires that the USB system contain its own acquisition aids.

The acquisition information is normally provided by the antenna programmer, which drives the antenna system so that the target stays within its one-degree beamwidth. The second acquisition device is the 3-foot antenna and its associated receiver which provide a 10-degree beamwidth for initial target acquisition. The acquisition antenna is mounted on the large antenna to simplify the overall system.

The USB system includes a number of self-checking features to assist in the checkout and maintenance of the system and to increase its overall operating reliability. The verification receiver monitors the transmitted data to the spacecraft to ensure the proper performance of the up-data transmission link. The system also contains built-in test equipment which will allow test data to be inserted into the data demodulator. The JPL equipment allows the signal from the transmitter/exciter to be injected directly into the receiver to provide an internal check of the RF system. In addition to these internal checks, the system also uses the transponder on a boresight tower for checkout of the RF system, the angle system, and the ranging system.

There are several variations to the basic USB system: the single and dual 30-foot systems, the primary and backup 85-foot antenna systems, and the single and dual instrumentation ships. The Collins Radio Company has the complete responsibility for the 30-foot antenna systems (Figure 7). They are provided with the JPL-supplied and other equipment to be integrated with Collins-supplied equipment to provide a complete system. Collins is required to erect, install, and checkout the systems on site. The facilities will be made available to the contractor and the network equipment will already be available on site. This equipment will be installed by others; however, Collins is responsible for the proper interface with this equipment.

For the primary 85-foot systems (Figure 8), Collins will be supplied Government-furnished equipment which he will integrate with the equipment he furnishes prior to shipment to the site. The 85-foot antenna structure is being provided under a separate contract. The contractor is responsible for integrating the USB on site.

At the JPL facilities (Figure 9), the contractor will supply components as indicated. These will be added to the other equipment at the facility to allow it to be used in support of the Apollo missions. These units will be checked out prior to shipment and installation of the equipment

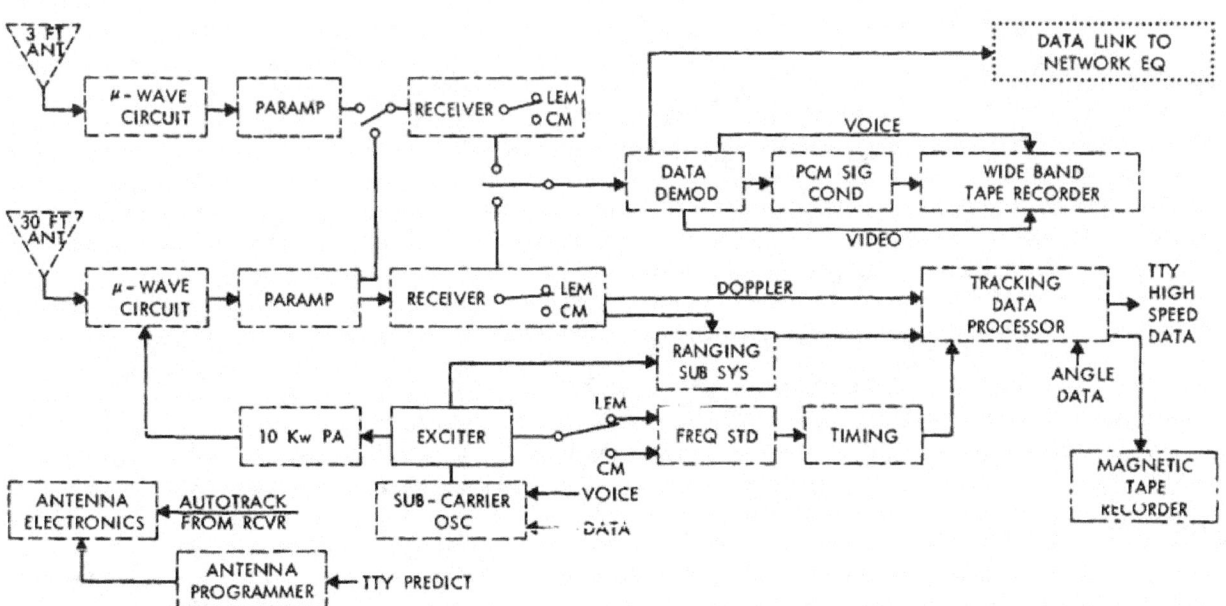

Figure 7—USB system 30-foot antenna facility.

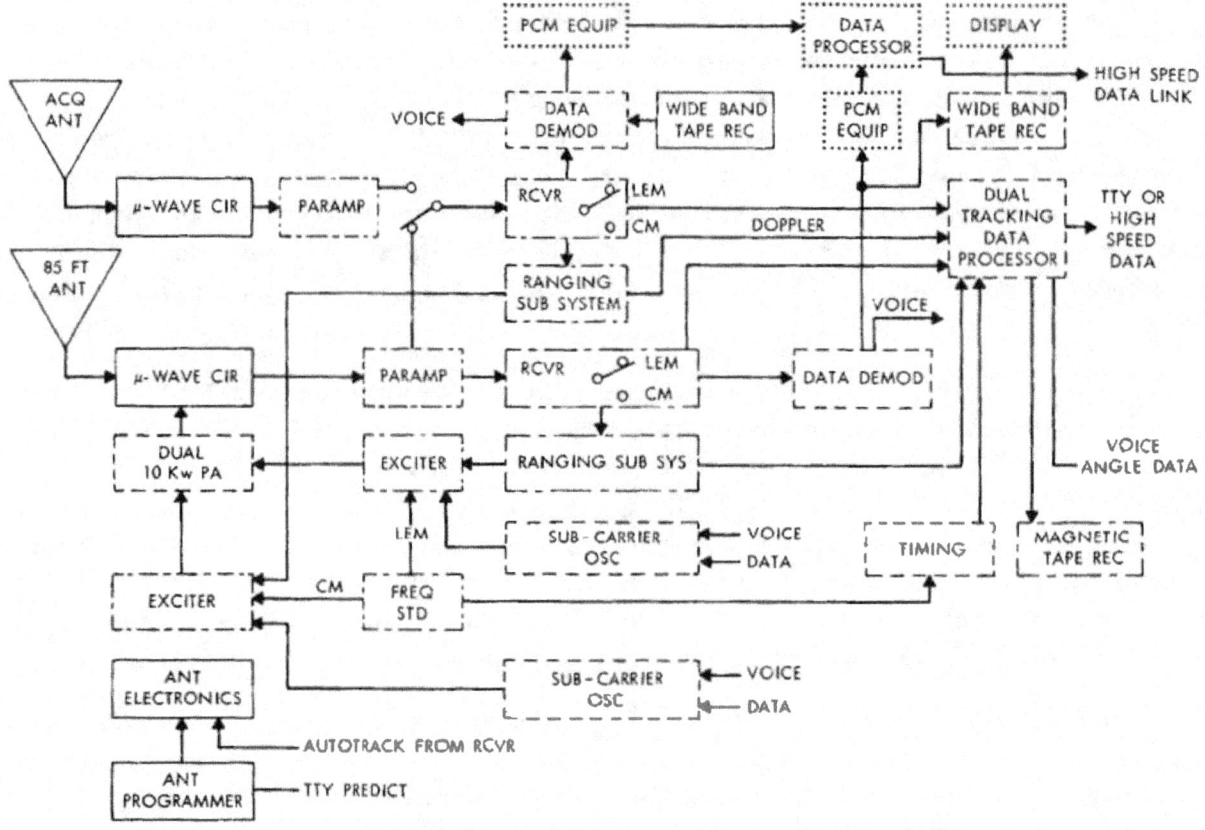

Figure 8—USB system 85-foot antenna facility primary stations.

and its integration into the system. A microwave link will be used to remote the data from the JPL site to the MSFN site, where both sites will have a common set of network equipment.

The contractor will be required to supply instrumentation for five ships. Three of these are the insertion and injection ships which will utilize the 30-foot antenna. The contractor will supply the equipment indicated in Figure 10. Reeves Instrument Company will supply the antenna system and be responsible for the overall integration of the system.

The contractor will also supply test and training units as a part of the systems contract. These consist of essential elements of the system and will be used for checkout of the spacecraft and as aids in the test and training program. These units are shown in Figure 11.

GSFC awarded the USB system contract to the Collins Radio Company on July 14, 1964. Blaw-Knox is building the 30-foot antenna structures under a subcontract to Collins. The antenna structure was shipped to Guam on June 7, 1965 and to Carnarvon on June 18, 1965. The first USB (for Guam) has been delivered and the remainder of the 30-foot systems will be delivered at a rate of one per month. The 85-foot antenna structures were built by Blaw-Knox under separate contract.

The USB system will be installed in the network during the next year. They will be checked out on the SA-202 through SA-206 and on SA-501 and SA-502 missions and will be used to provide primary mission support data beginning with SA-207 and SA-503.

FUNCTIONAL DESCRIPTION OF UNIFIED S-BAND SYSTEM AND INTEGRATION INTO THE MANNED SPACE FLIGHT NETWORK 11

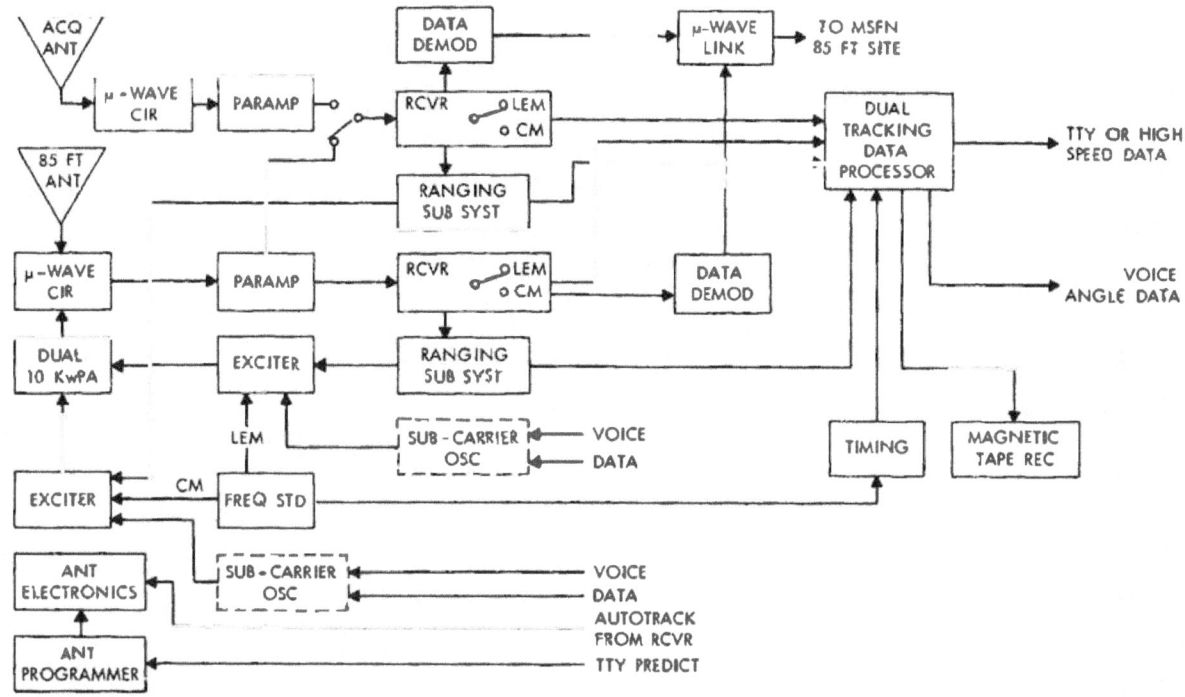

Figure 9—USB system 85-foot antenna facility backup station.

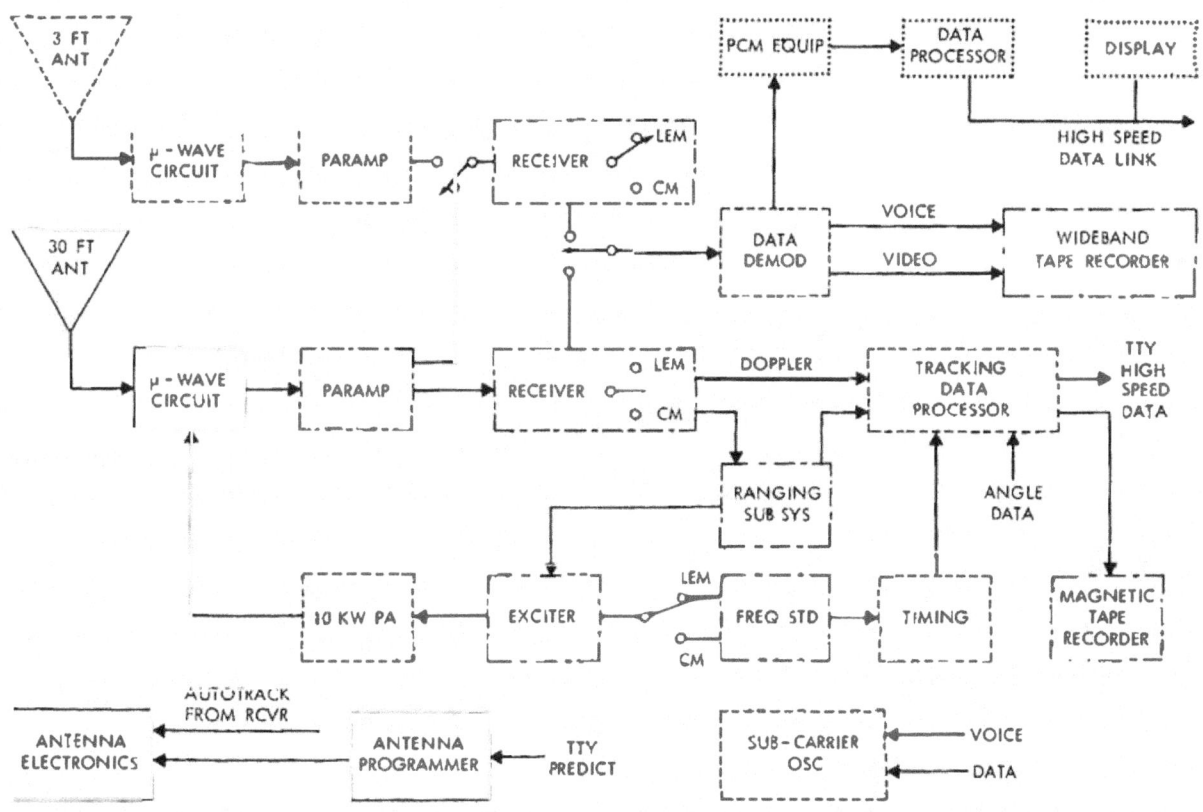

Figure 10—USB system antenna facility primary stations.

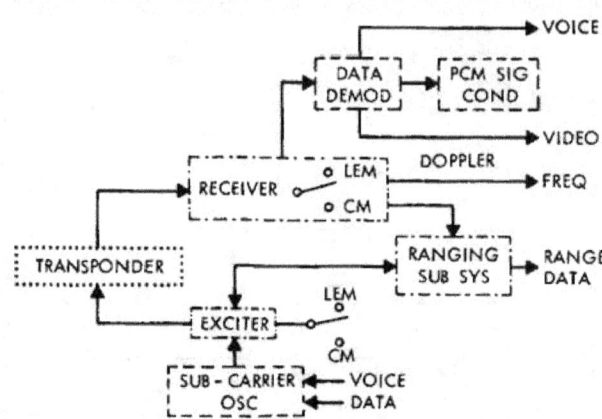

Figure 11—USB system test unit.

TRACKING STUDIES FOR PROJECT APOLLO

by

W. D. Kahn

Goddard Space Flight Center

ABSTRACT

A tracking error analysis study is presented in which the earth orbital, translunar, lunar, and transearth phases of the Apollo Mission are considered. Principal error sources, such as measurement random errors, measurement bias errors, and errors in tracking station location, are analyzed to determine how well a tracking network such as the Manned Space Flight Network (MSFN) can determine the orbit of the spacecraft.

Results of an error analysis study considering tracking of the spacecraft during all the different phases of the Apollo Mission are described. Specifically, tracking during the earth orbital, translunar, lunar, and transearth phases are simulated in this study.

INTRODUCTION

Tracking radars of the Manned Space Flight Network (MSFN) will be used to determine the trajectory of the Apollo spacecraft during the earth orbital, translunar, lunar, and transearth phases of the Apollo Mission. Because of the effects of errors in the measurements as well as errors in the equations of motion, it is possible to obtain only an estimate of the spacecraft's true trajectory. How good the estimate will be is dependent on the volume and quality of the tracking data, as well as the adequacy of the mathematical model used in the data reduction process. Error analysis studies, incorporating the effects of the principal error sources, simulate the data acquisition and data reduction processes. Such studies, subject to the initial assumptions, provide information on the capability of the MSFN tracking radars to determine the trajectory of the Apollo spacecraft.

THE EARTH ORBITAL PHASE

The Apollo spacecraft is inserted into a 100 nautical miles (185 km) earth parking orbit. Tracking of the spacecraft is assumed in this study to commence immediately after insertion. The trajectory profile from insertion through translunar injection is given in Figure 1. Tracking coverage during the first orbit by stations of the MSFN is given in Figure 2.

Propagation of the rms errors in the state vector (a vector composed of the components of the position and velocity vectors) resulting from random errors in the measurements, bias errors in the measurements, and errors in tracking station location is analyzed during the first parking orbit of the Apollo spacecraft (Figures 3 and 4).

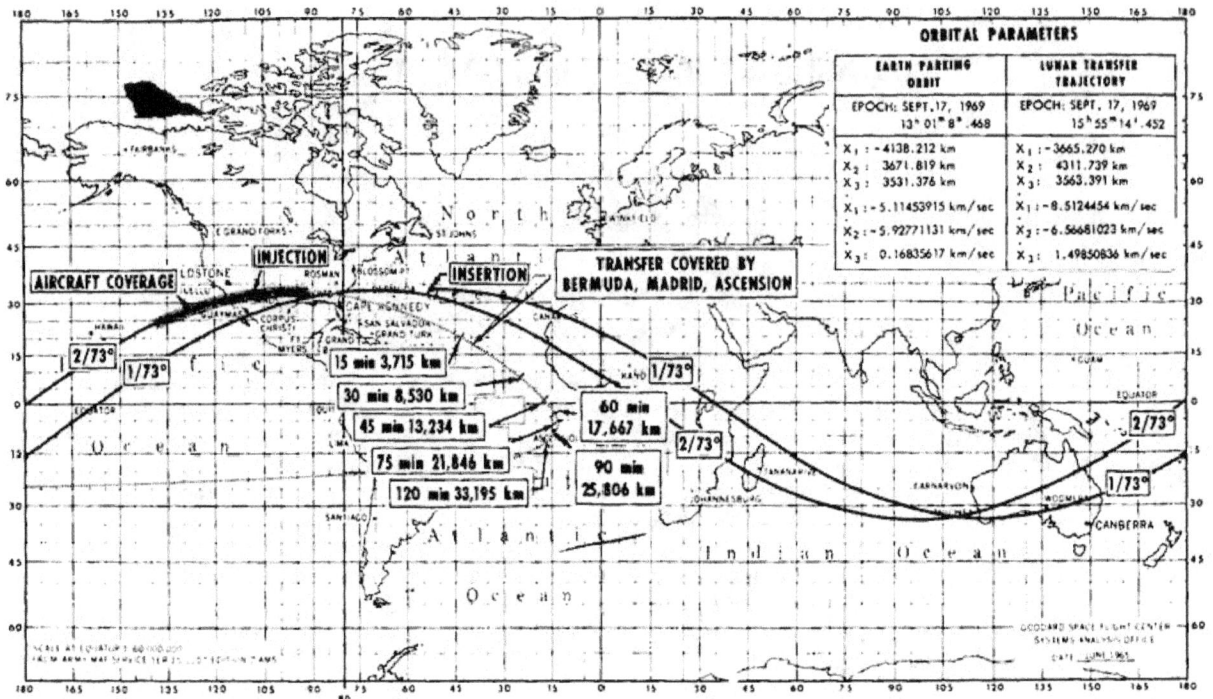

Figure 1—Lunar transfer from the second parking orbit, $\alpha = 73°$.

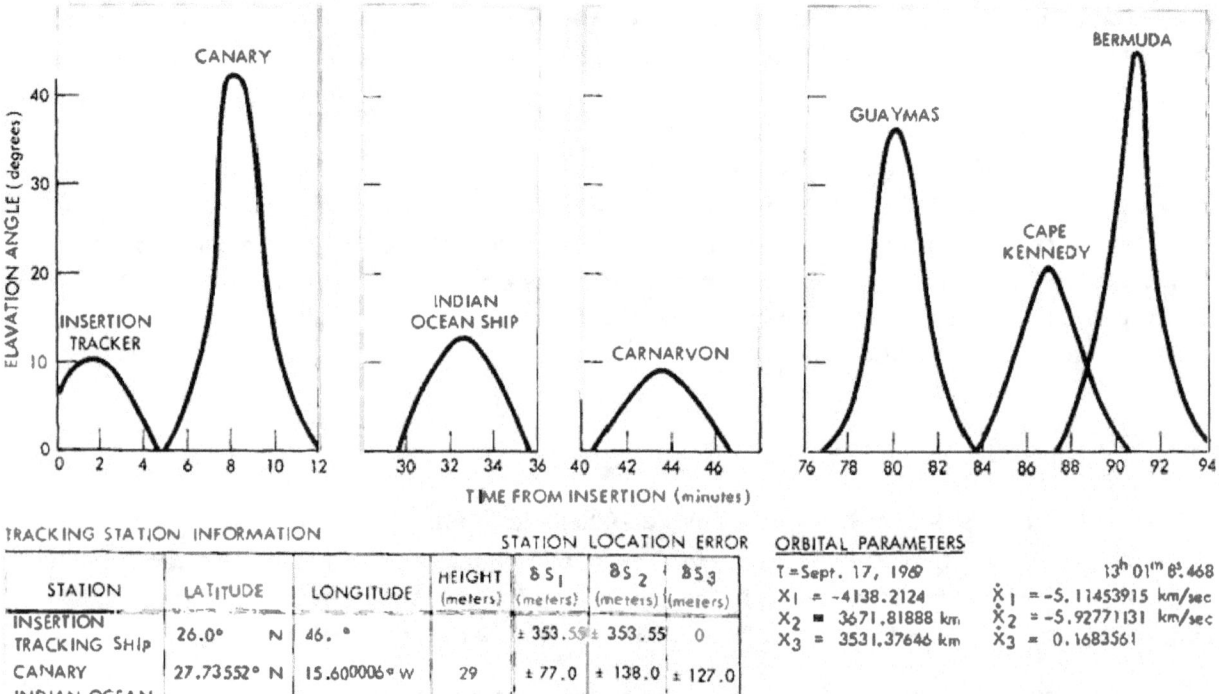

Figure 2—Tracking station coverage during first Apollo parking orbit.

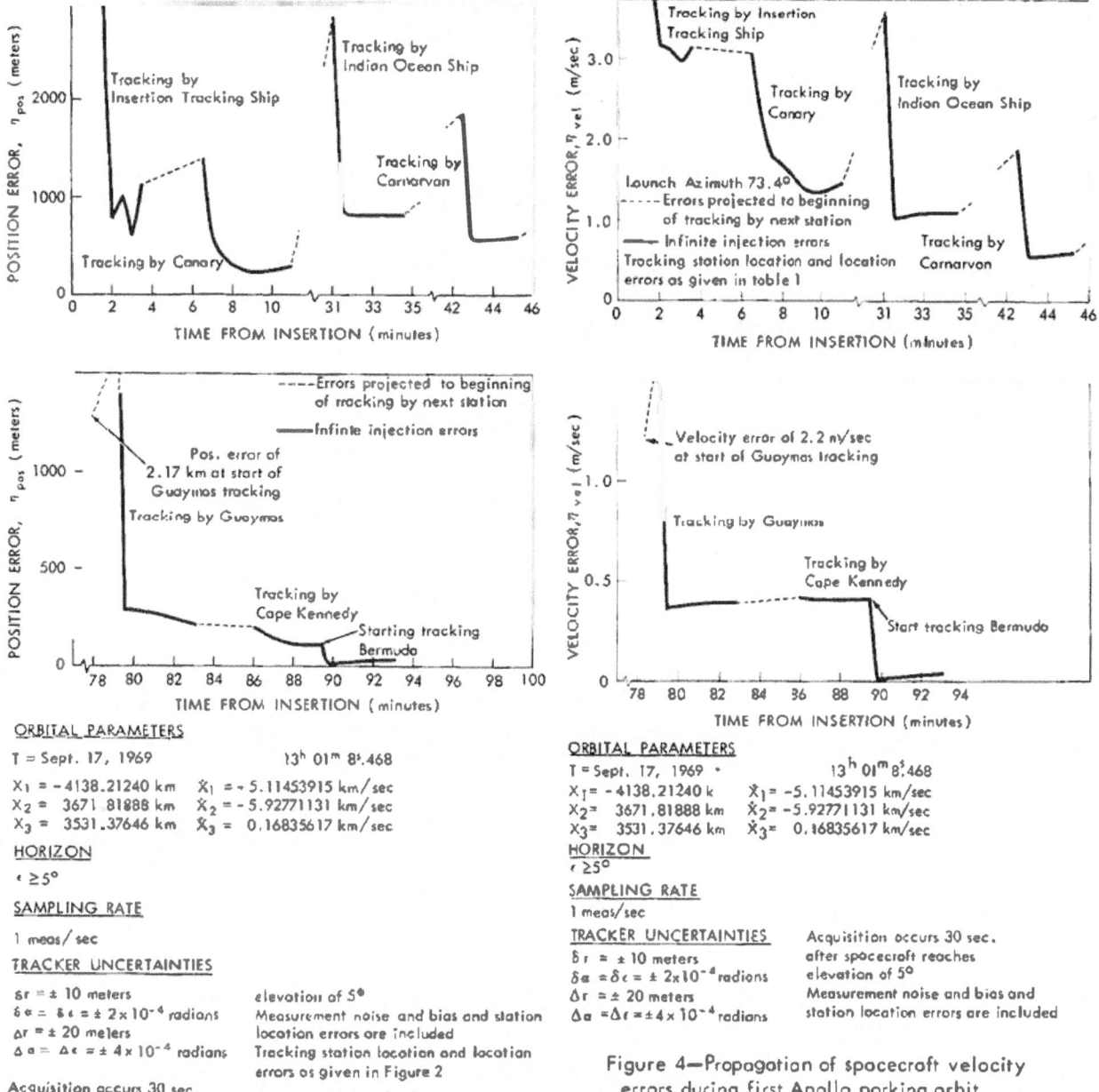

Figure 3—Propagation of spacecraft position errors during first Apollo parking orbit.

Figure 4—Propagation of spacecraft velocity errors during first Apollo parking orbit.

During periods when observational data are obtained, the rms errors in the state vector slowly decrease until such time as no tracking data are available. As the rms errors in the state vector are propagated through a region where there is no tracking coverage, their magnitudes increase or decrease, depending on where along the orbit they are evaluated. In the study presented, the rms errors in the state vector are increased when propagated without tracking data. As soon as tracking data are added to the projected rms errors in the state vector, these errors decrease very rapidly.

Despite the assumption of no initial knowledge about the state (position and velocity), as well as the inclusion of the principal error sources in the tracking data, it is shown in

Figures 3 and 4 that the rms errors in position and velocity will be ±40 meters and ±4 centimeters per second at the end of the first parking orbit. Therefore, it is safe to conclude that the spacecraft's orbit can be determined very accurately from tracking data.

THE TRANSLUNAR PHASE

The spacecraft is injected into a lunar transfer orbit between the second and third earth parking orbits. In Figure 1, a profile of the transfer trajectory is given from injection up to several hours beyond injection.

Figures 5 and 6 show the propagation of rms errors in the state vector up to three hours after injection. A comparative analysis is made of the effects of random measurement errors,

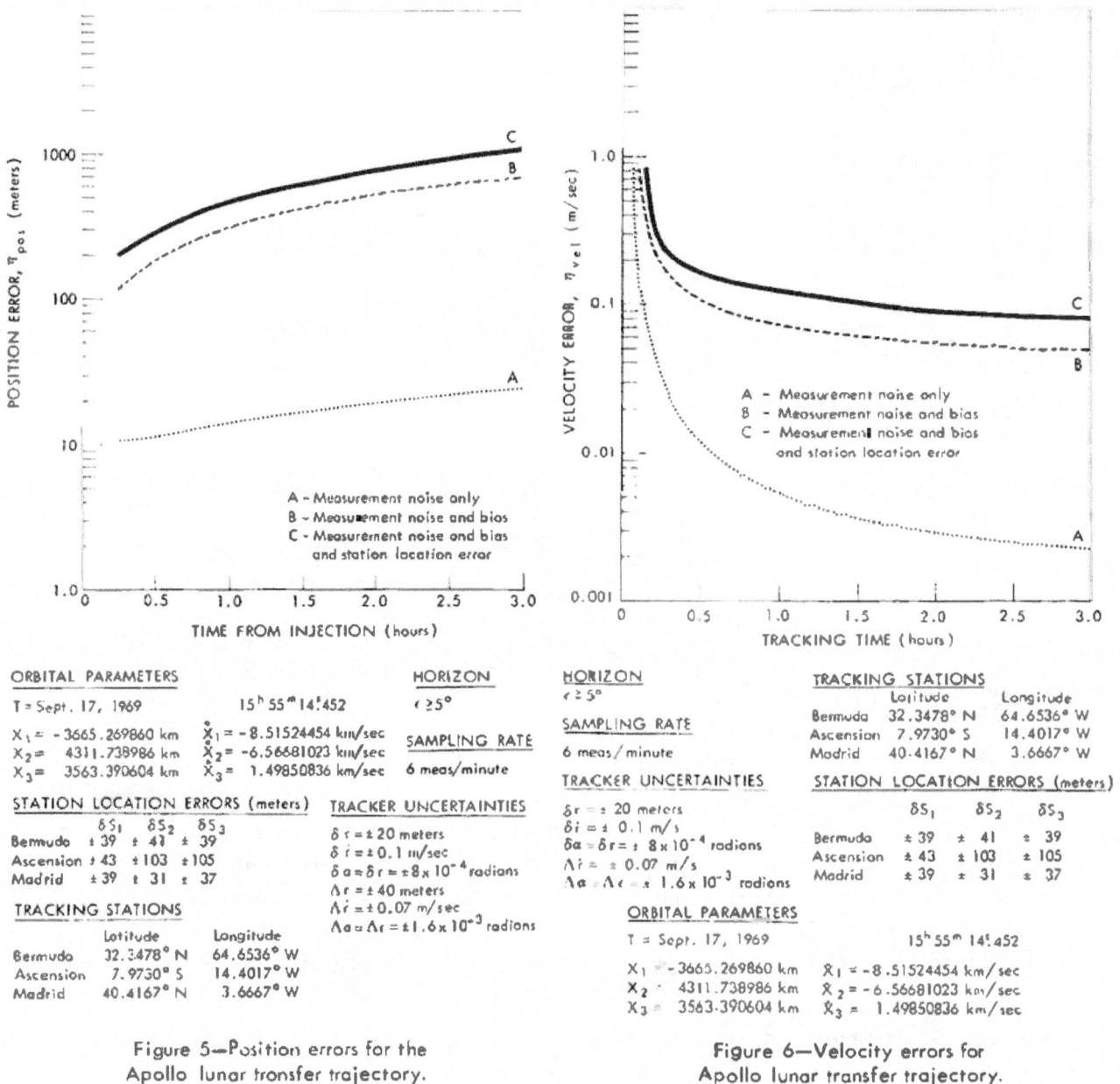

Figure 5—Position errors for the Apollo lunar transfer trajectory.

Figure 6—Velocity errors for Apollo lunar transfer trajectory.

measurement bias errors, and errors in tracking station location, when these errors are considered alone and in combination. The results of the analysis indicate that measurement bias errors produce the most significant effect on the rms errors in the state vector. Because of the continuously increasing distance between the spacecraft and the earth, the contribution of the errors in the station location on the rms errors in the state vector is not too significant.

For this error analysis study, it was assumed that the spacecraft is tracked by the 85-foot unified S-band (USB) antennas at Madrid, and by two 30-foot USB antennas at Bermuda and Ascension. The 85-foot dish tracks the spacecraft in the two-way doppler mode, and the two 30-foot dishes track in the three-way doppler mode (passive doppler).

Because the tracker-spacecraft geometry weakens as the spacecraft recedes from the earth, an increase in the rms position error of the spacecraft results. However, the spacecraft's velocity relative to the earth decreases as the spacecraft's distance from the earth increases. Therefore, the rms error in spacecraft velocity also decreases. This is due to the relative decrease in the trajectory's sensitivity to velocity.

THE LUNAR PHASE

The Command and Service Module/Lunar Excursion Module (CSM/LEM) is inserted into an (80 ± 5) nautical mile parking orbit around the moon, with orbital insertion occurring on the backside of the moon. For purposes of this study, 22 minutes are required before the spacecraft becomes visible to the tracking stations on earth. At this time the 85-foot USB antennas at Canberra, and two 30-foot USB antennas at Carnarvon and Hawaii, will track the spacecraft. The 85-foot dish will track in the two-way doppler

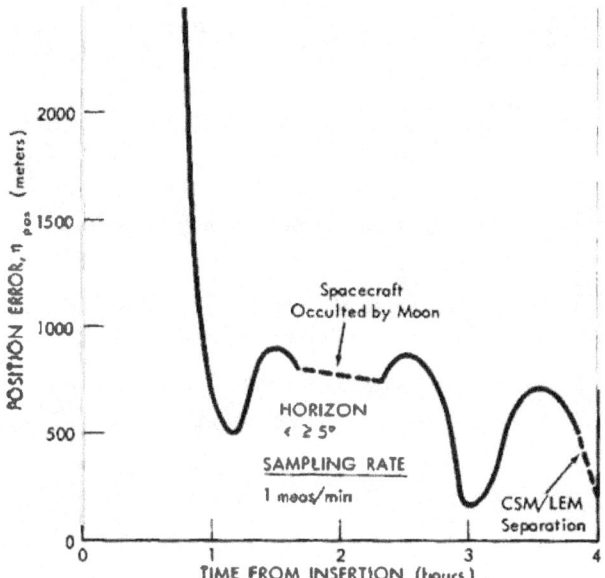

Figure 7—Propagation of spacecraft position and velocity errors during CSM lunar parking orbits.

mode, and the two 30-foot dishes will track in the three-way doppler mode. To insure good tracker-spacecraft geometry, the tracking station configuration on earth was selected for maximum north-south separation.

Approximately 3.8 hours after insertion, CSM/LEM separation occurs. In order to provide *a priori* knowledge of the state to the LEM before the LEM descent maneuvers, very good knowledge of the state must be determined by tracking the CSM/LEM from the earth. Figure 7 shows the rms errors in spacecraft position and velocity to be ±500 meters and ±24 centimeters per second at the time of CSM/LEM separation. Because the effects of measurement-bias errors are not included in this study, the results in Figure 7 are on the optimistic side.

The elapsed time from the initiation of the LEM descent maneuvers up to CSM/LEM docking maneuvers is approximately 36 hours. After this time period, the astronauts will have abandoned the LEM to re-enter the CSM. Upon re-entering the CSM, the LEM is jettisoned. Shortly thereafter, the CSM is injected into its earth transfer trajectory.

TRANSEARTH PHASE

Twenty hours after transearth injection, the first midcourse correction is made. Two other midcourse corrections are made at 65 hours and at 88 hours after transearth injection. The last midcourse correction is made one hour before re-entry.

Error analysis studies during this phase of the Apollo Mission are made from transearth injection up to the first midcourse correction (Figure 8), after the first midcourse correction up to the second midcourse correction (Figure 9), and eight hours before re-entry up to re-entry (Figure 10). The last error analysis includes the time at which the third midcourse correction is made. For all these tracking-error analysis studies, the best tracking complex-spacecraft geometry configuration was chosen.

Attention is called to the increase in the rms velocity error as the spacecraft approaches the re-entry altitude of 122 kilometers (Figure 10). This increase in the rms error in spacecraft velocity is due to the return trajectory's increased sensitivity to velocity as it approaches the earth. A corresponding decrease in the rms error in spacecraft position results.

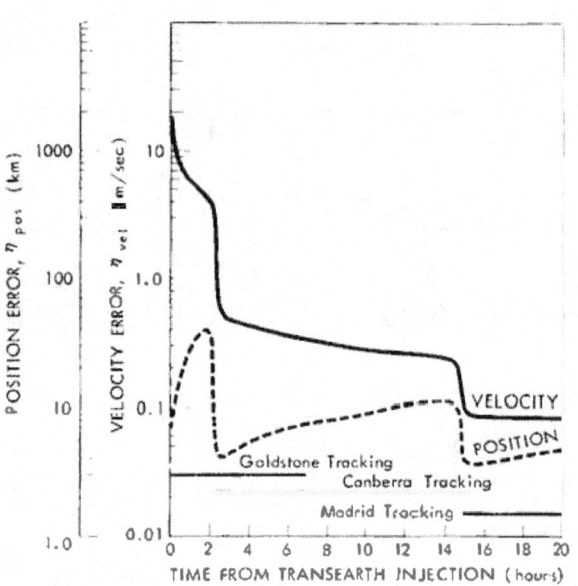

Figure 8—Errors in spacecraft position and velocity for Apollo return trajectory (first 20 hours).

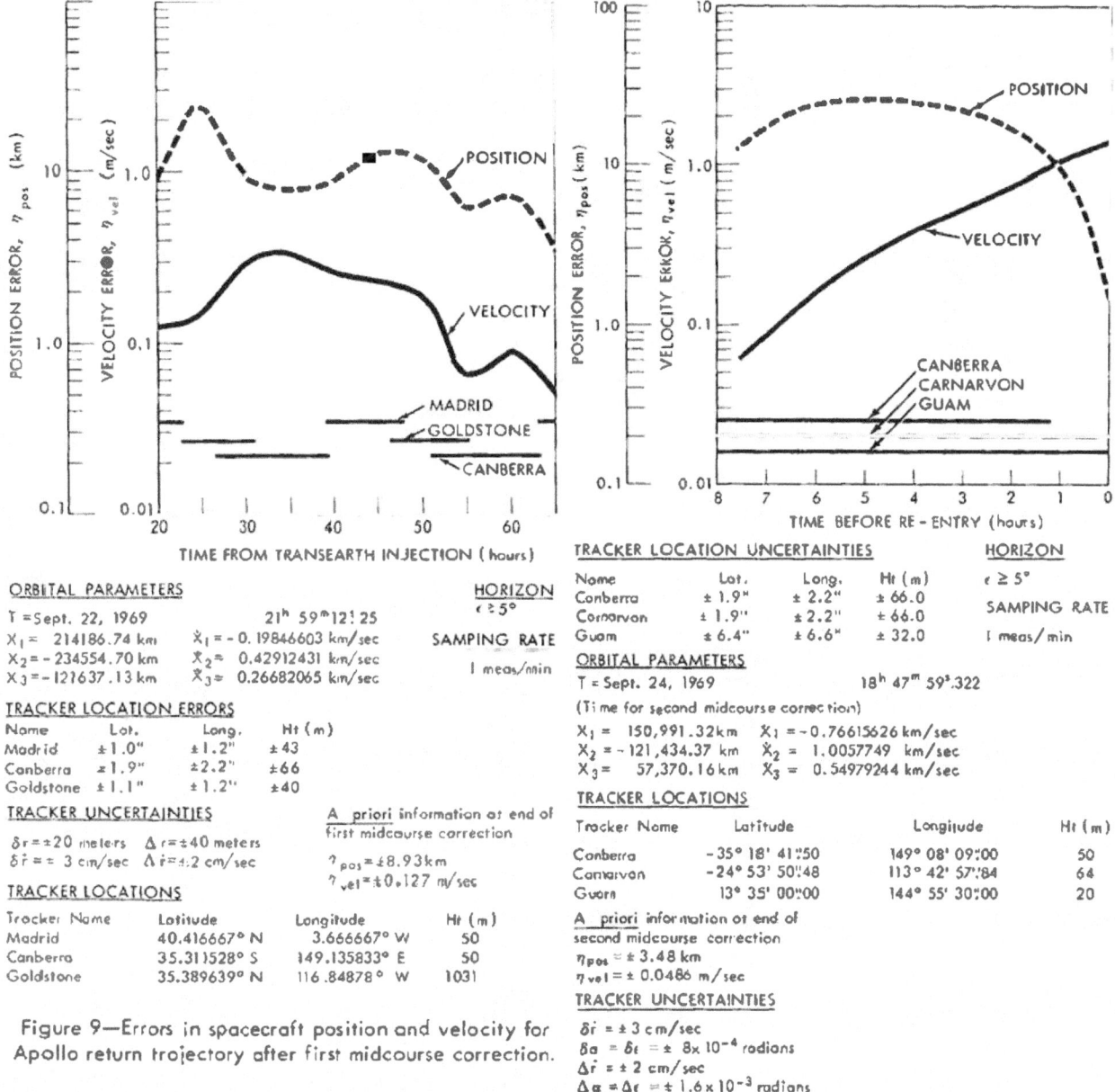

Figure 9—Errors in spacecraft position and velocity for Apollo return trajectory after first midcourse correction.

Figure 10—Errors in spacecraft position and velocity for Apollo return trajectory during final eight hours prior to re-entry.

CONCLUSION

The error analysis studies contained in this paper demonstrate that the bias errors in the measurements and in tracking station location most significantly influence the errors in the state vector. These error sources, unlike the random errors in the measurement, *do not* decrease as the amount of observational data is increased.

ACKNOWLEDGMENTS

The author expresses his appreciation to Mrs. A. Marlow and Mr. J. L. Cooley for their help in preparing the data used in this paper.

REFERENCES

1. Kahn, W. D., and Vonbun, F. O., "Tracking Systems, Their Mathematical Models and Their Errors, Part II — Least Square Treatment," to be published soon as NASA Technical Note.

2. Philco Corporation, "User's and Programmer's Manual for Interplanetary Error Propagation Program," prepared for contract NAS 5-3342.

3. Schmidt, S. F., "The Application of State Space Methods to Navigation Problems," Philco WDL Guidance and Control System Engineering Department, Technical Report No. 4, July, 1964.

4. Vonbun, F.O., and Kahn, W. D., "Tracking Systems, Their Mathematical Models and Their Errors, Part I — Theory," NASA Technical Note D-1471, October, 1962.

USB ANTENNA STRUCTURES

by
L. E. Hightower
Goddard Space Flight Center

ABSTRACT

This paper describes the main features of the unified S-band antennas, their design considerations, characteristics, parameters, functions, and modes of operation. Description covers both land-based and shipboard antennas, specifications, and maintenance problems. Both Rosman and Apollo antennas are treated.

INTRODUCTION

This discussion will treat the main features of the unified S-band antennas and will give some of the reasons behind these features. The presently planned Apollo ground system network will employ ten 30-foot and three 85-foot diameter steerable antennas. These antennas use what is known as X-Y mounts. The network also uses ships instrumented with azimuth-elevation antennas.

TYPES OF ANTENNA MOUNTS

Most steerable antennas have two mutually perpendicular rotational axes as shown in Figure 1. The main difference between antenna mounts is the orientation of the lower of these two axes. There are three main types of antenna mounts.

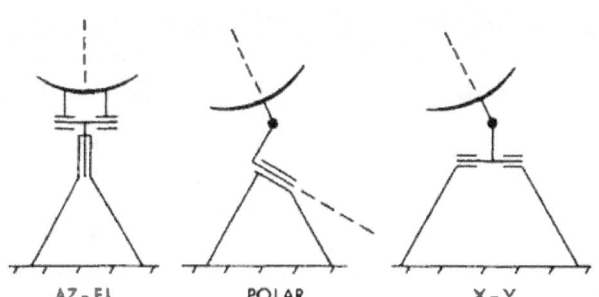

Figure 1—Antenna mounts.

Azimuth-Elevation Mount

In the azimuth-elevation mount, the lower axis is vertical. This axis arrangement permits compactness of design and rigidity and it is the logical choice when tracking through zenith is not a requirement. It is by far the most popular mount. The problem is that extremely fast azimuth rates are required to track through zenith with the azimuth-elevation mount.

Polar (Hour Angle - Declination) Mount

In the polar mount the lower axis is parallel to the earth's rotational axis. This arrangement facilitates ease of tracking of celestial objects. It is used for radio telescopes and for antennas in the NASA deep space effort.

X-Y Mount

Both axes of the X-Y mount are horizontal at zenith. The mount was designed especially for tracking earth-satellites. Its main advantage is that it can track through zenith. All three types of mounts have gimbal lock positions at the ends of the lower axis. For the X-Y mount, these positions are half cones (10° wide) just above the horizon (areas which are not greatly significant).

LAND-BASED ANTENNAS

It has been established that all the Apollo land-based antennas must be capable of maintaining contact with the spacecraft through zenith (orbital transfer could occur at zenith) and essentially complete sky coverage is required. These requirements dictated selection of the X-Y mount for both the 30-foot and the 85-foot land-based antennas.

30-Foot Antenna

The first of the 30-foot S-band antennas has been erected at Collins Radio Company's Dallas, Texas facility. The following features can be seen in Figure 2 (starting at top):

1. Three-foot diameter acquisition antenna at apex of quadripod (note radome cover).

2. Secondary reflector for Cassegrain feed system located just below acquisition antenna.

3. Y axis (upper).

4. Y-wheel house (so-called cement mixer) which houses boresight package and equipment such as parametric amplifiers that should be near the feeds. This room is air conditioned by circulating a chilled glycol solution.

5. X axis and X-wheel assembly which houses Y-axis drive assembly and provides access way for personnel to Y-wheel house.

6. Lower platform which provides mounting for X-drive units.

7. Room beneath antenna which houses power amplifier units and motor starters for drives.

Figure 3 shows the following:

1. The main reflector, which is made up of 36 solid surface panels. These panels are individually adjustable. Paint on the surface panels scatters solar radiation to prevent overheating feeds.

2. The feed cone mounted at center of dish.

3. Lights on rim of dish to warn when antenna may be transmitting. An audio warning is also used.

Figure 4 shows the axis movements. Figure 5 shows the optical boresight package mounted in the Y-wheel house. The package looks through a window (optical flat). (Note foam insulation on inside walls of wheel house.)

85-Foot Antenna

Much of the structure for the three 85-foot Apollo antennas has been fabricated by the Blaw-Knox Company, and an antenna bought under the same contract as the Apollo antennas has been

Figure 2—Thirty-foot antenna, side view of reflector.

Figure 3—Thirty-foot antenna, front view of reflector.

erected at Rosman, North Carolina. Figure 6 shows the two 85-foot antennas at the Rosman Data Acquisition Facility. The antennas are used for tracking earth satellites. The antenna in the foreground has been in essentially continuous operation for some time and a free period longer than 30 minutes is unusual. GSFC personnel have learned a lot about the maintenance of large antennas from this and other big dish facilities. The antenna in the background is shown in Figure 7. The Rosman antenna in Figure 7 is the same as the Apollo antennas except for some slight changes (mainly different quadripod support for acquisition antenna and secondary reflector) which were made to meet special S-band requirements. The main features of this antenna are enlarged versions of those we have seen on the 30-foot antenna. Some features worthy of note are:

1. Axis wheel structures. In this antenna these become large space frames.

2. Y-wheel house. This becomes a building mounted in the structure. In this case it houses the power amplifier units.

3. Optical boresight room (just left of the ladder suspended from dish structure).

Figure 8 gives a better view of the X-wheel structure. Note large counterweight box which is filled with lead. Note also bridging beneath lower platform to hold drive pinion teeth in full contact with the bull gear.

Figure 9 shows the Rosman 85-foot antenna reflector. This is the same as the S-band antennas except for the quadripod feed support. The feed cone is not attached, and one can see the three-foot diameter opening through the center of the dish structure. This provides a conduit for connecting feed components with the parametric amplifiers and transmitter power amplifiers in the room mounted in the Y-wheel structure. The ring to which the feed cone will be

Figure 4--Thirty-foot antenna, axis movements.

attached is also shown. Four reflector panels have been removed to uncover the alignment datum points. These points define the reference plane from which all reflector panel adjustments are made. The opening in the surface for the optical boresight equipment is also shown.

SHIPBOARD ANTENNAS

Although the X-Y antenna can track through zenith, it has some drawbacks. When it is designed for essentially complete sky coverage, the rotational axes are separated by a considerable distance. As a result, both axes must be counterweighted. This means that the design lacks compactness and the lower axis has a high moment of inertia. For shipboard application these disadvantages were judged to overshadow the advantage of being able to track through zenith (considering the fact that the ship location could possibly be changed to avoid an overhead pass). Therefore, azimuth-elevation mounts are used for the Apollo ships' antennas.

Figure 5—Thirty-foot antenna, optical boresight package.

Figure 6—Eighty-five-foot antennas at Rosman.

SPECIFICATIONS

The use of higher RF frequencies is the obvious trend in spacecraft communications. This, plus the fact that updating a completed antenna is generally impractical, was the reason why most main features of the Apollo antennas were specified at the highest practical level considering the present state of the art. Many major parts of the antenna are specified to greater accuracy than necessary for operation at S-band. Some examples are:

1. Reflector surface accuracy is held to 0.030 and 0.040 rms for the 30-foot and 85-foot antennas, respectively. This should permit satisfactory operation in the 10,000 megacycle region.

2. Alignment of rotational axes is held to five seconds of arc. This is as close as is feasible with existing field alignment equipment.

3. Pointing accuracy is specified to be within 40 seconds of arc.

Figure 7—Eighty-five-foot antenna, similar to Apollo antenna.

MAINTENANCE

Experience at GSFC has shown that steerable antennas, particulary 85-foot diameter antennas, require considerable maintenance. Maintenance of drive systems and gears is expected, but one of our most troublesome problems has been the loosening of bolted joints.

The problem is partly peculiar to X-Y antennas. As the X-Y antenna tracks from horizon to horizon, the gravity loading on practically every joint in the structure is completely reversed. For an azimuth-elevation antenna, only that part of the structure below the azimuth axis experiences reversal of stresses from gravity loading. Simple calculations show that gravity loadings are, for most joints, higher than drive and brake loadings.

Figure 8—Eighty-five-foot antenna X-wheel structure.

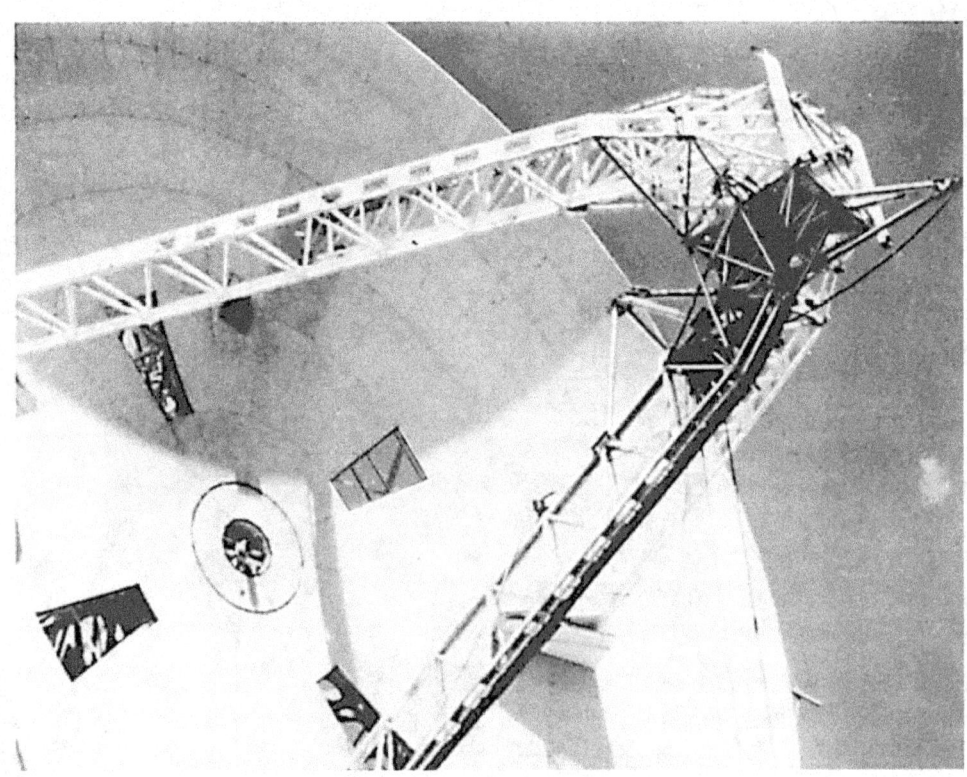

Figure 9—Eighty-five-foot antenna reflector.

Why does a bolted joint designed for a relatively high factor of safety loosen under normal antenna usage? First, most load values given for strengths of bolted joints will allow a very small amount of slippage, so that the margin of safety for a joint subjected to reversal of loading may not be as high as calculated. Second, the actual loading of a particular joint can be higher than calculated because of slight inaccuracies in lengths of members. If a joint slips ever so slightly as the antenna is exercised, the bolt threads will unscrew because mechanical rectification is inherent in a bolted joint which is slipped. Even if the threads are locked, the joint will still loosen at a slower rate because of wear. This loosening of joints can progress throughout the structure.

A related problem is that bolts torqued to known values and then subjected to very light loads have shown an appreciable relaxation of torque. There are at least three possible causes:

1. Cold flow of paint films.

2. Some extrusion of washers into bolt holes.

3. Some yielding of "pinnacles" of irregular mating surfaces.

Because of GSFC's experience with loosening of structural joints, a program of periodic bolt checking and torqueing is being included in the preventive maintenance program. It is expected that a similar program will be necessary for antennas in the Apollo network.

USB SERVO SYSTEM

by

N. Raumann

Goddard Space Flight Center

ABSTRACT

A unified S-band (USB) servo control and drive subsystem has been designed and is being presently developed. The intent of this discussion is to present an overall view of the subsystem and its anticipated capabilities. The text includes, as an introduction, a general description of the servo control and drive subsystem, the system's principal modes of operation and its required sky coverage. More specific discussion on the dynamic behavior of the system follows the introductory portion. Finally, some preliminary results are given and the present status of the subsystem is discussed.

INTRODUCTION

The servo and drive system is the portion of the overall antenna system that permits the accurate positioning of the gimbal axes in response to various input signals. First, the land-based antennas will be discussed and then a few comments on shipboard systems will be made. An X-Y mount is used for this application because zenith coverage is accomplished which is not possible with a more conventional Az-El mount. Even though a two-axis mount could have been designed mechanically to cover the whole hemisphere, a cone exists in which tracking is impossible due to excessive drive rate requirements. This cone of silence, or keyhole, is always centered around the major axis of the antenna and its size is proportional to maximum rates that the drive system can deliver. For the X-Y mount, the keyhole appears along its major axis, the X axis, which is parallel to the surface of the earth and has a north-south orientation for 30-foot systems; thus, only targets on the horizon appearing in a northern or southern direction are affected by keyhole considerations. The Az-El mount, which has its keyhole at zenith, usually requires a larger keyhole for a given maximum drive rate because satellite dynamics, as seen from the antenna, approach maximum values at zenith and minimum values at the horizon.

The antenna gimbal axes are positioned by means of a hydraulic drive system. A hydraulic system has been chosen rather than an electrical one because of inherent advantageous characteristics, such as high torque to inertia ratio, large dynamic range, lack of radio interference, and lack of predominant time constants within the servo bandwidth. Nevertheless, in selecting a hydraulic drive, certain possible problem areas have to be considered, and these are mainly concerned with hydraulic leaks and contamination of the fluid. Careful design and preventative maintenance will, however, minimize these problems. The rating of the drive system has been chosen to provide for obtaining maximum velocities and accelerations under maximum wind conditions.

The servo and structural interface has been adequately covered for this application, which requires that the lowest natural frequency of the structure is sufficiently high to realize the required servo bandwidth. The servo bandwidth is mainly determined by satellite dynamics and system noise considerations and is in the order of 1 cycle per second. The natural frequency of the 30-foot antenna is specified to be 4 cycles per second and that of the 85-foot antenna, 3 cycles per second.

Table 1 shows the performance specification of the two antenna types, the 30-foot and 85-foot systems. It can be seen that the maximum tracking velocity is 4 and 3 degrees/second respectively and an acceleration capability of 5 degrees/second^2 is provided. These rates are adequate to track a satellite in a low earth orbit, of about 100 miles. The antenna will be able to track in winds up to 45 miles per hour. Full tracking accuracy will be realized in winds up to 20 miles per hour. Tolerances will be doubled for winds between 20 and 30 miles per hour and quadrupled for winds between 30 and 45 miles per hour. The drive system will be powerful enough to move the antenna to a stow position in 60 miles-per-hour winds.

Table 1

System Performance.

Criteria	30 Ft	85 Ft	Units
Velocity	4	3	Degrees/Second
Acceleration	5	5	Degrees/Second^2
Winds: Operating	20	20	MPH
Operating (reduced accuracy)	45	45	MPH
Stowing	60	60	MPH
Survival	140	120	MPH
Sky Coverage	2	2	Degrees Above Horizon
Keyhole Cone	20	20	Degrees
Keyhole Orientation	North-South Axis	East-West Axis	
Accuracy: Pointing	±0.6	±0.6	Minutes
Tracking	1.5 max	1.5 max	Minutes

Due to the particular arrangement of the axes of this mount, the keyhole will be oriented along the X axis. The keyhole will describe a 20-degree cone at each end of the X axis. Except for the keyhole, the antenna will be capable of tracking in all directions above a horizon of 2 degrees. Pointing accuracy can be defined as the closeness to which the antenna can be directed to a given coordinate position. Pointing accuracy could be determined in the program mode, for example, by introducing a fixed position into the programmer and measuring the error between this commanded position and the actual. Pointing accuracy of this system will be ±0.6 minutes of arc. Tracking accuracy is determined by measuring the overall angular error between the axis of the RF beam of the antenna and a line drawn between the antenna and the target. In particular, this measurement could be performed by autotracking a calibration plane and observing the position of the plane on an optical monitor mounted on the antenna. The 3-sigma tracking error will not exceed 1.5 minutes of arc.

DRIVE SYSTEM

Each antenna axis is driven by two fixed displacement hydraulic motors which are connected to the bull gear through individual gear boxes. This configuration has been chosen to eliminate backlash in the drive system. Figure 1 shows the X bull gear, a pinion extending from a gearbox, and the hydraulic motor. The hydraulic pump unit is in the background. Figure 2 shows a closer picture of the gearbox with the motor. Figure 3 shows the pump unit with its hydraulics. On one end of the pump unit, the variable displacement pump which drives both motors is shown. On top of the pump is a servo valve. This controls a ram which in turn positions the yoke of the pump. Yoke angle, for feedback purposes, is derived from a potentiometer in front of the pump. The pump is driven by a squirrel cage motor barely visible behind the structure. The motor also drives two auxiliary fixed-displacement pumps which are not visible. On top of the pump unit the reservoir is visible. Next to it is a box which houses the brake control unit. This device permits a gradual application of the brakes for normal shutdowns. Only during an emergency stop is sudden complete application of the brakes used. In front of the pump unit are various filters, valves, and gauges required for the satisfactory operation of the system.

A very simplified schematic of the hydraulic drive system is shown in Figure 4. There are the two motors which are connected through gear reducers to the antenna axis bull gear.

Hydraulically, the motors are connected in series and are energized by the main pump. When the yoke of the pump is in its neutral, or

Figure 1—Photograph of USB antenna drive system showing X-bull gear and hydraulic motor.

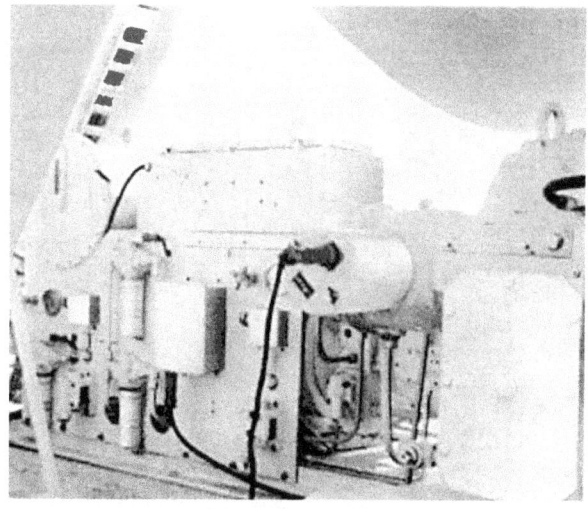

Figure 2—Gearbox and motor.

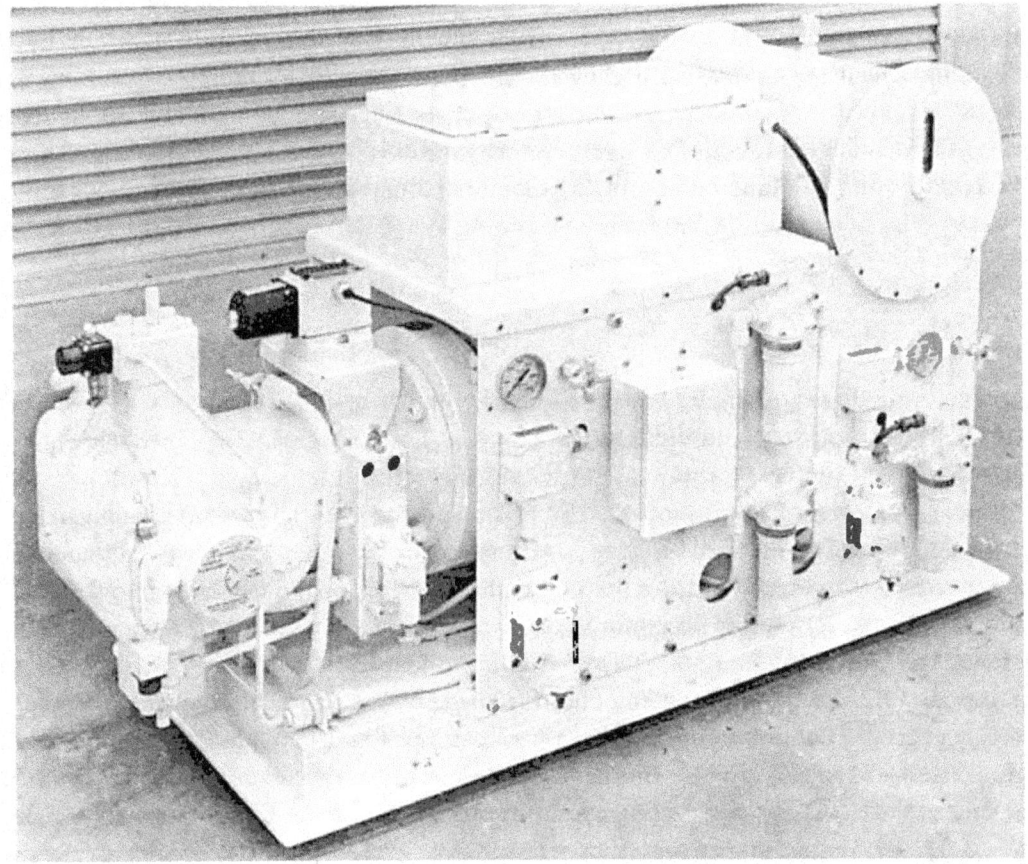

Figure 3—Pump unit with hydraulics.

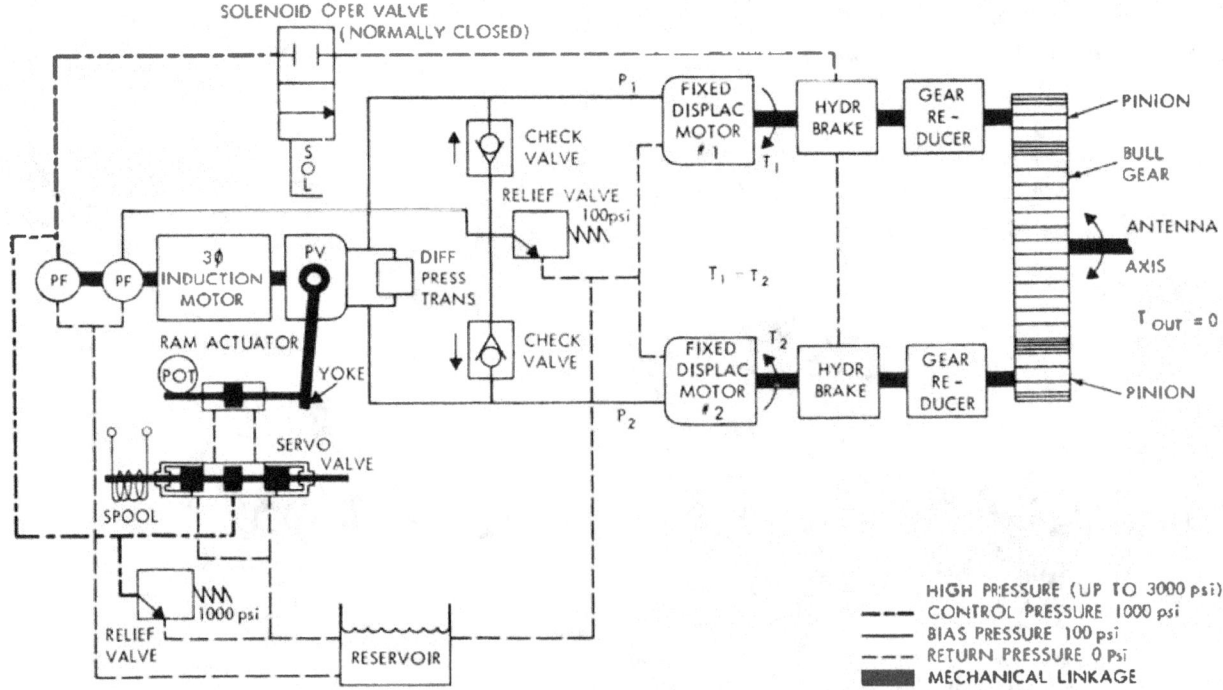

Figure 4—Schematic of hydraulic drive system with no excitation and brakes applied.

central, position, as shown in this slide, no differential pressure is produced across the pump. The anti-backlash feature of this drive system is accomplished by the first auxiliary pump. The output of this pump is held to 100 pounds per square inch by means of a relief valve. This 100 pounds per square inch, or bias pressure, is applied through check valves to opposite ports of the hydraulic motors, thus producing equal and opposite torques on the bull gear. Even though no motion results, backlash in both gearboxes and between pinion and bull gear will be taken up. The second auxiliary pump produces a 1000 pounds per square inch control pressure which is used to power the yoke servo and to lift the brakes whenever the solenoid operated valve is energized. Having to rely on control pressure and current in the solenoid makes the brakes fail-safe. Failure, electrical or hydraulic, will apply the brakes automatically.

Figure 5 shows the same schematic, only now the brakes are lifted and a signal has been applied to the servo valve causing its spool to be displaced to the right. This action raises the pressure on one side of the piston of the ram actuator and moves it to the right. This in turn moves the yoke to the right and causes a pressure increase at motor #1. Maximum pressure could be as high as 3000 pounds per square inch. The check valve at motor #1 closes because pressure on upper port of the valve is higher than on the lower port. The other check valve remains open, thus maintaining the 100 pounds per square inch bias pressure on motor #2 which is required for the anti-backlash feature. Because pressure P_1 is larger than P_2, motor #1 will develop a torque T_1 that is larger than that of the other motor, and consequently a net torque will be applied to the bull gear which is proportional to the difference of T_1 and T_2. This net torque, if sufficiently high, will cause motion of the antenna, say in a clockwise direction.

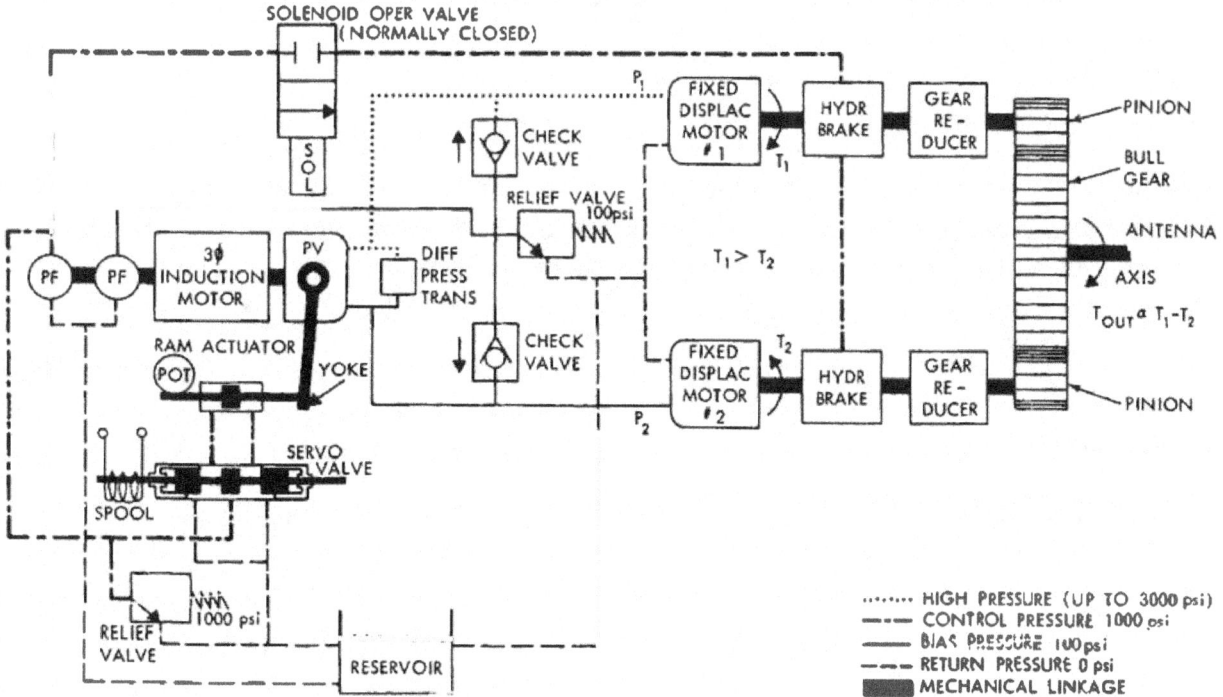

Figure 5—Schematic of hydraulic drive system with brakes lifted and signal applied to servo valve causing its spool to be displaced to right.

Figure 6 shows a similar condition only now the drive signal to the servo valve has been reversed. This will produce a torque T_2 at motor #2 which is higher than that of motor #1 and motion of the bull gear will result in the opposite direction to that in the previous case.

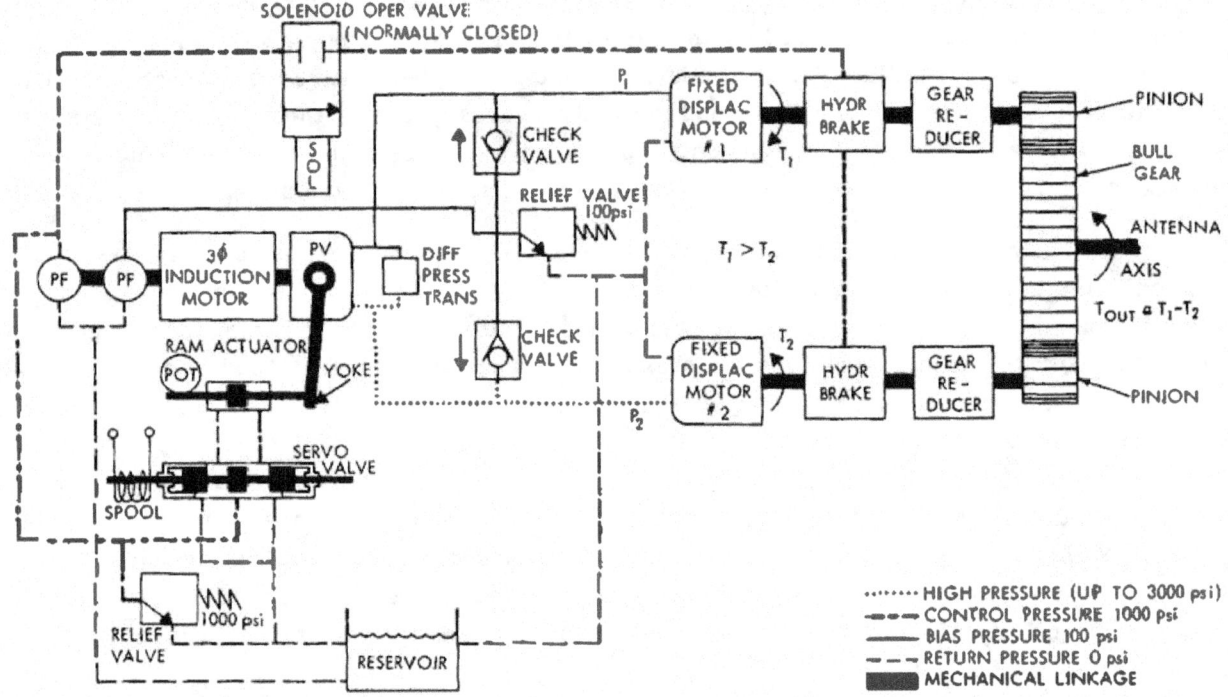

Figure 6—Schematic of hydraulic drive system with drive signal to servo valve reversed.

Note that several feedback transducers have been provided in the hydraulic drive system, namely: a potentiometer measuring yoke angle, a differential pressure transducer across the pump, and a tachometer at each motor shaft, which, however, is not shown. Also not shown are many other hydraulic components and circuits, for example: a heat exchanger is provided to cool the oil, several filters in the ranges between 1-1/2 - 25 microns are provided to keep contamination to a minimum, an oil path through the pump housing is provided to prevent overheating of the pump (especially at no flow conditions), and several relief valves are provided in case excess pressures appear.

SERVO SYSTEM

The servo system is capable of operating in any of the following modes.

1. manual
2. slew
3. programmer
4. slave

5. scan

6. acquisition track

7. automatic track

8. auto-program

9. test

A manual mode is provided which permits an operator to position the antenna to any desired coordinate position by means of a ball tracker.

The ball tracker can also be used in the slew mode, in which the antenna can be operated at various constant velocities. In the program mode, the antenna follows a command which is generated from a prediction tape in the programmer. The antenna is also capable of following any other antenna in the slave mode. A scan function generator has been incorporated to permit superposition of a search pattern on most other modes of operation. Scan functions available are spiral, circle, raster, and sector search patterns. The acquisition track mode permits automatic tracking of targets with the acquisition monopulse system and the automatic track mode permits tracking with the narrow beam, high-gain, unified S-band monopulse system.

A new mode of operation has been added to this system which has not appeared on previous GSFC antennas and this is the autoprogram mode, which will be explained a little later. Finally, there is a test mode which permits testing of the various operational modes prior to a satellite pass.

Figure 7 shows the servo control panel. The various mode switches are arranged in the center of the panel. Several other switches required for operation of the antenna are provided below, such as power on-off switch, hydraulic on-off switches, and disable switches. Indicator lights are provided for each axis monitoring oil temperatures, oil filter conditions, and antenna limit conditions. Servo error meters for each axis are also provided. The operator can, at his discretion, adjust the servo loop bandwidth by means of a switch between the error meters.

Above the servo control panel is the error monitor and slave selector panel. By depressing any of the buttons, the antenna can be slaved to any of six external sources. The additional servo error meters permit measurement of errors in any mode not selected by the servo control unit. Figure 8 again shows the servo control unit and the error monitor and slave selector panel, but it shows it in relation to the ball tracker which is just in front of the operator.

This ball tracker, as previously mentioned, permits simultaneous positioning of

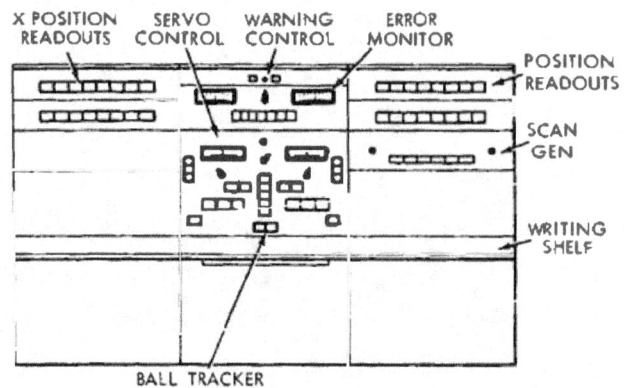

Figure 7—Servo control panel.

both axes of the antenna in the manual mode of operation. In the slew mode, the ball tracker permits variations in axis speeds. Figure 9 shows the complete operator's station. The servo control unit is in the center. X and Y position readouts are on either side of the control unit. Just below the Y position readout is the scan generator. Next to the servo console is the TV monitor and camera equipment. Next to it is the servo rack housing the various amplifiers and other electronic components.

SYSTEM PERFORMANCE

Figure 10 shows a rough schematic of the servo system. The antenna mount is represented by the right-hand block. The antenna axis is driven from the hydraulic system through a gear box. Each antenna axis is provided with a synchro transmitter — actually, there are two, a coarse and a fine transmitter, but for simplicity only one has been shown. Furthermore, RF electronics are provided which generate servo error

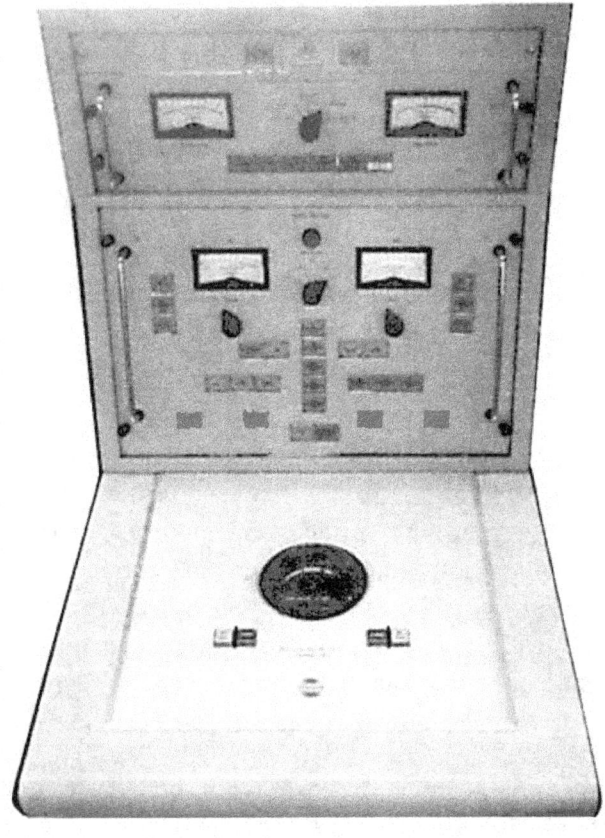

Figure 8—Servo control unit in relation to ball tracker.

Figure 9—Complete operator's station.

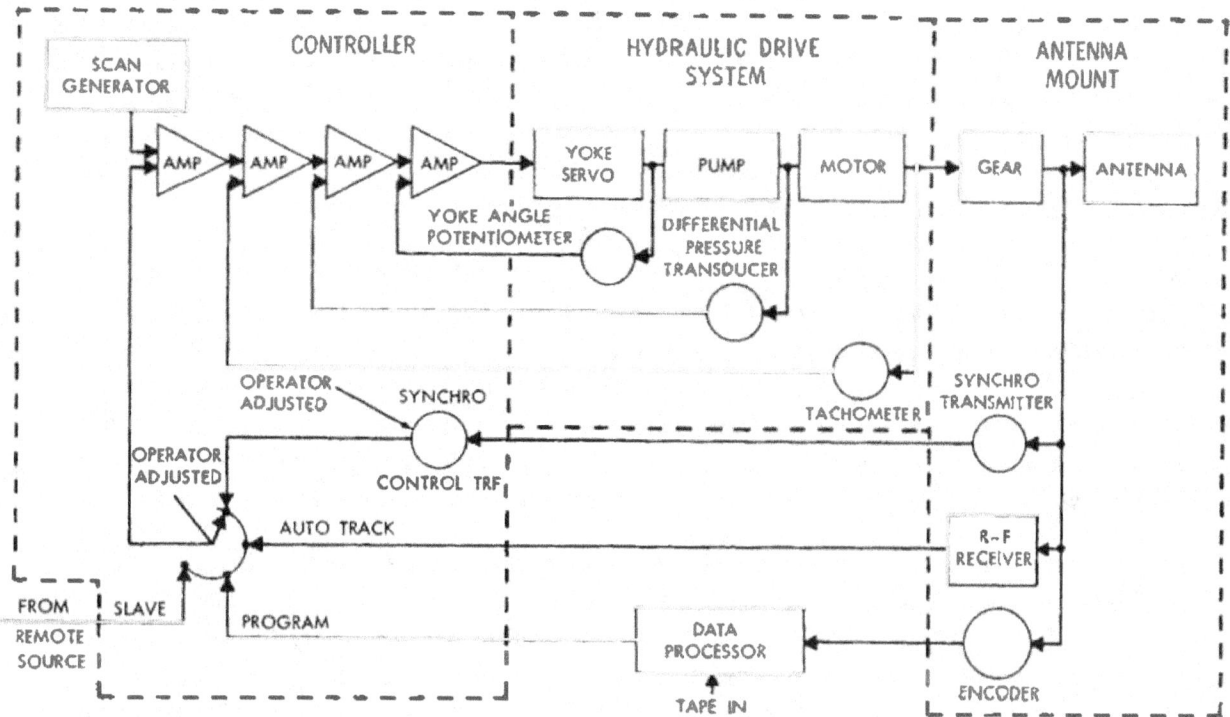

Figure 10—Schematic of servo system.

signals in the tracking modes, and each axis has a digital position encoder. The hydraulic system consists, as previously stated, of the motors, the pump, and the yoke servo with their respective feedback elements. The controller is made up of several operational amplifiers and switching relays.

The system utilizes three minor servo loops for stability and linearization purposes. These are the yoke loop, pressure feedback loop, and the velocity loop. The main, or position, loop is closed around equipment that depends on the various modes of operation. In the manual mode, the operator adjusts the ball tracker, which is coupled to a control transformer. This transformer compares the position of the mount to its shaft position and generates an error signal which will move the drive system and therefore the mount in a direction to null this error signal. During autotrack, the RF receiver acts as a position transducer and is used to close the loop. In the program mode, the encoder output is compared to coordinates on a prediction tape, and an error signal is generated which again is fed to the servo system. Similarly, in the slaved mode, an error signal is applied to the system.

The various amplifiers in the controller have been compensated to give the proper frequency response for the various loop gain conditions and the servo bandwidth requirements. System accuracy is a direct function of loop gain, bandwidth, and the type of servo system used. To realize the tracking accuracy in the autotrack mode, a Type II servomechanism is used. This type of system reduces velocity errors to zero, and one has to contend with acceleration errors only; however, this type of a system is more difficult to stabilize than a Type I system, which is utilized in all other modes of operation. Bandwidth is dictated by target dynamics, wind spectra, and noise considerations. Normally tracking low altitude satellites in windy environment requires a bandwidth of about 1 cycle per second. Satellites far out in space have a very slow apparent motion and therefore can use lower bandwidth. Lower bandwidth is especially desirable from a noise standpoint because RF thermal noise increases with satellite distance due to poorer signal-to-noise ratios. To accommodate these conditions, a variable bandwidth switch has been provided, permitting servo bandwidth selection by the operator. Bandwidths between 0.12 cycle per second and 1.0 cycle per second are available.

Switching to a lower bandwidth may, however, not be justified for a distant satellite target when tracking occurs during windy conditions. Even though target dynamics could use a narrow servo bandwidth, varying winds require a wide bandwidth. To accommodate these contradictory requirements, the autoprogram mode has been provided. This combination mode uses narrow bandwidth tracking information for following target dynamics and uses wide bandwidth program information to reduce wind effects. Computer results have indicated that definite improvement in operation can be expected; however, this type of operation has not been field-tested as yet.

The shipboard antennas, the 30-foot dishes on injection ships, and the 12-foot dishes on re-entry ships, have basically the same type of servo and drive system. The mount has an Az-El configuration. The drive system must have the capability of not only following a target but also of stabilizing the mount against roll and pitch of the ship. This requires antenna velocities of 50 degrees/second and accelerations of 50 degrees/second^2. Stabilization against ship's motion is accomplished by use of rate gyros on the mount and by utilizing information derived from the ship's inertial navigation system. Contrary to the land-based antennas, an electric drive system is provided which consists of a torque motor and an amplidyne connected in a Ward-Leonard loop. The advantages of torque motors are that they do not require gearing and consequently eliminate backlash. Also, these motors display a large dynamic range which cannot be duplicated with an ordinary dc motor. Torque motors permit a compact design which influences favorably the nautral frequency of the structure. At present, a natural frequency of 10 cycles per second is anticipated. These systems have essentially the same modes of operation as their land based counterparts. Their tracking accuracy will also be 1.5 minutes of arc.

ANTENNA FEEDS AND ACQUISITION ANTENNAS

by
J. Flowers
Goddard Space Flight Center

ABSTRACT

This paper presents the history and a technical description of the Apollo Cassegrainian Feed System and Acquisition Antenna. Characteristics of the feed and acquisition antenna systems are discussed, including design considerations, configuration, constraints, parameters, and interfaces with the Cassegrainian feed system and the more conventional focal-point feed system. Discussion includes the 30-foot parabolic dish antenna, the shipboard 30-foot antennas, and the feed and acquisition antenna systems for the projected 85-foot dishes. Also described are typical receiving patterns, the method of TE_{12}/TM_{12} mode excitation, efficiency factors, and various packaging problems.

INTRODUCTION

The Cassegrainian feed system configuration was decided upon because it is better able to carry the complexity of equipment required to be located in proximity to the feed. A simple adaptation of already proved designs was made a firm requirement early in the design of the Cassegrainian feed. The feed design is a basic four-horn monopulse in which the communications channel is formed by summing the four horns. The E-plane distribution is altered by higher order modes, generated in side-wall launchers, to produce equal E- and H-plane illuminations. Simplicity of the feed design is further enhanced by limiting available polarizations to right-and-left circular, remotely selectable. A diversity communications channel is available but unused.

An early decision in the design of the acquisition antenna placed this unit on the periphery of the 30-foot dish; with further study it was determined that for an X-Y mounted antenna the apex of the quadrapod was the more desirable location from both an RF and a mechanical standpoint. The acquisition antenna has a simple four-horn receive only feed. Any of four polarizations is available, remotely selectable.

CASSEGRAINIAN FEED SYSTEM

The Cassegrainian feed assembly serves as the illuminating system for the 30-foot parabolic dish antenna, and in due course, will serve likewise for the 85-foot model. In this discussion such terms as "illuminating" and "radiator" are used for the sake of simplicity in describing functions of the feed, irrespective of its use as a transmitting or receiving feed, even though a receiving feed does not illuminate a reflector antenna, in a true sense.

This assembly consists of the hyperboloidal subreflector and the feed cone or feed housing, which contains the feed, composed of the aperture horn, mode control sections circular polarizer, orthogonal mode junctions, comparators, polarization switches, and the filters and diplexer.

The feed system is the connecting link between the physical antenna structure and the unified S-band transmitters and receivers. By the very uniqueness of its position in the overall system, the feed system, like the antenna structure, must have near 100 percent reliability, as it is virtually impossible to provide redundant circuits which may be quickly switched into the system.

To the end of achieving reliability we chose a design approach which combined good microwave engineering techniques and proved, established principles to create a simple and dependable feed system. At the same time we felt constrained, by the complexity and weight of receiving equipment which was required to be located in the immediate vicinity of the feed, to introduce a small element of additional unreliability by the use of a Cassegrainian configuration.

The advantage of the Cassegrainian system over the more conventional focal-point system is apparent in Figure 1, which illustrates in simplified sketches the principal difference between the two. The focal-point feed illuminates the main reflector surface directly, and its total weight plus signal and control cabling must be supported by the quadrapod legs, whereas the Cassegrainian system uses a passive, secondary reflector in front of the focal point, permitting the active feed components to be mounted close to the main reflector surface. The author places the "unreliable" label on the Cassegrainian system only because there is an additional item of equipment to sustain damage, or to become misadjusted; in reality it is highly probable that a Cassegrainian system will prove more reliable, as environmental protection capability is enhanced by the closed, weather-tight wheel-house and cable runs for RF, power, and controls are shortened and simplified.

DESIGN PARAMETERS (30-FOOT DISH)

To summarize the pertinent basic design parameters of the 30-foot dish microwave subsystem: The feed assembly will receive data in the 2270- to 2300-megacycle band with a minimum gain of 44.0db, corresponding to an overall efficiency of 53%, including I^2R losses of less than 0.5db. Monopulse sum and error signals of comparable gain are provided to the tracking receiver in this band as well. Transmission of up to 20-kilowatt RF power, at a minimum of 43.0db gain, is possible over the 2090- to 2120-megacycle band. The feed system receives and transmits only circular polarization with remote switching capability. Receiving and transmitting circuits are switched simultaneously, with the primary data-receiving output, as well as the monopulse tracking signals, being of the same polarization sense as the transmitted

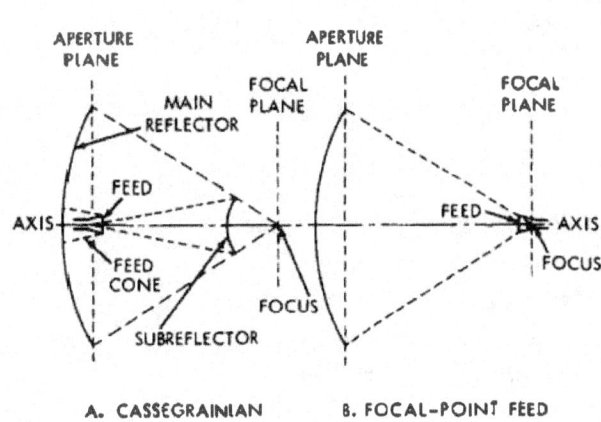

Figure 1—Antenna feed systems.

signals. Very close attention has been paid to the distribution and symmetry of phase and amplitude in the feed system, providing a clean sidelobe structure and a sharp, stable tracking-signal output.

DESIGN PARAMETERS (85-FOOT DISH)

The feed and acquisition antenna systems for the 85-foot dishes have not yet been developed. They will be electrically similar to those of the 30-foot dish; with less of a packaging problem being posed by the larger feed cone of the 85-foot structure, the feed system can be further simplified, particularly with respect to ease of assembly and disassembly. The primary difference between the two subsystems, of course, is that in the 85-foot dish the feed will develop gains of 50.5db receiving and 50.0db transmitting. Better sidelobe control will also be possible. Except for these differences, what is said here applies as well to the 85-foot feeds and acquisition antennas as to the 30-foot equipment.

SHIPBOARD ANTENNAS (30-FOOT DISH)

Also, in a general sense, the basic parameters given here for the feed system of the ground station 30-foot unified S-band antennas apply as well to the shipboard 30-foot antennas, built for the prime contractor, Reeves Instrument Company, by Hughes Aircraft Company in Fullerton, California. The basic technical difference is in the method of injection of higher order modes which are described later. The designer of the shipboard 30-foot feed system had an easier packaging problem than did the Rantec engineers, due to more space being available in the back-up structure of the Az-El mounted shipboard antennas than in the ground station structure.

UNIFIED S-BAND FEED SYSTEM

The basic aperture components of the unified S-band feed system (produced by Rantec Corporation in Calabasas, California) are the four square waveguide horns, disposed two opposing two to develop the error signals in each plane (Figure 2). These are represented by the four parallel lines. From this point back the system is a simple, classical four channel-waveguide monopulse system. Problems were encountered in packaging the components within the feed cone

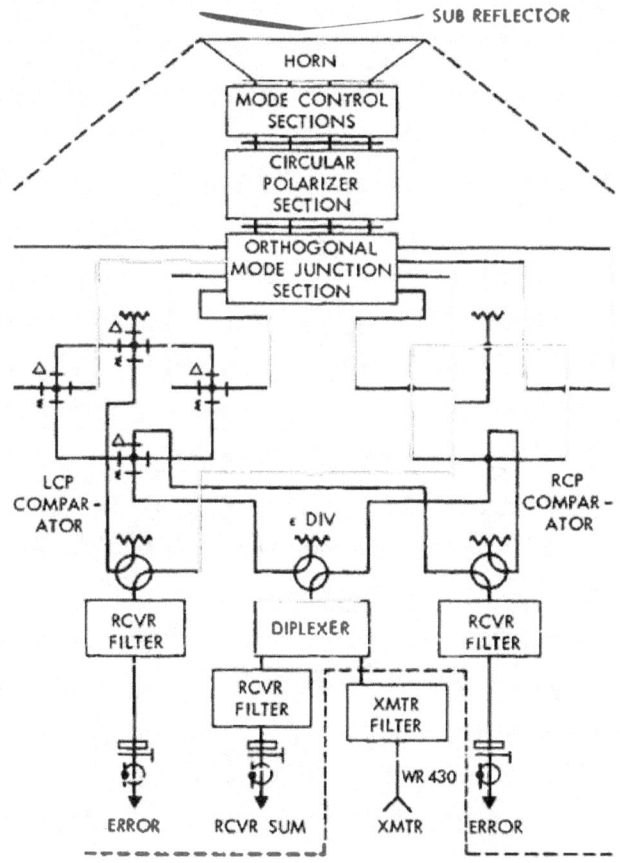

Figure 2—Cassegrainian feed system, block diagram.

(shown in outline here by the dashed lines), approximately 7 feet tall by 3-1/2 feet in diameter at the base, to the extent that the original design goal of packaging entirely within the feed cone area was not achieved. However, observing the packaging design as it now stands would lead one to the conclusion that the spilling-out of microwave equipment from the base of the feed cone is intentional, as the input filters of the feed extend down or back into the Y-wheel house to a very convenient point for short-cable connection to the preamplifiers.

The four square waveguide horn outputs lead into the orthomode junction section in which the orthogonal circular polarizations are extracted. These are operated upon by the proper combination of magic-T hybrids and transfer-switch positions to give sum and error channels of remotely selectable right or left circular polarization. The transmitter input is diplexed from the primary sum output, hence is of the same sense circular polarization as the primary receive-sum output, which is also the tracking-reference channel, and the same sense as the tracking-error channels. A diversity receiving-sum output, of the opposite sense circular polarization to the primary, is available but unused in the present system. Better than 190db isolation is obtained between the transmit and receive ports in the transmitting frequency band; and better than 165db isolation to any spurious signals generated by the transmitter in the receiving frequency band.

Outward from the basic four-horn monopulse aperture the feed system becomes what has come to be called a "multimode" horn. The multimode portion of the feed comprises the sections at the top of Figure 2 and is shown schematically in Figure 3.

The effect of the multimode action is to operate upon the amplitude distributions (shown at the four-horn aperture to the left in Figure 3) to produce the distribution illustrated at the right. The final result is that the H-plane distribution is unaffected, and the E-plane is modified to be essentially identical to the H-plane, leading to higher efficiency and improved side lobe control.

RANTEC FEED

The Rantec feed differs from others in this class which have been described in the published literature in the method of TE_{12}/TM_{12} mode excitation. Section B-B is a section through the main square waveguide beyond the four waveguide aperture, and includes a section through one of the sets of higher mode exciters, which are essentially four auxiliary waveguides, shorted at their far ends. When the electric field is as shown, the upper and lower auxiliary waveguides are excited. Being of very small axial dimension, the mouths of these auxiliary

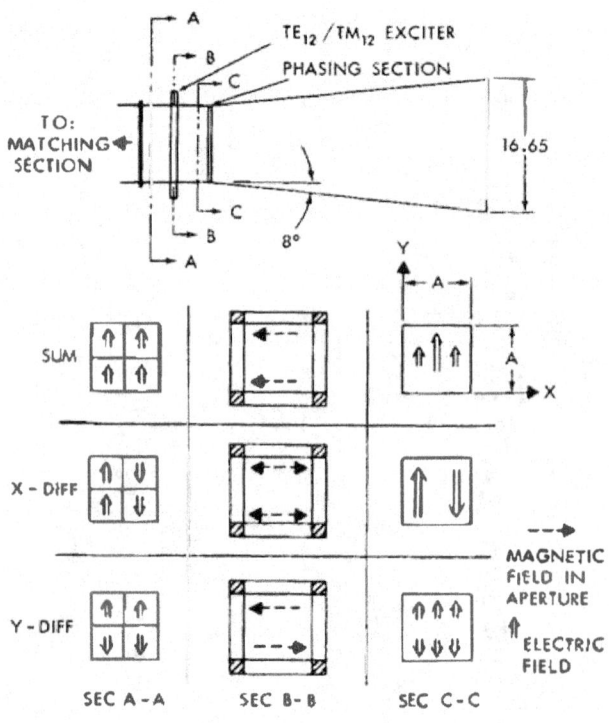

Figure 3—TE_{12}/TM_{12} exciter.

waveguides act as magnetic line sources. For the sum mode, these line sources couple to the TE_{10}, TE_{12}, and TM_{12} modes, plus higher modes which are prevented from propagating by choice of the main waveguide dimension A. Essentially only the TE_{10} and the TE_{12}/TM_{12} modes are present in the throat of the horn. Two independent and essentially non-interacting exciter sections are used in series, the parameters of each being chosen so that one optimizes the phase and amplitude relations for the 2270-2300-megacycle band, and the other functions likewise for the transmit band.

When the feed is operated in the X-difference mode, the higher mode exciter couples to the higher order modes. However, these modes do not propagate in the main square waveguide size chosen. In the Y-difference mode, the exciter section couples to the TE_{11}/TM_{11} modes, which are the desired modes already launched by the phasing of the four-waveguide sections. Other modes are again cut off by the choice of the main square waveguide size.

The resulting aperture distributions approach the ideal, with the exception which is common to all orthodox monopulse systems; for an optimum amplitude taper across the dish in the sum modes the difference mode tapers are too low, resulting in high difference pattern sidelobes (in the order of -15db). The phasing section and horn are designed to cause the TE_{20} and the TE_{11}/TM_{11} modes to be phased to maintain the orthogonal phasing generated by the circular polarizer.

RECEIVING PATTERNS

A set of representative sum and error receiving patterns of the 30-foot dish is shown in Figure 4. These are hand transcribed from data recorded at the experimental site at Dallas, Texas; our instrumentation was not the best and the site is far from ideal for an exhaustive evaluation of a large aperture antenna. Ground reflections were an obvious problem. Nevertheless, a sufficient number of our patterns recorded on this poor range were similar to those recorded on the shipboard 30-foot dish at the relatively ideal Carbon Canyon range to give us a measure of confidence in the results. It must be pointed out, however, that the sidelobes in all planes about the antenna axis will not be as good as shown here; in some instances the first sidelobes in the sum pattern are as high as -17.5db below the sum pattern peak.

TRANSMITTING PATTERNS

The transmitting patterns of the 30-foot dish have not yet been evaluated satisfactorily at Dallas due to instrumentation problems. Indications are, in the preliminary data taken, that these will show somewhat higher close-in sidelobes than the receiving patterns. This high energy content out to some 5 degrees will actually be used to advantage as the

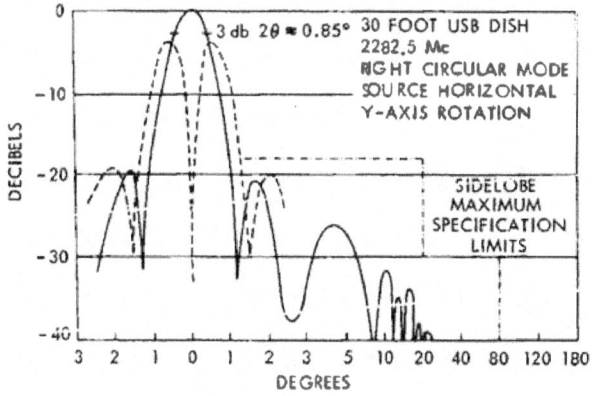

Figure 4—Typical receiving patterns.

acquisition-transmission mode, which will be discussed in this symposium by Mr. R. H. Newman.

ACQUISITION ANTENNA SUBSYSTEM

The acquisition antenna subsystem was developed and built by the prime contractor of the unified S-band system, Collins Radio Company. It was our intention at the outset of the program to mount the acquisition antenna on the periphery of the main dish, but a study made by Collins Radio in cooperation with Blaw-Knox convinced us that for an X-Y mounted antenna, the apex of the quadrapod was the more desirable location, from both an RF and a mechanical standpoint. From the RF viewpoint, the apex-mounted antenna maintains a more uniform earth-to-antenna relationship, irrespective of the direction of pointing, than does a peripherally mounted antenna on an X-Y mount. Isolation between the main dish, transmitting, and the receiving acquisition antenna is essentially equal in either location. The mechanical analysis showed that the deflection of the subreflector remained within the specified limits of 0.05 inches; this has been substantiated by measurements at Dallas. Furthermore, the deflections are more uniform than the off-center deflections produced by the periphery location.

The acquisition antenna has a simple four square waveguide horn feed, as is shown in Figure 5. Orthogonal linear polarizations are extracted from the square waveguides through probes, and are carried through the rest of the circuitry in type-n/coaxial components. The switch-hybrid packages are modularized components manufactured by Ramcor, and are sandwiched, with the comparator package, in the space between the acquisition dish and the subreflector. From this network one sum and two error channels, of any linear or circular polarization remotely selectable, are fed back through coaxial lines to the acquisition preamplifiers. Three waveguide filters identical to the receive filters used in the main feed system are inserted into these lines, and are mounted in the back-up structure of the main dish.

For the 30-foot dish the acquisition antenna is a 42-inch diameter paraboloidal dish of 0.4 f/d, with a beam width of approximately 10 degrees and a minimum of 22db gain over the receiving band of 2270-2300 megacycles. For the 85-foot system, the acquisition antenna will be proportionately scaled.

ANTENNA FEEDS AND ACQUISITION ANTENNAS

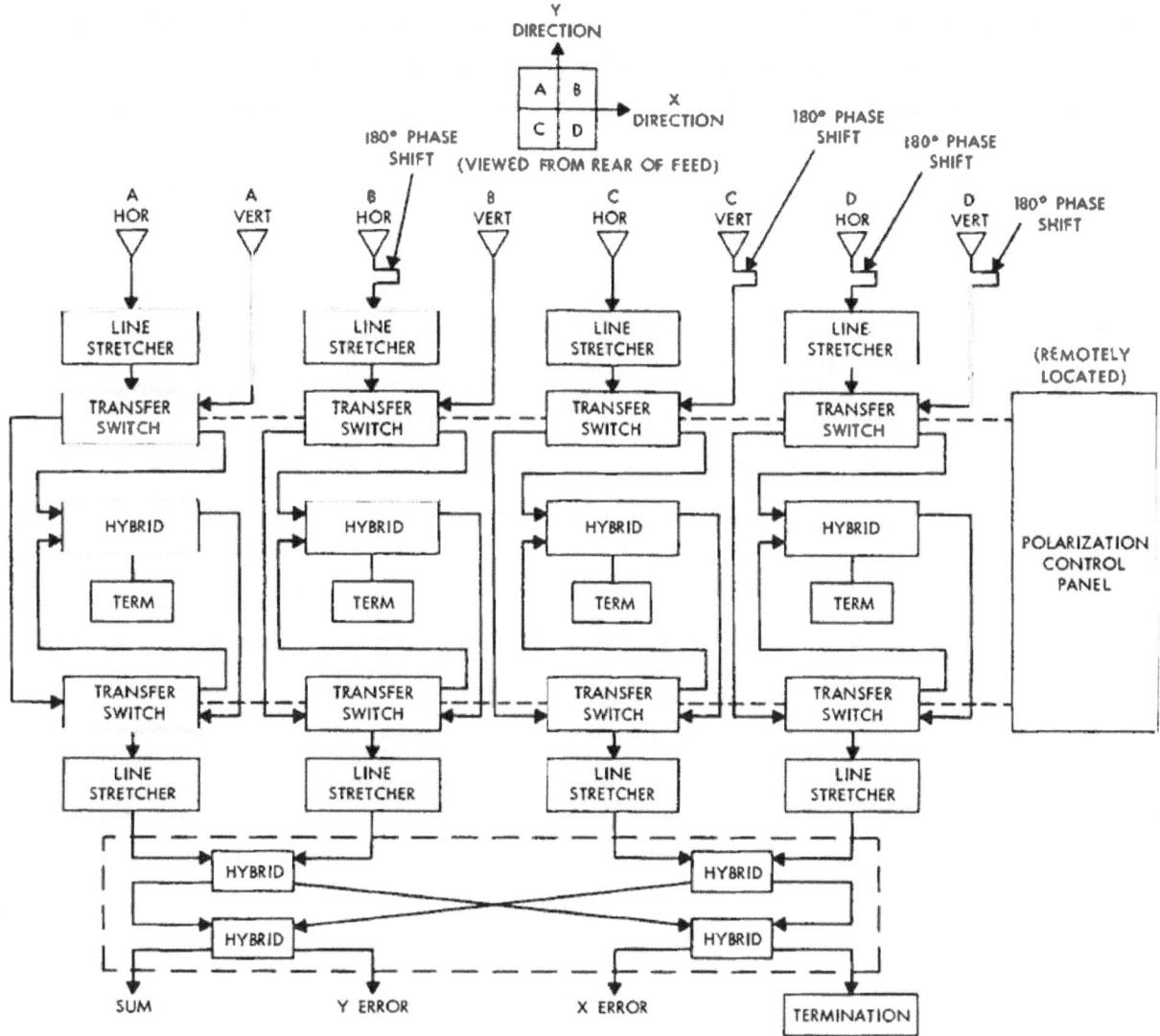

Figure 5—Acquisition antenna, block diagram.

PARAMETRIC AMPLIFIER, AND NOISE FIGURE AND TEST SIGNAL NETWORK

by
J. B. Martin
Goddard Space Flight Center

ABSTRACT

Parametric amplifiers are used to provide a low system noise temperature for both the main tracking and acquisition antennas. Identical units are used to simplify maintenance and allow substitution in an emergency. A noise temperature of 170°K degrees is achieved without cooling. The units are housed in weatherproof, pressurized containers to allow mounting without weather protection.

It is important to determine system readiness for operation without disabling the equipment. The Noise Figure and Test Signal network is designed to measure noise figure or inject test signals while the receivers are connected to the antenna terminals. This arrangement also allows more realistic tests of signal threshold since the effects of antenna temperature and sky noise are included in the measurement.

INTRODUCTION

A large variety of techniques and devices is used to calibrate a tracking system prior to its operational use. Typical examples which have already been discussed are the method of aligning surface panels and the use of airplanes to calibrate all signal-receiving subsystems. It is necessary, in a complex system such as this, to perform daily preoperational tests to assure the operator that the equipment has been set up properly and is working to expectations.

Two subsystems—the Boresight Equipment, and the Noise Figure and Test Signal network—are used in performance of these tests. The Boresight Equipment is located on a remote tower and is effective only when the dish can be pointed in the direction of the boresight tower. The Noise Figure and Test Signal network is located near the feed on the dish and may be used with the antenna in any position.

NOISE FIGURE AND TEST SIGNAL INJECTION SUBSYSTEM

The Noise Figure and Test Signal Injection subsystem is used to measure receiver noise figure (which includes the parametric amplifier) or to inject a variety of test signals into the receiving equipment. This enables the operator to verify that the system has the proper sensitivity and that the data demodulator and data handling equipment will operate properly with the receiver. A primary goal in the design of this subsystem has been flexibility: it may be used with the antenna in any position and it allows the use of a wide variety of static or dynamic test signals.

Figure 1 shows the functional block diagram of the Noise Figure and Test Signal Injection subsystem. The receivers are included, since this subsystem connects into the receiver both at the input and output. Note the division drawn between the antenna-mounted equipment on the left and the control room equipment on the right. The control assembly located in the control room turns the network on, determines what signals will be injected and which receiver channels will be measured. We can, with the test transmitter, inject a CW or a phase-modulated signal through the network assembly to selected receiver inputs. Note that the test inputs are shown in parallel with the inputs from the feed. The test signals are injected into the receiver without disconnecting the receiver from the antenna. This is done for two purposes: first, we may be sure when the test signal is turned off that the receiver is ready to operate; second, this enables a test of system threshold and system noise figure to be made under conditions which are very realistic because all noise from the antenna is present in the system at the time the test is made. If we wish to inject a different type of signal, the test transmitter may be disconnected and a special signal inserted. An example of an alternate source is a sweep generator for checking the portion of the receiver which is mounted on the antenna.

It was mentioned previously that parametric amplifiers are mounted on the antenna. They are used only on the sum channel of the main and acquisition receivers, because the error channels in this application do not need the low noise figure. In addition, the first receiver

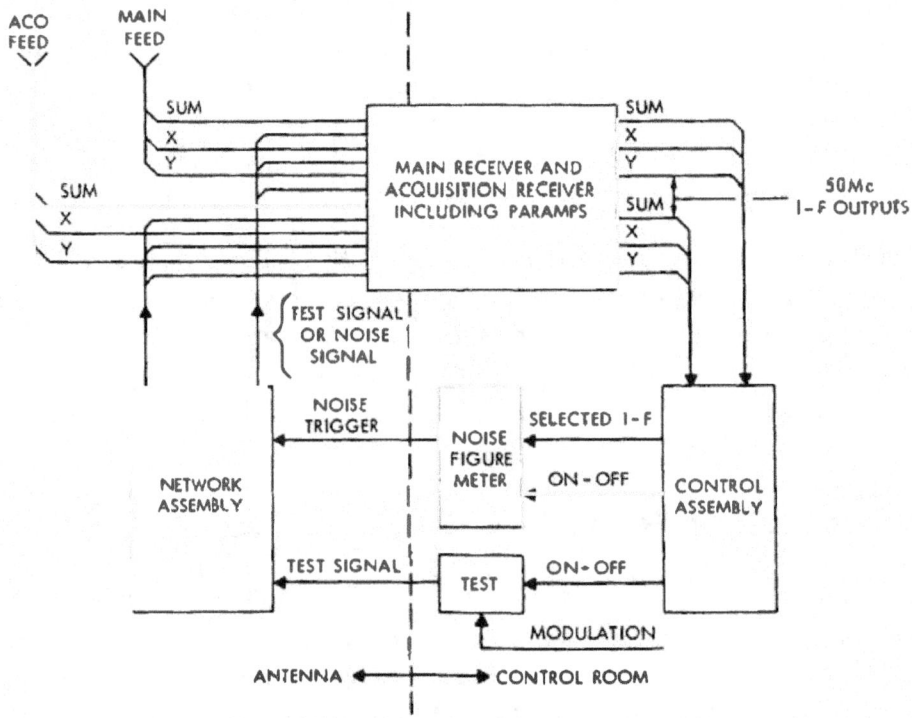

Figure 1—Functional block diagram of noise figure and test signal injection subsystem.

frequency conversion is performed on the antenna. The total amount of antenna-mounted equipment is quite significant. Thus the need for noise figure and test signal injection on the antenna becomes quite clear. This network measures noise figure to an accuracy of about 1/2db on the sum channel and about 1db on the error channels. This is not intended to be a precise measurement. The primary purpose in this case is a measurement of relative accuracy which can be repeated from day to day to obtain a trend of system deterioration.

Figure 2 shows the basic concept of the process of "automatic" noise figure measurement. At the input of the receiver three noise signals may be present: noise which is due to the antenna, noise which is due to the receiver, and noise from a noise source. For this purpose, a noise source located on the antenna is alternately fired on and off by the noise figure control circuitry in the operations room. When it is fired on, its output is added to the total noise present at the receiver input. This is amplified through the receiver circuits and fed to the noise figure indicator which adds gain to the signal to produce a constant amplitude (N_2) in the noise figure indicator.

When the noise source is turned off, the gain of the indicator is held constant and the amplitude of noise left (N_1) is a measure of the total noise present in the receiving system. This amplitude is displayed on the noise-figure meter as an indication of noise figure. Examples are shown for both a low and a high system noise figure. Both cases have the same amount of antenna noise and noise-source noises but a different amount of receiver noise. The amount of noise added to N_1 is the same for both cases. The magnitude of the ratio of N_2 to N_1 is inversely proportional to system noise figure. In other words, the larger the ratio of N_2 to N_1, the lower the system noise figure. It is also evident that the noise-figure meter not only displays the noise figure of the receiving channel but also includes the total antenna noise as a part of the measurement.

Figure 3 shows a view of the noise figure and test signal network equipment. The lower panel is the control panel with its single switch. With this switch we may measure noise figure or insert a test signal into either the main channel or the acquisition channel. Noise figure is measured individually for each receiver channel, but the test signal is inserted into the three channels (sum, X, and Y) or either receiver simultaneously. This panel is 3-1/2 inches tall by 19 inches deep. The upper panel is the noise-figure indicator. The reading of noise figure is displayed by the meter movement. The controls are for operation of the noise-figure meter. This panel is 5-1/4 inches tall by 16 inches deep. The network assembly, which is the part of this subsystem located on the antenna, will be shown later with a

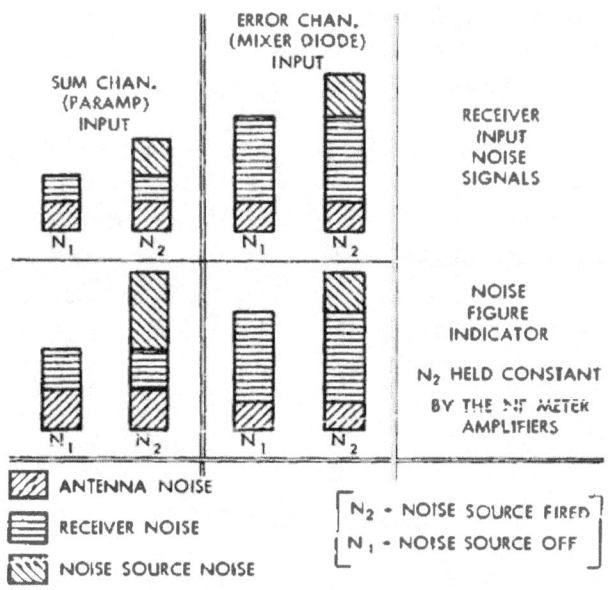

Figure 2—Basic concept of automatic noise figure measurement.

view of the parametric amplifier. That panel is 10-1/2 inches tall by 25 inches deep. It contains coaxial switches, directional couplers, signal equalizers, and the like and is actually the place where the noise-figure signals or the test signals are routed to their intended destinations. The test transmitter will be described as part of the JPL equipment.

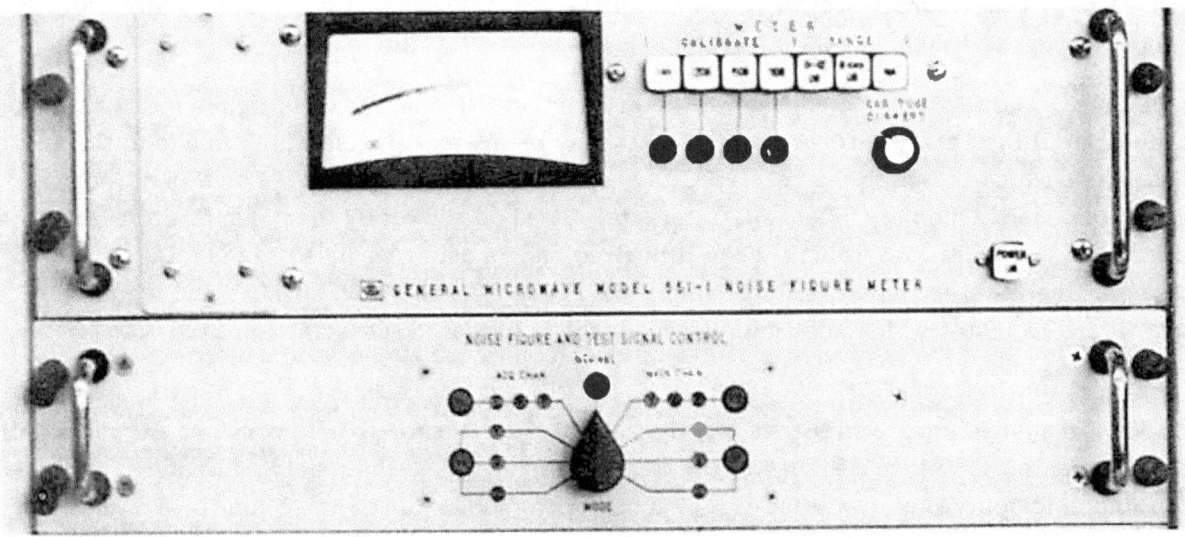

Figure 3—Noise figure and test signal network equipment.

PARAMETRIC AMPLIFIER

The purpose of the parametric amplifier (paramp) is to provide a low system noise temperature constrained by such things as the necessity to produce equipment that will be reliable under widely varying field conditions. This equipment will not be operated in a laboratory by engineering personnel but will be operated under field conditions which are not ideal and by people who perhaps are not ideal. The noise temperature must be as low as practical within the constraints of reliable performance, reasonable cost, and required sensitivity.

To provide a little background, a discussion of system noise temperature is presented. Although simplifications have been made for the sake of clarity, the conclusions are accurate to within a very few degrees. In this discussion, all noise present is considered as noise temperature. Noise temperature is a convenient means of stating the noise power present in a unit of bandwidth. The total noise power may then be determined from the expression:

$$P_n = T_n K B_n$$

where

P_n is the noise power,

T_n is the noise temperature (°K),

K is Boltzmann's constant,

and

B_n is the noise bandwidth.

System sensitivity may be easily determined by assessing the effect of each system component and adding all effects to obtain the total temperature. The concept of noise figure can be confusing in computing system performance because noise figure presupposes a source temperature of 290°K. The db number commonly used to express noise figure cannot be generally applied to comparisons of system sensitivities.

Figure 4 shows a representation of the system including the antenna, the parametric amplifier, and the receiver circuits. The noise temperature will be measured at the paramp input. If the system is entered at that point and a measurement taken toward the antenna feed, antenna temperature will be determined. That antenna temperature will include feed losses, as well as sky noise, side-lobe noise, and so on. If the signal enters again at the same point and looks toward the paramp, the total effect of parametric amplifier and receiver noise temperature will be seen. The receiver will have an effect on the noise temperature at the input of the system, but the contribution to input-noise temperature will be divided by the gain of the stages that precede it.

Antenna temperature will depend upon the position of the antenna. If the antenna is pointed toward zenith in a quiet section of the sky, the temperature will be lower than if it is pointed toward the horizon. A temperature of 30°K is expected for the 30-foot antenna when it is pointed near zenith. This temperature would increase to about 185°K when the antenna is pointed at the horizon. If there is a discrete source of noise that falls within the antenna beam width. This will tend to raise the antenna temperature. If the antenna is looking toward zenith in the quiet sky but with the moon in the field of view, the temperature is raised from 65°K to about 83°K. Under all of these conditions, the noise temperature of the equipment following would remain a constant. The total of paramp noise temperature plus receiver noise temperature would be 168°K in each case. This figure results from a paramp having a noise figure of about 1.7db followed by a receiver having a noise figure of about 10db when the paramp has a net gain of 20db. All temperatures may be added to get a total system temperature of 233°K for the quiet sky, 353°K at the horizon, and 251°K if the moon is in the field of view.

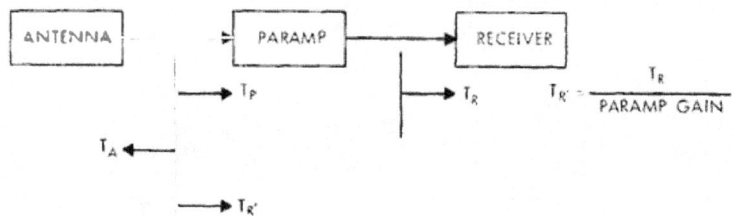

ALL LINE LOSSES ARE INCLUDED IN TEMPERATURE $T_S = T_A + T_P + T_{R'}$

ANTENNA POINTING	30 FOOT ANTENNA TEMP (°K)	PARAMP & RCVR TEMP (°K)	SYSTEM TEMP (°K)	COOLED PARAMP & RCVR TEMP (°K)	COOLED SYSTEM TEMP (°K)	IMPROVEMENT FACTOR (db)
ZENITH QUIET SKY	65	168	233	35	100	3.7
HORIZON	185	168	353	35	220	2.1
ZENITH QUIET SKY WITH MOON IN VIEW	83	168	251	35	118	3.3

Figure 4—Antenna, parametric amplifier, and receiver circuits.

When the system was originally designed it was expected that the system temperatures stated here would be sufficient to meet the need for the Apollo program. Lately it has come to light that the spacecraft in some attitudes will not be quite what we expected, and there will be conditions where system temperatures as originally specified will be marginal.

There are means that we can use to improve the system temperatures. It can be seen that the paramp temperature is considerably larger than the antenna temperature from a comparison of the two. We can lower the paramp noise temperature by employing a cooled paramp. Typical numbers on the right-hand part of the chart show a cooled-paramp-plus-receiver temperature of 35°K. This compares with the previous figure of 168°K. If the temperatures are added as before, a cooled paramp would provide system temperatures of 100°K, 220°K, and 118°K. Comparing the cooled-system temperature with the uncooled-system temperature indicates an improvement of 3.7, 2.1, or 3.3db. In other words, employing a cooled paramp in this application could improve system sensitivity by approximately 3db for the average condition. This improvement is bought for a price. That price is a complexity of the parametric amplifier which would be two to three times that of the present unit.

Figure 5 shows a block diagram of the parametric amplifier. The signal input is fed into a device which is referred to as a five-port circulator. It is really three, three-port circulators connected together and built as a single subassembly. A decoupled input is also included for injection of the test signal which was discussed a little bit earlier. This is a two-stage parametric amplifier which provides a total gain of 30db with a good degree of stability. A single-stage paramp could achieve 30db of gain, but at the cost of poor gain stability with change in time and change in temperature. Since this stability is important, the added complexity of the two-stage parametric amplifier is warranted.

The first stage of the parametric amplifier is shown on the left. The signal enters and exits the paramp through the same connection. The five-port calculator is the key to the

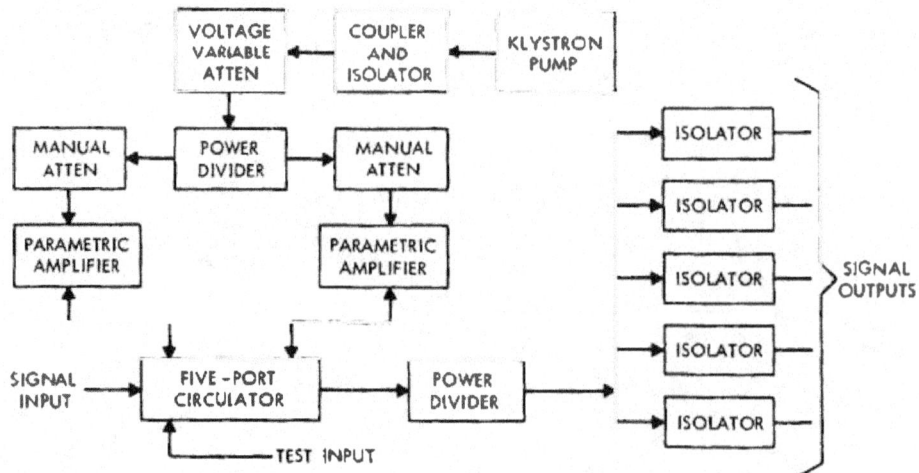

Figure 5—Block diagram of parametric amplifier.

successful operation of this paramp because energy that enters the circulator at the input will appear at the first output and at no other output (within reasonable limits, of course). The signal will appear at other outputs but will be attenuated by some 45 or 50db. Again, signals that enter at the first output port will exit at the second output port, enter the second parametric amplifier, experience gain, and be reinserted into the directional coupler to appear at the third output port. Energy fed into the test input will be combined with the normal signal input after being attenuated by the coupling loss of 20db.

The klystron pump provides the microwave energy needed to drive the parametric amplifiers. The output of the pump is fed through a coupler and an isolator and then through a voltage variable attenuator. The pump power that is actually inserted into each stage of the paramp may be adjusted from the control panel. The pump signal is then passed through a power divider and manual attenuators. These attenuators would be set as part of the alignment procedure. Bias controls for each of the parametric amplifier stages are also located on the remote control panel.

The signal that has passed through the two stages of the paramp is then fed through the power divider to five isolated outputs. These five outputs are provided so that more than one receiver at a time may be connected to the same parametric amplifier. The bandwidth of this paramp is about 30 megacycles. As such it can pass all the expected unified S-band signals. Isolation is provided so that the receivers will not interact with each other.

A typical noise figure which has been measured on the parametric amplifier is 1.68db, which would be an excess noise temperature of about 136°K. This measured noise temperature includes loss due to the input circulator. Further, this noise figure can be obtained with diodes which are of average quality. The diodes are tailored to the diode holders and the diode-plus-holder would be replaced as a unit in the field. The holders would then be returned to a central facility or to the manufacturer for outfitting with a replacement diode should the diode burn out. The gain stability of the parametric amplifier assembly has been measured to be 0.7db per day and this measurement was made while the environmental temperature outside of the paramp enclosure was varied from about 50° to 100°F.

Figure 6 shows a view of the parametric amplifier in its enclosure. This is a pressurized box. The top row of attachments includes the pressurizing connection, pressure relief valves which are set to prevent the pressure inside the box from exceeding twelve pounds per square inch, and a manual depressurization switch for use should it be necessary to disassemble the box for service. The bottom row shows the five isolated outputs, the test input, the signal input, the name-plate, and the power plug.

Figure 7 and 8 are views of the parametric amplifier with the cover off. Figure 9 is a view of the control panel. The two bias adjustments and the pump power adjustment which controls gain are visible. The writer was informed that this unit was photographed prior to acceptance testing, hence the running-time meter reading of 0000.0 hours. At the center of the panel are the diode current meters and the pump power monitor. The three enclosure-temperature

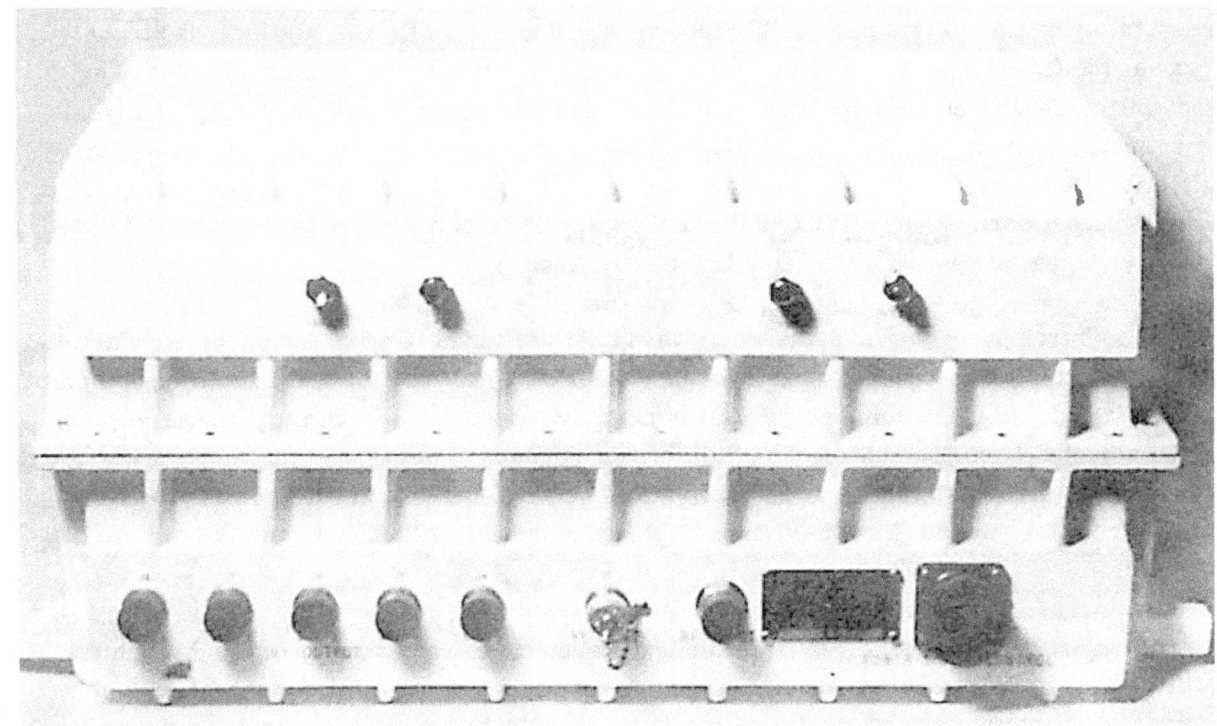

Figure 6—Parametric amplifier, enclosure attached.

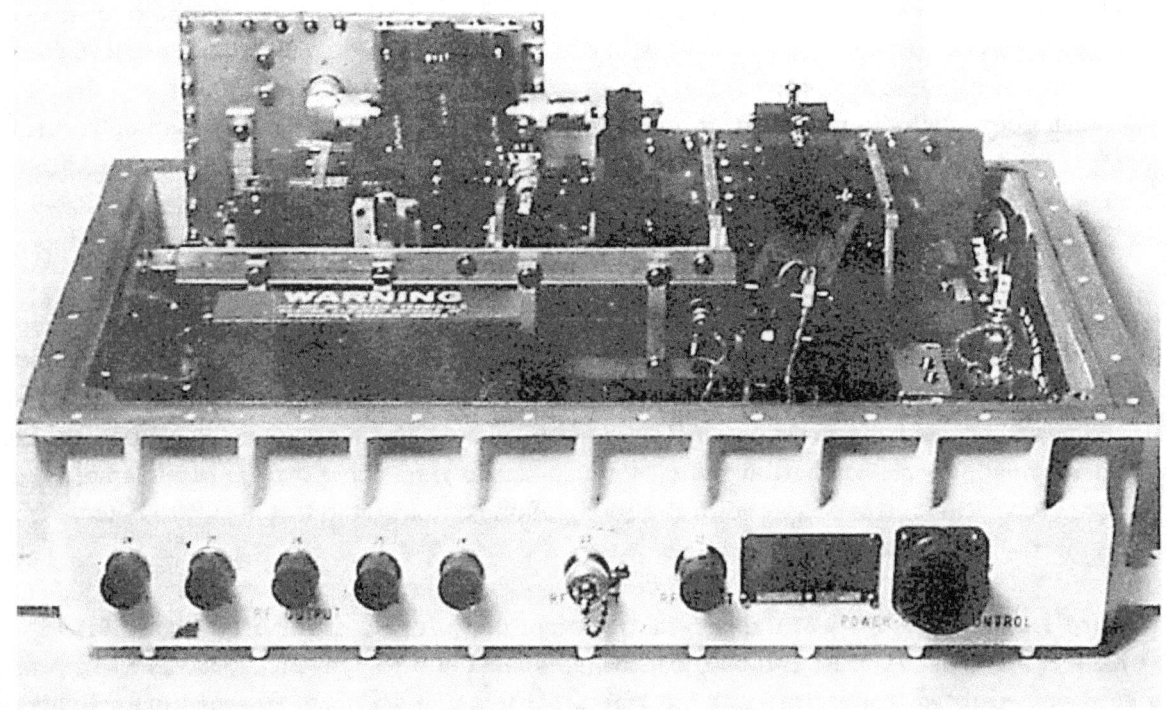

Figure 7—Parametric amplifier, enclosure removed.

Figure 8—Top view of parametric amplifier.

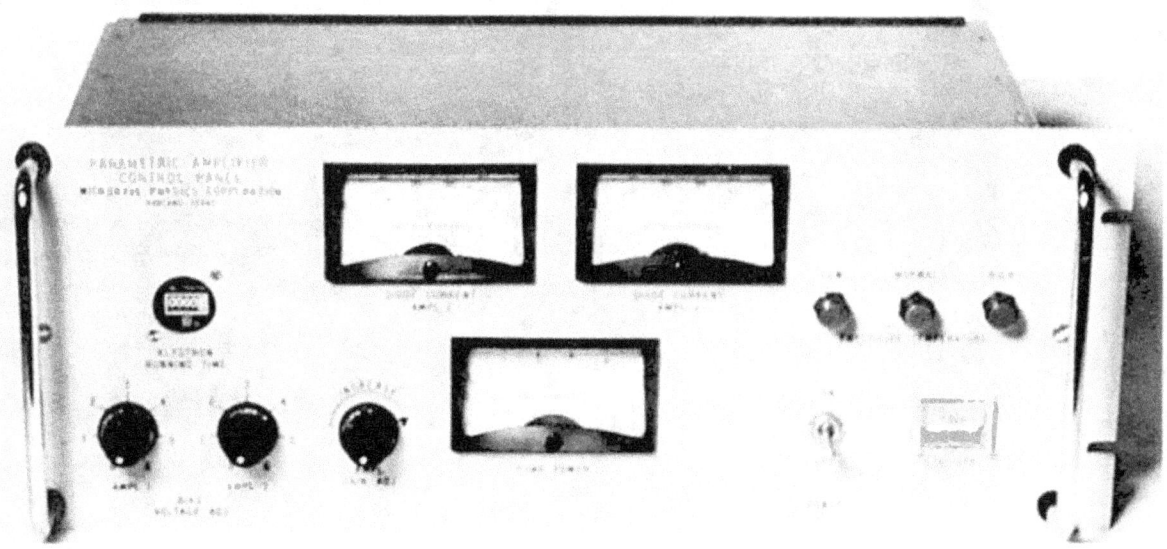

Figure 9—Parametric amplifier control panel.

lights show that the temperature in the antenna enclosure is either low, normal, or high. At the lower right are the ON/OFF switch and an OPERATE/STANDBY switch.

Figure 10 shows the parametric power supply, including the klystron bias monitors, reflector voltage adjustment, and fuses. Figure 11 shows the main channel paramp and the acquisition

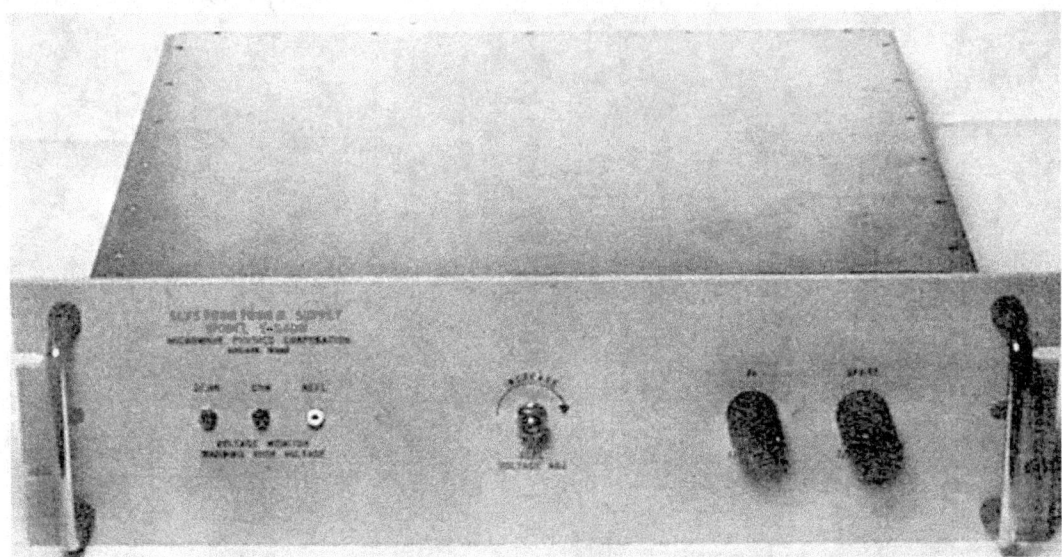

Figure 10—Parametric amplifier power supply.

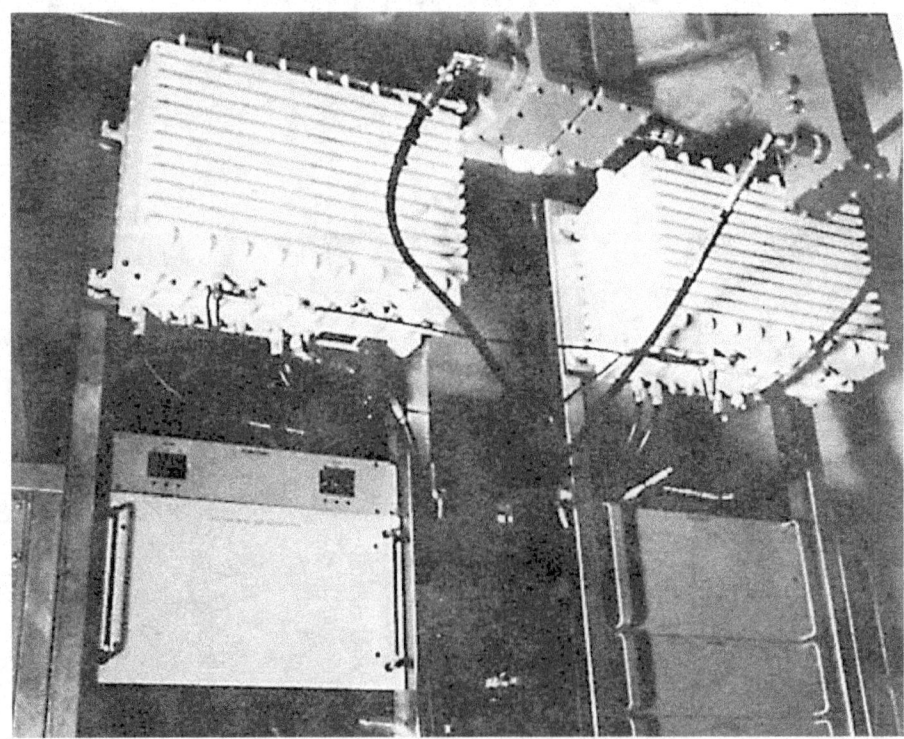

Figure 11—Main channel and acquisition parametric amplifiers mounted in wheelhouse of antenna.

paramp mounted in the wheelhouse of the antenna. The panel below the left paramp is the network panel. The panels to the lower right are part of the JPL receiver. The outputs from the feed which were discussed earlier can be seen at the top.

Figure 12 shows the antenna installation from the outside. Paramps, of course, are located in the top of the wheelhouse. Cables from the acquisition antenna come down the legs, enter the wheelhouse, and are fed to the paramp. Cables come out the side of the wheelhouse and pass over the axes of the antenna and down and into the operations building.

Figure 12—Antenna installation.

RECEIVER-EXCITER SUBSYSTEM

by

R. C. Bunce

Jet Propulsion Laboratories

ABSTRACT

The receiver-exciter subsystem, MSFN version, may be described as nine functional units interfacing with nine external subsystems. The prime interunit and interface signals, gross frequencies, levels, and functions are initially presented in diagram form.

Following the initial description, six functional block diagrams showing the mechanization within the functional units in greater detail are also presented.

Finally, photographs and diagrams showing equipment layout within the cabinets and views of the control panels are included, and functions of the important controls and indications are explained.

INTRODUCTION

In earlier NASA manned flight programs, several functionally independent systems using different frequency bands have been employed in the two-way spacecraft-ground links, resulting in highly complex facilities. However, in the Unified S-Band (USB) equipment for the Apollo program, most of the communications functions have been integrated, for the first time, into a single comprehensive capability. For example, all of the carrier frequencies in the two-way path are in S-band region (between 2100 and 2300 megacycles). Voice, television, telemetry data, range, range-rate, and antenna-tracking information may all be processed separately or simultaneously by the same radio frequency equipment.

Within the ground station facilities of the Manned Space Flight Network (MSFN), this unified concept is extremely evident in the receiver-exciter subsystem equipment. The subsystem acts as a link between the microwave equipment (such as the power amplifier and parametric amplifier) and the low-frequency RF, digital data processing, and dc actuated equipment. Information and reference signals from ten different external subsystems interface with the receiver-exciter equipment, which is, in essence, a focal point in the USB concept.

These basic interfaces are shown in Figure 1, together with the gross classifications of equipment within the subsystem. Only one of two identical receivers in the normal single configuration is indicated. The equipment is also supplied, for most stations, in the dual configuration. This configuration contains two complete receiver-exciter subsystems for redundancy and multiple-vehicle operation. For simplicity, only a single configuration will be discussed in this description.

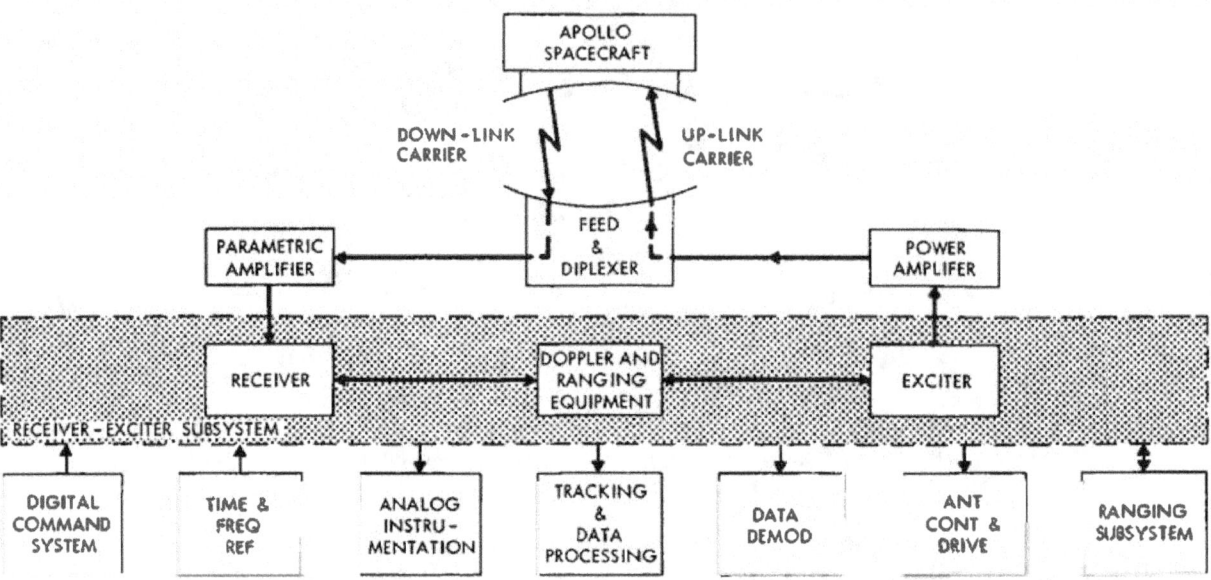

Figure 1—Receiver-exciter Unified S-Band system functions: doppler extraction, two-way communications, angle tracking and ranging.

Operation of the receiver-exciter subsystem within the Unified S-Band system can best be understood through a description of the following four major functional capabilities:

Doppler Extraction

The subsystem provides a signal whose frequency is proportional to the doppler shift occurring on the two-way transponded carrier. The doppler shift is a result of spacecraft motion with respect to the ground equipment.

Two-way Communications

The subsystem contains an S-band transmitter exciter that processes the up-data and voice modulation for the Apollo spacecraft, and also contains two functionally identical receivers that process the modulated received carriers from the Apollo spacecraft. The received modulation consists of spacecraft TV and data telemetry, as well as voice information.

Angle Tracking

The subsystem contains dual-channel angle receivers which operate in conjunction with the antenna feed and antenna control and drive equipment to form an antenna position tracking servo system.

Ranging

The subsystem contains a ranging receiver and other associated subassemblies that operate in conjunction with the digital ranging subsystem to provide data which, when properly reduced, yield the instantaneous range between the Apollo spacecraft and the ground station.

RECEIVER-EXCITER SUBSYSTEM

The fundamental S-band two-way carrier path is diagrammed in simplified form in Figure 1. Excitation from the exciter is applied to the power amplifier. The amplifier output is transmitted as the up-link carrier via the diplexer, antenna feed, and antenna. At the spacecraft, the up-link carrier is received, transponded and retransmitted as the down-link carrier. This carrier is received by the antenna and feed, passed through the diplexer and amplified by the parametric amplifier. The amplifier output is applied to the receiver.

The receivers and exciter interconnect with the doppler and ranging equipment to perform the listed functions. In the paragraphs that follow, the mechanization of these four major functional capabilities are discussed in greater detail.

DOPPLER EXTRACTION FUNCTION

Let the exciter output carrier frequency at S-band (between 2100 and 2110 megacycles) be designated F_T, as shown in Figure 2. The frequency F_T has a precision based upon the accuracy of a 1.0-megacycle reference supplied by the timing and frequency reference assembly.

The output frequency is amplified and transmitted to the spacecraft, where it is coherently transponded by the ratio 240/221, and then retransmitted to the ground station. On the ground, the received signal is preamplified by the parametric amplifier and appears at the receiver input as the frequency

$$(240/221) F_T + D$$

The quantity "D" is the two-way doppler-shift frequency, and has a maximum value of about 200 kilocycles at earth escape velocity.

The receiver reference loop is phase-locked to this received frequency, and receiver reference signals containing frequencies coherently related to the received frequency are applied to the doppler extractor.

Similarly, frequencies coherently related to the transmitted frequency are also applied to the extractor.

Within the doppler extractor, the transmitter references are suitably combined and shifted coherently to simulate the 240/221 ratio occurring in the spacecraft. The resulting signal is functionally differenced with the receiver references to yield the doppler

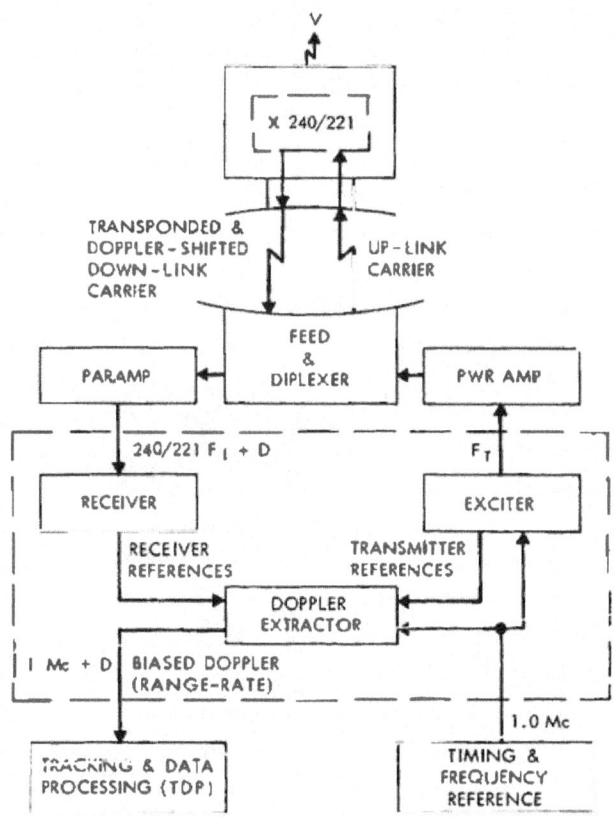

Figure 2—Doppler extraction function.

frequency D. Finally, this frequency is added to a 1.0-megacycle bias from the timing and frequency reference assembly, and the resulting biased doppler, or range rate signal, is supplied for further reduction to the tracking and data processing (TDP) subsystem. The biasing is done to supply the doppler signal in a form that is convenient for further reduction by a computer.

The frequency D is approximately related to the spacecraft radial velocity vector and transmitter frequency by the expression

$$D \approx \frac{240}{221} \times F_T \times 2 \frac{\vec{V}}{C}$$

where $\vec{V}$ is considered *positive* when the range is increasing. Thus, if the spacecraft is moving *away* from the ground station, the biased doppler frequency will be greater than one megacycle, while if the spacecraft is approaching the ground station, the biased doppler frequency will be less than one megacycle.

TWO-WAY COMMUNICATIONS FUNCTION

A typical operational configuration using both receivers is shown in Figure 3. Up-data and voice FM subcarriers from the subcarrier oscillators are applied to the exciter phase-modulator. The PM-modulated carrier from the exciter drives the power amplifier, which, in turn, feeds the PM-modulated up-link carrier to the spacecraft — Lunar Excursion Module (LEM) or Command and Service Module (CSM) — via the antenna and microwave equipment. Within the spacecraft, the carriers are suitably demodulated to provide up-link information for the in-flight equipment and personnel.

FM and PM-modulated carriers are generated within the spacecraft (LEM, CSM, or S-IV-B) and transmitted to the ground station.

In the configuration shown, the separate carriers are amplified through the multi-channel, parametric amplifier and applied to the separate receivers.

Receiver 1 operates as phase-lock, double-conversion equipment, and coherently detects the phase-modulated carrier. The resulting detected spectrum consists of information subcarriers, frequency-modulated by voice and data information. This spectrum is supplied to the data demodulator assembly for subcarrier demodulation.

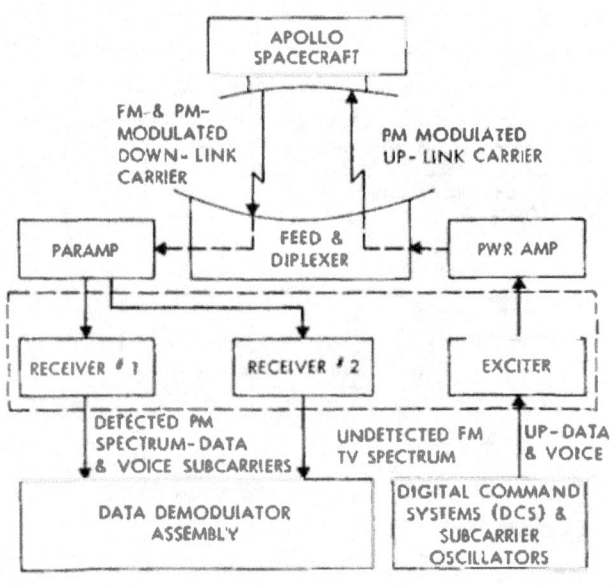

Figure 3—Two-way communications function.

Receiver 2, in the example shown, operates in an open-loop, single-conversion, wide-band mode. It supplies a gain-controlled FM spectrum (usually TV information) around a center frequency of 50 megacycles, the receiver first intermediate frequency (IF). This spectrum is also supplied to the data demodulator assembly for FM demodulation.

Receivers 1 and 2 are not limited to the modes of operation shown in Figure 3. Either or both receivers may be simultaneously operated in either the open-loop or closed-loop configurations on any one of four received channel frequencies in the 2270- to 2290-megacycle band. The receiver internal configuration is identical, except that only one source of reference signals is required, and this is included in receiver No. 1, for use by both receivers.

ANGLE TRACKING FUNCTION

The received carrier from the spacecraft is split by the antenna feed equipment into three channels, as shown in Figure 4: the sum channel ("Σ"), the "X" channel, and the "Y" channel.

The sum channel signal is amplified by the parametric amplifier, and is the main received carrier for the reference loop of the receiver.

The X and Y channel signals are not preamplified, but are applied directly to the dual-channel, angle receiver. Using reference signals generated by the receiver reference loop, the angle channels operate as dual-conversion receivers. They produce dc outputs (E_x and E_y) with magnitude proportional to the amplitude of the channel input signal.

The antenna pattern associated with each channel is such that, when the radial axis of the antenna is perpendicular to the plane of the incoming wavefront, the sum channel amplitude is maximum, but the angle channel inputs are minimum, or "null" inputs. Under this condition, the error signal dc outputs E_x and E_y are also at a minimum.

When the antenna is slightly displaced from radial alignment in either the X or Y tracking planes, as occurs during angular tracking, the angle channel input amplitude increases. The detected error voltages then take on dc values proportional to the angular displacement, or tracking error. The polarity of the error voltage is a function of the phase of the channel input signal, which in turn is dependent on the direction of the angular tracking error. The antenna pattern associated with the angle channels is essentially biphase; that is, the phase goes through a 180° reversal at the null (alignment) position of the antenna.

The error signals thus contain information as to the direction and magnitude of the

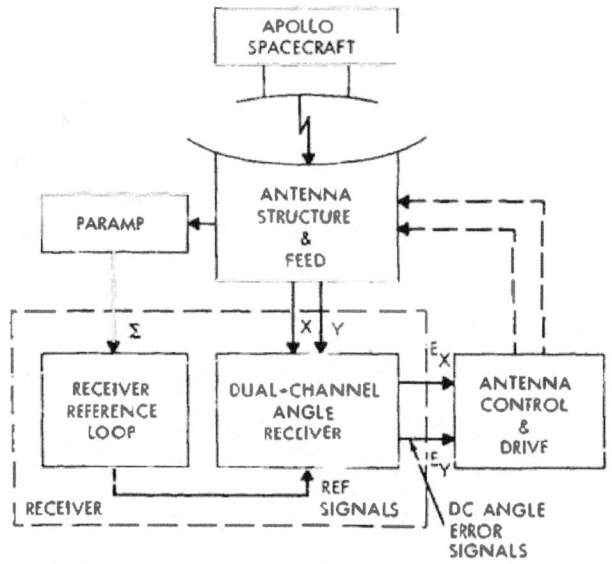

Figure 4—Angle tracking function.

angular tracking error, and the angle channels function as the amplifiers and detectors in the antenna tracking servo loop. The other elements of the loop are the antenna feed, which performs the *sensing* function, and the antenna control and drive equipment, which actuate the motions of the antenna structure.

The standard single configuration contains two complete angle channel receivers, one associated with each of the reference loops. Receiver No. 1 is ordinarily used with the main (30-foot or 85-foot) antenna, while receiver No. 2 is ordinarily associated with the small, widebeam acquisition antenna. When the acquisition antenna is not in use, receiver No. 2 reference loop is ordinarily switched to receive via the large antenna through the multi-channel parametric amplifier.

RANGING FUNCTION

The major signal paths associated with the ranging function are shown in Figure 5. The digital ranging equipment, known as the ranging subsystem, although not a part of the receiver-exciter subsystem, is shown in the diagram to simplify the description.

A pseudo-random noise code spectrum containing a "clock" component is applied from the ranging subsystem as phase modulation (code x clock) to the exciter. The resulting modulated carrier is transmitted to the spacecraft, "turned around", and retransmitted to the ground receiver. Within the receiver reference loop, the carrier containing the received code x clock modulation is translated to an IF of 10 megacycles and applied to the ranging receiver.

Within this receiver, the received code x clock is correlated with a locally generated code from the ranging subsystem. The correlation process is functionally subtractive, yielding an output of clock signal alone, whose amplitude is proportional to the degree of correlation. This signal is tracked by a receiver phase-lock loop, and its amplitude is detected to appear as a dc *correlation indication*. This indication is routed back to the ranging subsystem as a primary information input.

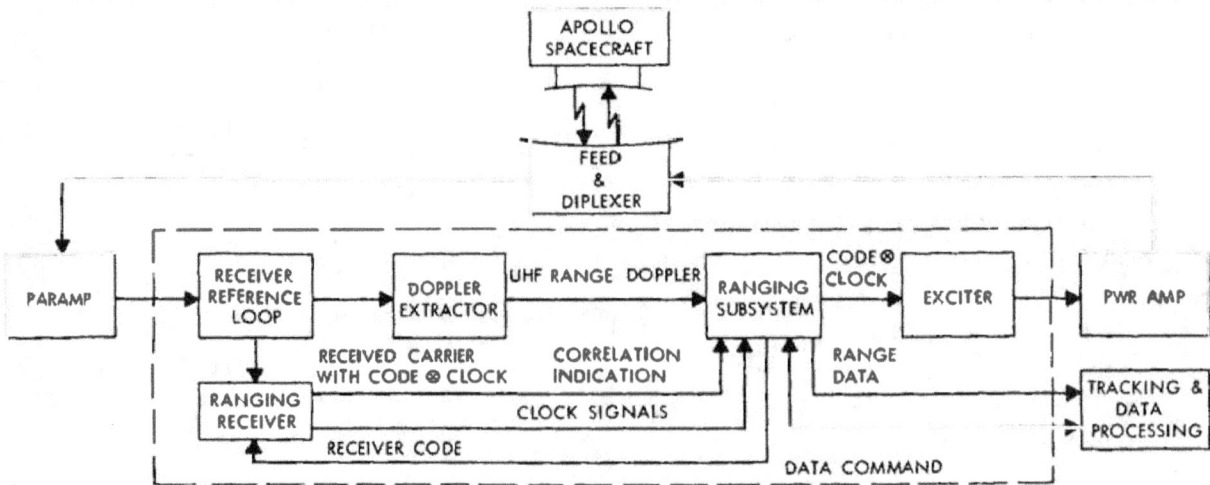

Figure 5—Ranging function.

The ranging receiver also supplied clock frequency reference and clock doppler signals, while the reference loop supplies a UHF range doppler signal (at one-fourth the S-band doppler value or D/4), for use by the ranging subsystem.

Using these various inputs, the ranging subsystem programs an acquisition sequence from which data proportional to the range of the spacecraft is obtained.

Upon completion of the acquisition program, the ranging subsystem delivers updated range information to the tracking and data processing subsystem upon command from that subsystem.

THE RECEIVER REFERENCE LOOP

The reference loop of a typical receiver is particularly important as an element of the subsystem, as it contains equipment that is operational in all four of the major functions. The loop and its associated branches are shown in detail in Figure 6.

S-band RF input, at one of four carrier center frequencies in the 2270- to 2290-megacycle range, is applied to the first mixer and preamplifier. At the mixer, the signal is differenced with the local oscillator (LO) chain injection signal, which is 50 megacycles lower in frequency than the received signal. The resulting 50-megacycle IF signal is preamplified and applied to the automatic gain control (AGC) 50-megacycle IF amplifier.

During open-loop operation, when the carrier is frequency-modulated by television information, the 50-megacycle spectrum is branched off at this point, passed through a gain-controlled, wide-band, 50-megacycle IF amplifier, and supplied as an undetected spectrum to the data demodulator assembly.

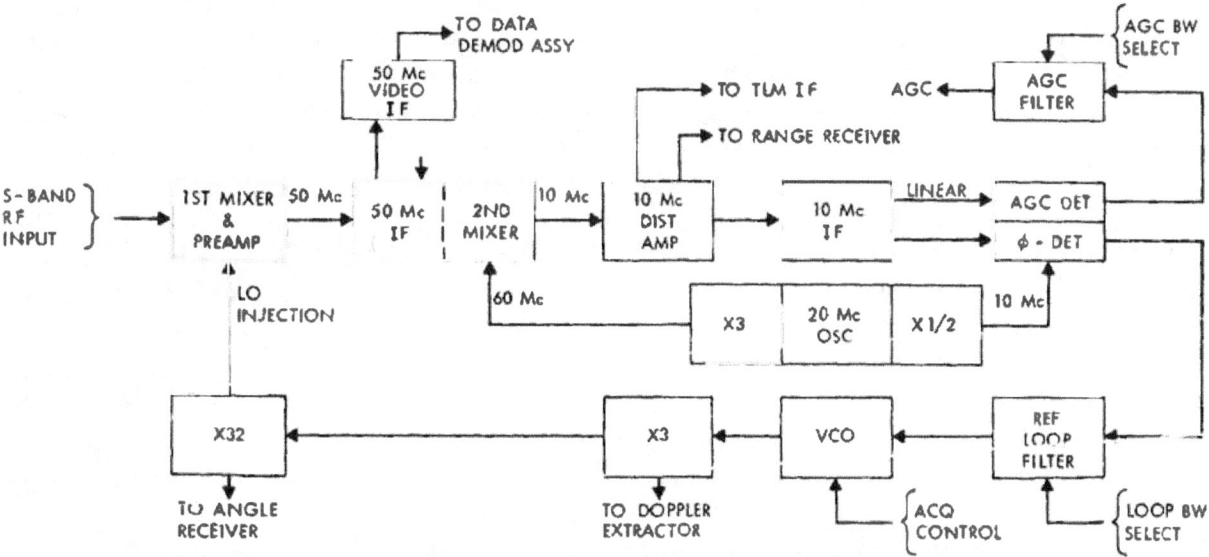

Figure 6—Receiver reference loop (typical).

In closed-loop operation, the signal is next gain controlled through a series of 50-megacycle AGC IF amplifier stages, and then differenced with a 60-megacycle reference signal in the second mixer to produce the second IF of 10 megacycles. The IF amplifier is capable of a total gain control range of 91db, operating at an overall gain between +51db and -40db. The phase and gain changes across this range must be carefully controlled during manufacture to assure compatible operation with parallel units in the angle receiver channels.

The 10-megacycle output is applied to a distribution amplifier, where telemetry channel IF and range receiver channel input signals are branched off. Operation of these channels are covered in greater detail later in this paper.

The reference loop signal is next applied to a 10-megacycle IF amplifier, where a crystal filter establishes the loop predetection noise bandwidth of about 7.0 kilocycles. After filtering, the signal is split into two channels. The first operates at high gain and contains a limiter whose output is applied to the loop phase detector. The second channel operates at lower gain without limiting, and this channel output is applied to the loop AGC detector.

Within the loop phase detector, and assuming loop phase lock, the limited output signal frequency is differenced with a 10-megacycle reference signal. The resulting output is a small dc voltage proportional to the angular phase error in the loop. This dc output is applied to the reference loop filter, within which time constants are selected manually to control the overall loop-noise bandwidth ($2B_t$). Threshold values for this bandwidth ($2B_{L_0}$) of 50, 200, and 700 cycles per second may be selected.

The loop filter output, known as the loop "static phase error" (SPE), is a small and relatively noise-free dc voltage. This voltage is applied to the voltage-controlled oscillator (VCO) where, during phase lock, it automatically adjusts the VCO frequency to maintain lock during input signal frequency variations.

An acquisition input voltage to the VCO is applied manually by the operator to obtain initial lock (acquisition), and then to balance out the residual phase error when acquisition has been accomplished. This latter function is indicated by a reduction of the SPE to zero.

The VCO output is next multiplied by three, and a coherent reference signal for the doppler extractor is branched off from the multiplier.

Finally, the VCO signal is multiplied by 32 for a total multiplication of 96, and applied as the local oscillator injection signal to the first mixer, thus closing the loop. Local oscillator injection signals for the angle channel receiver are also branched off at the x32 multiplier output.

Returning to the AGC path, the detector output is applied to the AGC loop filter. Within the filter, the AGC loop bandwidth is selected by the control operator for one of three values, grossly designated narrow, medium, or wide.

These values are ordinarily paired with the corresponding reference loop $2B_{L_0}$ settings, although this is not a necessity for proper operation.

The filter output is the dc AGC voltage, with a dynamic range of 10 volts. This voltage is applied to the gain-controlled stages in the 50-megacycle IF amplifiers in the reference loop, and to the parallel angle receiver channels. It is also displayed and recorded by the analog instrumentation equipment, as it varies with, and is a measure of, the input signal level.

The 60- and 10-megacycle reference frequencies are both derived from a 20-megacycle crystal oscillator. The 60-megacycle signal is obtained through a x3 multiplier, while the 10-megacycle signal is derived from a x 1/2 multiplier. This reference generation equipment is present only in one of the receivers. Reference signals for the second receiver, the angle channels, the telemetry channels, the range receiver, and the doppler extractor are all branched off of the x3 and x 1/2 multiplier outputs.

In summary, the reference phase-lock loop is of second order, with the dual-phase integration occurring through the loop filter and VCO, while the AGC loop is of first order with single integration occurring through the AGC filter.

VARIATION IN LOOP NOISE BANDWIDTH

The reference loop gain varies with the input signal level, primarily because of the suppression of signal by noise within the limiter preceding the phase detector. The increased loop gain at high signal levels results in an increased damping and widening of the bandwidth. The values of 50, 200, and 700 cycles per second mentioned earlier are values occurring at the system signal threshold; the strong signal bandwidths are much wider.

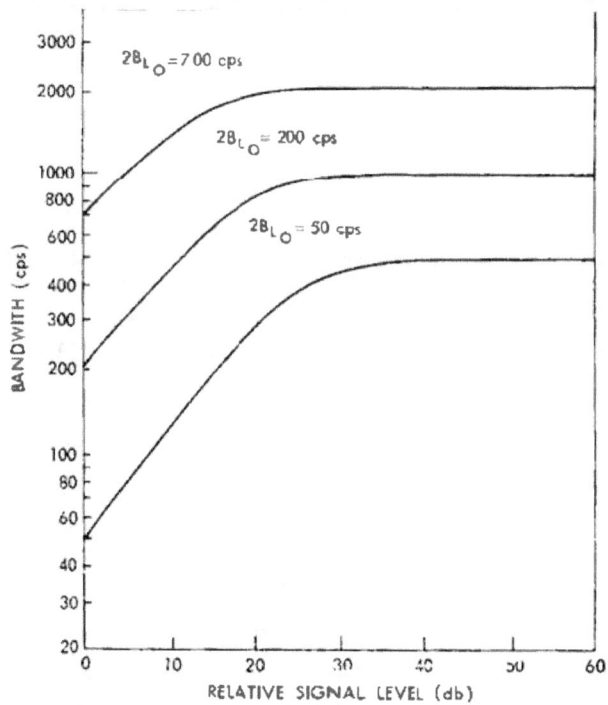

This effect is shown in Figure 7. Note that in the 700-cycles-per-second position, the bandwidth rises to about 2 kilocycles when the signal exceeds the threshold value by about 30db. This wide bandwidth is desirable for tracking the high doppler rates encountered during the earth orbital phase of the Apollo missions, and will ordinarily be used during these passes. Carrier phase modulation within the loop bandwidth cannot be properly demodulated because the loop "tracks out" these frequencies. This is of little concern for the Apollo program, however, as all modulation except the emergency voice information is on subcarriers at

Figure 7—RF loop noise bandwidth as a function of relative signal level above RF design frequency threshold.

frequencies greater than 1.0 megacycle, well beyond the low frequency cut-off of the loop.

The 50-cycles-per-second position, reaching a maximum bandwidth of 500 cycles per second, is intended for use during the lunar phases of the mission. Doppler rates will be low during these phases, and the increased sensitivity and narrow bandwidth will assure an adequate communications margin for the expected received signal levels, even if the emergency modes must be used.

The FM television spectrum will contain energy within the tracking bandwidths shown. However, the receivers are in open-loop condition during FM reception, and no attenuation occurs, as the tracking loop is inoperative.

THE RANGING RECEIVER AND DETECTED TELEMETRY CHANNELS

The 10-megacycle IF distribution amplifier in the receiver channel branches off signals for the ranging receiver and the detected telemetry channel. As these two signal paths are important to the basic functions of the subsystem, they are shown in greater detail in Figure 8.

The ranging receiver input, from either receiver as selected by the control operator, consists of code x clock modulation on the 10-megacycle IF. This modulation occupies a wide spectrum containing significant sideband components as far as 2 megacycles from the carrier. This spectrum is applied to a wideband phase detector which is referenced by "code x IF." The code x IF is a modulated spectrum centered at the IF frequency of 10 megacycles. The spectrum is derived from a phase switch, within which the 10-megacycle IF reference signal is

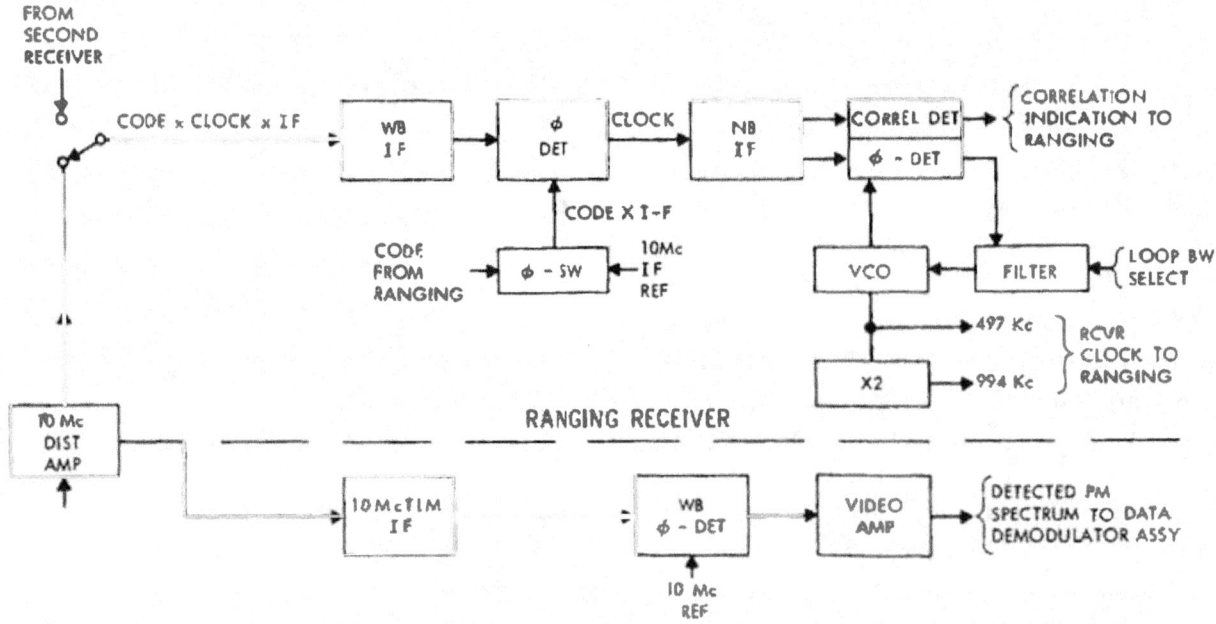

Figure 8—Detected telemetry channel (typical).

periodically switched ±90° by the code signal, also known as receiver code. This code is supplied by the ranging subsystem.

The phase detector differences the two signals, producing an output spectrum which always contains some energy at the clock frequency. The amplitude of this energy is directly proportional to the degree of correlation between the received code and the receiver code.

The energy at the clock frequency, known as the clock signal, is filtered and amplified through a dual-channel IF amplifier. The channel outputs are applied to a loop phase detector (limited output), and a correlation detector (linear output). The correlation detector develops the dc correlation indication for the ranging subsystem.

The phase detector output drives a loop filter and VCO, which in turn references the two detectors. These units together define the ranging receiver phase-lock loop. The loop bandwidth, as in the main receiver, is established by manual selection of the time constants in the loop filter. This bandwidth has threshold values of 4, 16, and 40 cycles per second. These are considerably narrower than the bandwidths of the main loop; therefore ranging threshold is not ordinarily reached during operation.

The receiver loop acts as a narrow-band tracking filter, providing relatively noise-free frequency references at the clock frequency and its second harmonic. These are supplied to the ranging subsystem. The frequencies are used to drive the receiver coder within that subsystem.

The detected telemetry channel is a simple series arrangement of IF amplifier, wideband phase detector, and video amplifier. The phase detector is referenced with a 10-megacycle signal from the reference signal generator in the receiver.

The detected signal is supplied at a level of 0dbm and a -1db bandwidth of 1.25 megacycles to the data demodulator assembly.

RECEIVER-EXCITER SUBSYSTEM EQUIPMENT LAYOUT

Control room cabinets containing the receiver-exciter equipment are shown in Figure 9. The first three cabinets on the left contain subsystem control panels and monitoring equipment, tilted and arranged for convenience by a seated control operator. Continuing from left to right, the fourth, fifth, and eighth cabinets each contain two roll-out frames which mount subassemblies of the subsystem. Over eighty different types of subassemblies are used, and the total count exceeds 200.

One frame of cabinet one is rolled out to show the subassembly packaging and mounting methods. All of the interconnecting coaxial cabling is routed on the outer surface of the mounting plates, while the power, dc, and low-frequency signal paths are all wired with shielded leads on the inner surface of the plates within the frame. Each subassembly is individually removable for quick replacement. Connections to the wiring within the frames are made through multipin connectors mounted at the ends of the subassemblies. Intracabinet cabling is routed

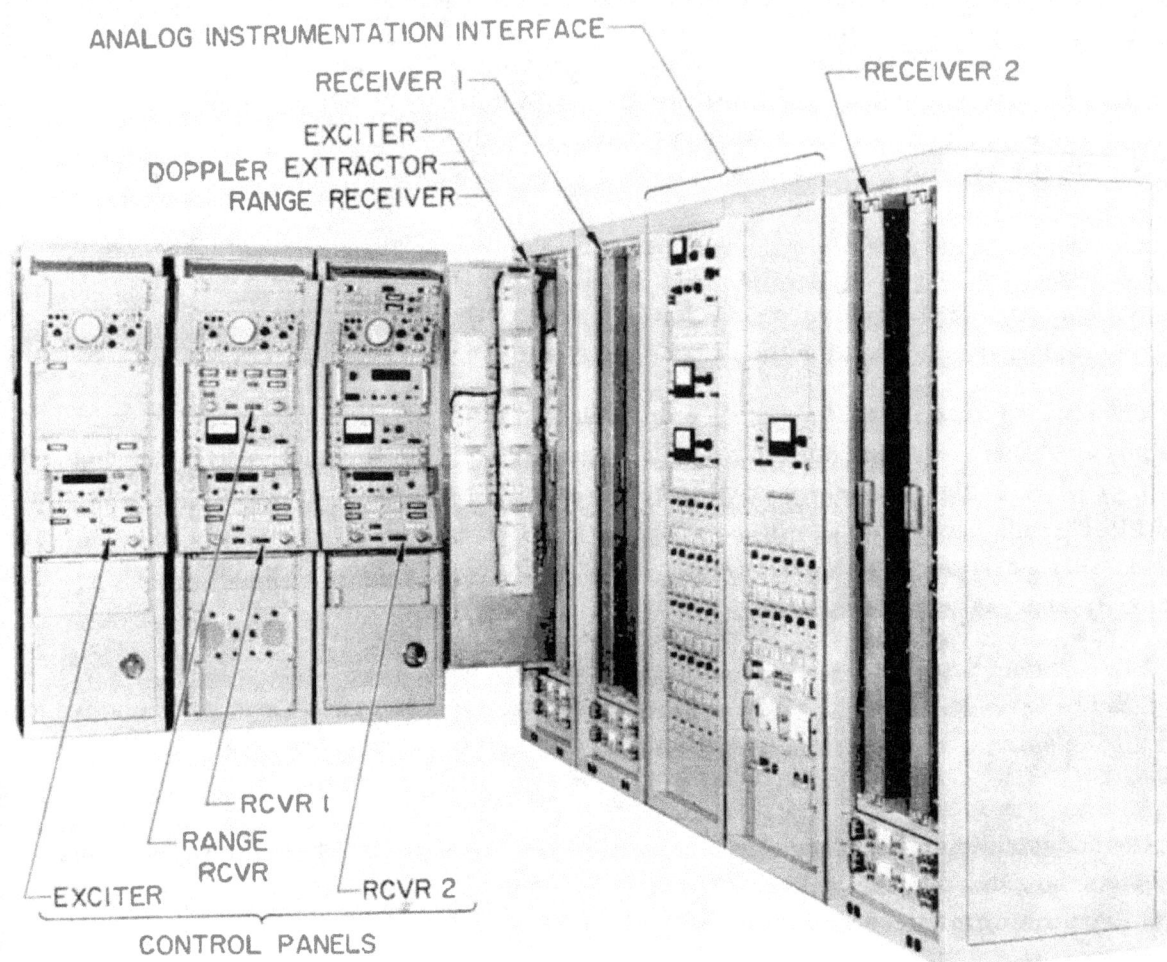

Figure 9—Receiver-exciter subsystem.

through floor channels beneath the cabinets, and all connections to these cables are made through connector plates at the base of the cabinets.

Subassembly power supplies are rack-mounted beneath the roll-out frames, and ac convenience outlets are placed on the cabinet lower lips.

Cabinets two and five contain the subassemblies for receiver one and receiver two, respectively. Each receiver thus housed consists of the reference loop, the angle channels, and the telemetry channels.

Cabinet one contains subassemblies of the exciter, the doppler extractor, and the range receiver, as well as other minor equipment used with the ranging receiver during the ranging program. The exposed plate contains subassemblies of the doppler extractor.

Additional subassemblies containing equipment operable in the S-band region are normally mounted near the antenna, and do not appear in this picture.

Cabinets three and four of the right-hand group contain isolation amplifiers and power supplies which preprocess monitoring signals before they are fed to the analog instrumentation subsystem. All of these signals are normalized for a peak-to-peak level of 10 volts from the low impedance output of the isolation amplifiers. The cabinets also contain instrumentation used while testing and evaluating the performance of the subsystem equipment.

The location of the system control panels for the exciter, the two S-band receivers, and the ranging receiver are indicated on racks one, two, and three. Figures 10, 11 and 12 show these control panels in greater detail.

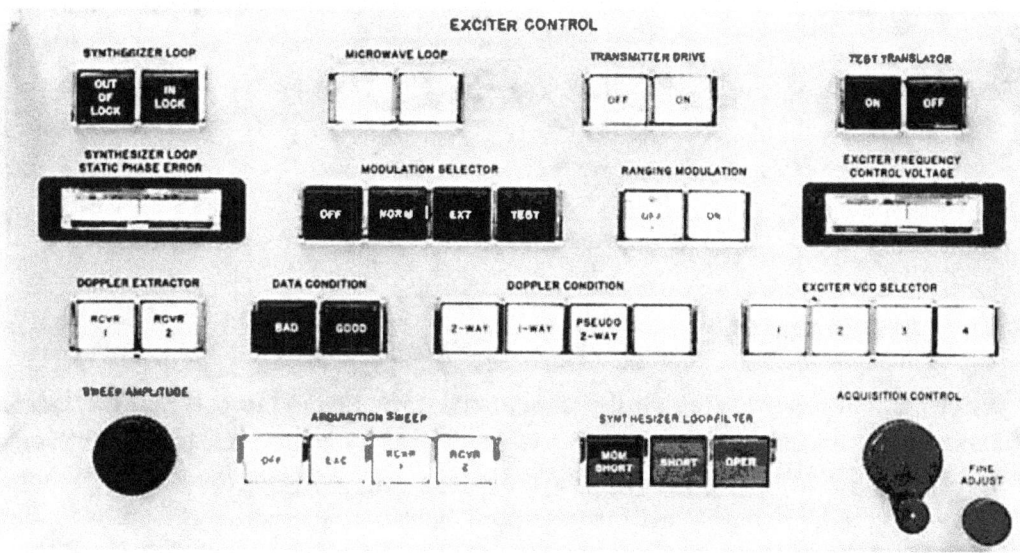

Figure 10—Exciter control.

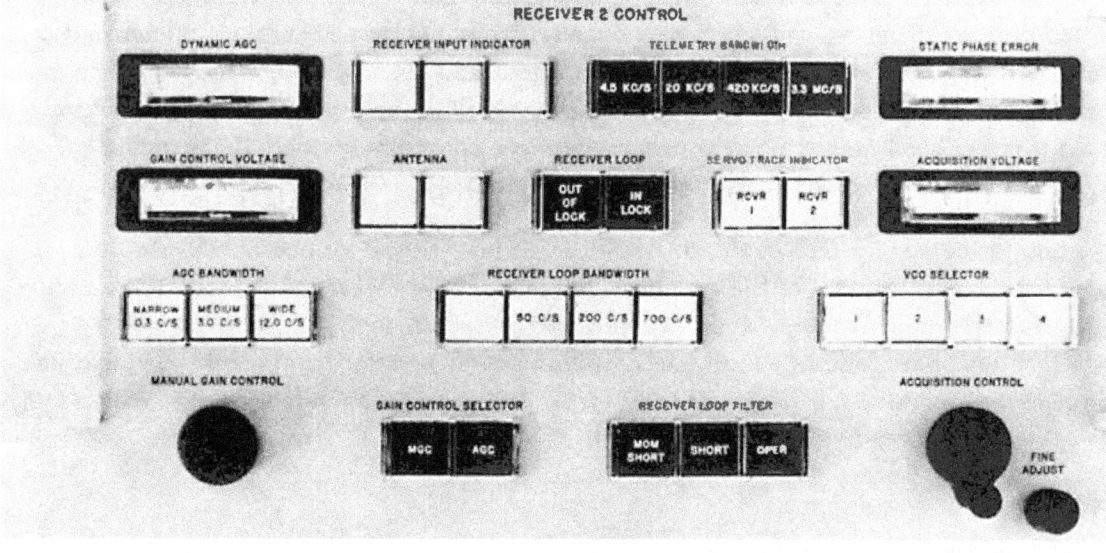

Figure 11—Receiver 2 control.

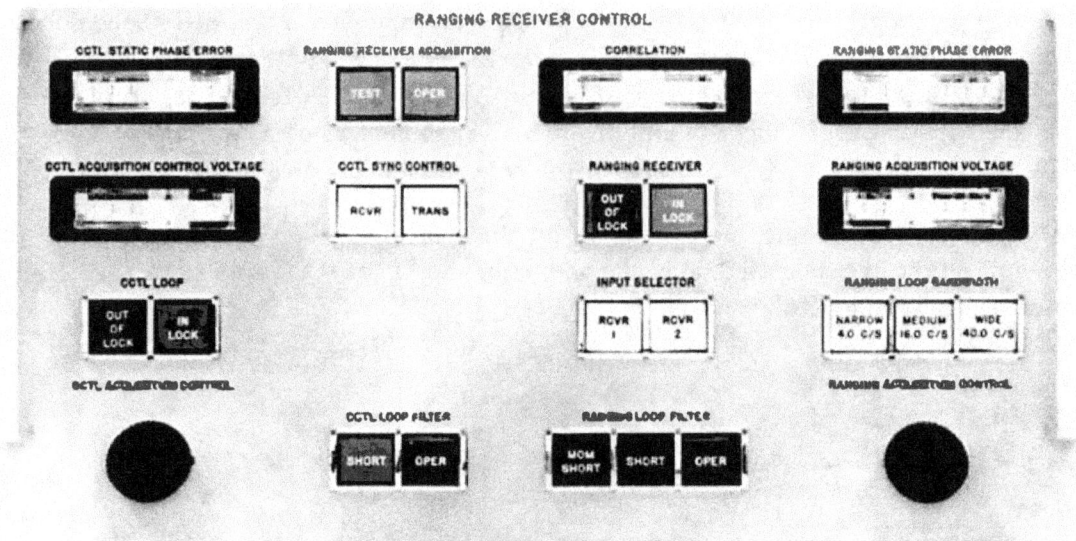

Figure 12—Ranging receiver control.

THE SYSTEM CONTROL PANELS

The exciter control panel contains all operational controls and indications for the exciter and doppler extractor, as shown in Figure 10. Controls for a phase-lock loop which locks one of four exciter VCO's to a system frequency synthesizer are included, together with controls for selecting the modulation source and the receiver input to the doppler extractor. The panel also contains controls for an automatic sweep generator that acts as an aid during acquisition of down-link carriers by the receivers. The exciter VCO's may also be automatically frequency-swept to aid in the two-way carrier acquisition process.

The receiver control panel contains all operational controls and indications for the receiver reference loop, angle channels, and telemetry channels, as shown in Figure 11. Push-button controls for selecting the reference loop noise bandwidth and the AGC loop bandwidth are included, as well as controls for selecting one of four VCO's for the corresponding four received carrier frequencies. Coarse and fine manual adjustment controls for the VCO acquisition voltage are located conveniently in the lower right-hand corner of the panel.

The ranging receiver control panel, as shown in Figure 12, is not ordinarily used during system operation, as all control of the ranging equipment is transferred to the digital ranging subsystem. However, during test of the receiver-exciter equipment, this panel is used to control and monitor operation of the ranging receiver and associated equipment. Typical controls are those for selecting the ranging receiver bandwidth, and for selecting the main receiver, from which the ranging receiver input is derived.

SUMMARY

The receiver-exciter subsystem interfaces with many of the other station subsystems to aid in performing four major functions.

1. Doppler extraction

2. Two-way communication

3. Angle tracking

4. Ranging

The subsystem embodies, in many ways, the heart of the unified S-band concept, as it:

1. Receives and generates the S-band carriers which define the single-system approach.

2. Operates upon modulation and frequency information contained in these carriers to aid in giving the ground station a total communications capability with the spacecraft of the NASA Apollo program.

VERIFICATION RECEIVER, SCO OSCILLATOR AND UP-DATA MODEMS

by

J. Jacobi

Goddard Space Flight Center

ABSTRACT

The verification receiver is a solid-state, S-band telemetry receiver with special demodulators. It is a fixed-tuned superheterodyne, multiple-conversion receiver of standard design. The purpose of the verification receiver is to sample the up-link signal at the power amplifier output and provide demodulated voice and data signals. The voice signal is recorded and the data is utilized by the command system as an input to the verification loop.

The subcarrier oscillator (SCO) subsystem comprises a 30-kilocycle voltage controlled oscillator (VCO) and appropriate mixing networks. The 30-kilocycle SCO is modulated with voice and the 70-kilocycle SCO is modulated with data. The two resulting signals are added linearly and modulated onto the S-band carrier.

The up-data modem (modulator-demodulator) accepts a command message from a modified Univac 642B computer and converts it to a form suitable for modulating onto the 70-kilocycle SCO.

INTRODUCTION

The equipment which will be discussed includes the up-data buffer modem, the subcarrier oscillator subsystem and the verification receiver. Together, these items comprise a significant portion of the up-link communications system. Figure 1 depicts the relationship between these subsystems.

The up-data buffer modem accepts data data from a computer and operates on the data to put it into a form suitable for modulation onto a subcarrier. The subcarrier oscillator subsystem accepts data and voice signals and modulates these signals onto their respective subcarriers. The verification receiver samples the output of the S-band power amplifier and demodulates the S-band carrier. The output of the verifica-

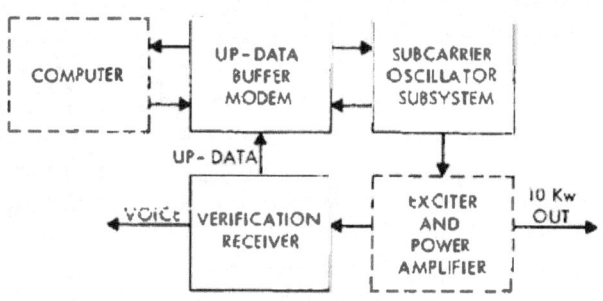

Figure 1—Relationship between up-data buffer modem, subcarrier oscillator subsystem, and verification receiver.

tion receiver is the original up-data and up-voice signals. The data is returned to the buffer modem as an input to the system verification process. The voice output is recorded.

UP-DATA BUFFER MODEM

The function of the up-data buffer modem (Figure 2) in the Unified S-Band system is to provide interface between a modified Univac CP-642B computer and the up-data subcarrier oscillator. The computer provides digital data to the buffer modem. The buffer modem stores this data and at the proper time, modulates the data onto an audio tone. This tone is mixed with a synchronization tone and filtered, and the resultant is applied to the 70-kilocycle subcarrier oscillator. The buffer modem also accepts phase-modulated audio from the verification receiver, demodulates this audio, and provides the demodulated data to the computer. The computer uses this information as a part of a complex verification process.

The buffer modem may be divided into four main sections: a transmit section, a receive section, the audio switching system, and control circuits. In the transmit section, the buffer processes digital data and provides a phase shift-keyed audio signal to the subcarrier for up-data transmission (the phase shift key will be subsequently referred to as PSK). The receive section processes the output of the verification receiver. The audio switching circuits automatically provide normal and emergency connections between modulators, demodulators and RF equipment. The control section generates control and timing waveforms required by the up-data buffer modem.

Transmit Section

The transmit section can be subdivided into two parts, the data input circuits and a PSK modulator. The data input circuits consist of a 27-bit shift register and a 5-bit storage register.

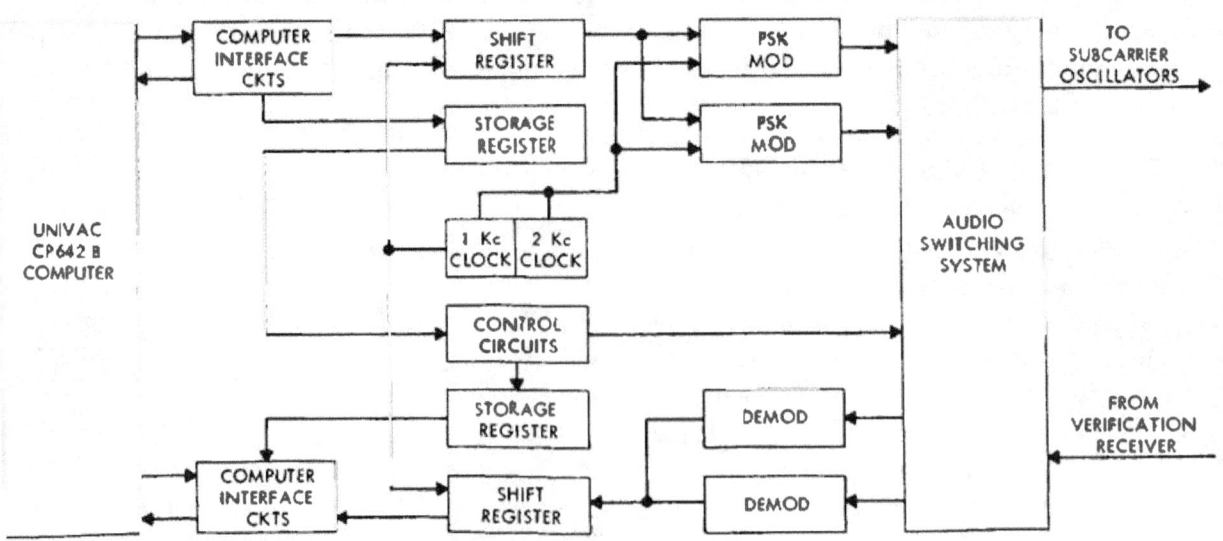

Figure 2—Up-data buffer modem.

When the buffer is ready to accept a word from the computer, it places a request signal on a line to the computer. The computer responds with a 30-bit parallel word, consisting of 25 data bits and 5 control bits. The 25-bit data portion of the word is entered into the shift register and the 5-bit control information is set into the storage register. The control information selects the modulator, demodulator and transmitting equipment to be used during the transmission of the data bits. If the computer does not respond to the request for a word, logic 1 data bits are transmitted each bit time.

When the data is completely entered into the shift register, it is automatically dumped to the PSK modulator at a 1-kilobit rate. After the data is transferred to the PSK modulator, a signal is generated to initiate another word transfer from the computer.

PSK Modulator

Two identical phase modulator circuits (Figure 3) are employed in the buffer, one acting as a standby to increase operational reliability. Digital signals from the shift register are applied to the modulator and shift the phase of a 2-kilocycle tone at a 1-kilobit rate. The total phase shift between a logic "0" and a logic "1" is 180 degrees.

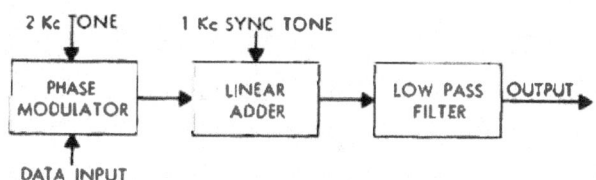

Figure 3—PSK modulator.

The phase-shifted signal is filtered and added linearly to the 1-kilocycle synchronization tone. The combined signal is filtered through a 3-kilocycle low-pass filter, amplified and supplied through two balanced outputs to the RF equipment.

Receive Section

The receive section consists of a pair of phase demodulators and data output circuits. The phase demodulators are depicted in Figure 4.

Phase-modulated audio is obtained from the verification receiver and applied to one of the two demodulation circuits. The 1-kilocycle synchronization tone and the 2-kilocycle PSK audio are separated by filtering, the 2-kilocycle PSK audio is applied to a phase detector, and the 1-kilocycle synchronization tone is doubled to 2 kilocycles, and is used as the phase detector reference. The demodulated output is then "squared up" to provide the digital data to the data output circuits.

The data output circuits consist of a 26-bit shift register and a 5-bit storage register. The storage register receives information from the audio switching section and drives a display unit for visual presentation of equipment configuration. The demodulated

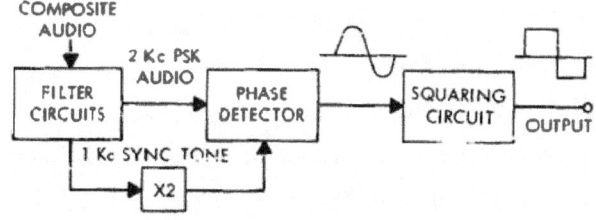

Figure 4—PSK demodulator.

audio output is entered into the shift register at a 1-kilobit rate. When a full word is stored the buffer requests the computer to accept the word. The computer responds by accepting the word in parallel readout and acknowledges to the buffer that it has accepted the word.

SUBCARRIER OSCILLATOR SUBSYSTEM

The purpose of the subcarrier oscillators (Figure 5) in the system is to convert baseband voice and data signals to frequency-modulated subcarriers. The subsystem also linearly adds these subcarriers and adjusts their respective levels so as to produce the proper up-link modulation index at S-band for the mode of transmission selected. For purposes of discussion, this subsystem will be broken down into three components: the voice subcarrier oscillator, the data subcarrier oscillator, and the mode selection circuit.

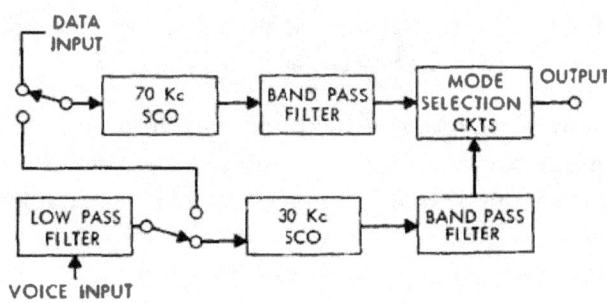

Figure 5—Subcarrier oscillator subsystem.

Voice Subcarrier

Voice signals are received from the station intercom and applied to the appropriate input of the subcarrier oscillator subsystem. The input voice is passed through a low pass filter which restricts the voice spectrum to a maximum frequency of 3 kilocycles. This filter has relatively sharp cutoff characteristics attenuating 6-kilocycle components by 60db which reduces the spreading of the voice subcarrier spectrum and eliminates a certain amount of noise.

The voice subcarrier oscillator is a voltage-controlled multivibrator which operates at a nominal center frequency of 30 kilocycles. The frequency deviation of the subcarrier oscillator is a linear function of the modulation voltage and has a maximum value of plus and minus 7.5 kilocycles about the center frequency. The linearity of the frequency deviation versus voltage is 1 percent or better over the full range of plus and minus 7.5 kilocycles.

The output of the voltage-controlled multivibrator is filtered by a band-pass filter to remove harmonics of the 30 kilocycles and also to remove undesirable voice components which might occur at frequencies of 3 kilocycles and less. The output of the band-pass filter is then supplied to the mode selection circuitry.

Data Subcarrier

The data subcarrier oscillator is also a voltage-controlled multivibrator. The nominal center frequency of this subcarrier is 70 kilocycles, and the maximum deviation of the data subcarrier is plus and minus 5 kilocycles about the center frequency. The linearity properties are the same as those of the voice subcarrier. Filtering of the data prior to application to the subcarrier is accomplished in the up-data buffer modem.

The output of the 70-kilocycle multivibrator is filtered by a band-pass filter for reasons mentioned in the discussion of the voice subcarrier. The output of the filter is supplied to the mode selection circuitry.

Mode Selection Circuitry

In the present Unified S-band system there are eight possible modes of up-link operation, designated 1A through 1H. The basic mode structure is given in Table 1.

Table 1

Basic Mode Structure of Unified S-Band System.

Mode	Operation
1A	No subcarrier outputs
1B	Voice subcarrier only
1C	Up-data subcarrier only
1D	Voice subcarrier only at a voltage level different from that in mode 1B
1E	Up-data subcarrier only at a voltage level different from that in mode 1C
1F	Both subcarriers linearly added
1G	Both subcarriers linearly added at voltage levels different from those in mode 1F
1H	Backup voice. This mode permits modulation of up-voice on the 70-kilocycle subcarrier in the event of certain failures.

The purpose of the mode selection circuitry is to make the proper subcarriers available at the voltage levels required by simply setting a selector switch to the mode desired.

To accomplish its purpose, the mode selection circuitry accepts the outputs of the two subcarrier oscillators and applies them to two banks of variable attenuator networks. Each bank of networks may be considered to contain eight variable attenuators corresponding to the eight modes of operation. One bank of attenuators adjusts the voice subcarrier level and the other, the data subcarrier level. The output of the two banks of attenuators are combined in a linear fashion and provided to the transmitter-exciter through a line driver. It should be noted that since each attenuator is variable, the level of either subcarrier may be adjusted independently of the mode selected and independently of the level of the other subcarrier. The range of output levels is sufficient to accommodate any present or future modulation index requirements.

VERIFICATION RECEIVER

The purpose of the verification receiver (Figure 6) in the Unified S-band system is to provide a means of demodulating a sample of the up-link signal as far upstream as possible. The

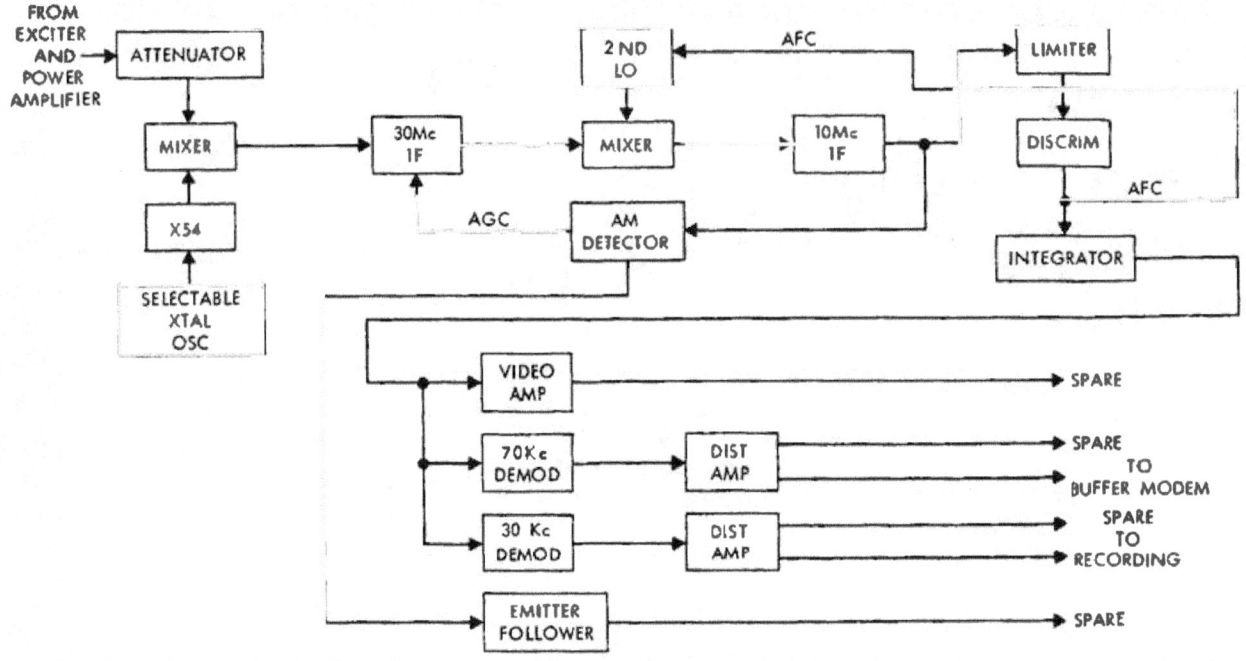

Figure 6—Verification receiver.

demodulated outputs of the verification receiver consist of up-data and up-voice. The up-data is returned to the buffer modem for further processing and the voice is recorded.

The verification receiver is a version of a commercial, solid-state, S-band telemetry receiver. It is a fixed-tuned, superheterodyne, multiple-conversion receiver of standard design. The unique items in this receiver are the phase demodulators and the subcarrier demodulators.

A sample of the up-link signal is obtained from a directional coupler located at the output of the S-band power amplifier. The power level out of the directional coupler is approximately +20dbm. Therefore it is necessary to reduce the signal to an acceptable level with the attenuator shown in the diagram.

After reducing its power level the signal is converted to the first IF frequency of 30 megacycles. The converter utilizes crystal-controlled oscillators that operate at approximately 40 megacycles, which requires a multiplication factor of 54. The output bandwidth of the converter is approximately 3 megacycles.

The 30 megacycle first IF signal is supplied to the second mixer where it is heterodyned down to 10.035 megacycles. The signal is amplified, limited and applied to the phase demodulator.

The phase demodulator consists of a conventional Foster-Seeley discriminator and an integrating network. With a phase-modulated input a Foster-Seeley discriminator provides a demodulated output which is a differentiated replica of the video intelligence. By integrating this output, a true replica of the video is obtained. This type of phase demodulator gives adequate performance at high signal-to-noise ratios. It has the advantage that it does not have the acquisition problems associated with a phase lock demodulator.

It should be noted that the receiver employs an automatic frequency control (AFC) loop. This eliminates the problem of having to retune the receiver because of local oscillator drift or because of slight changes in up-link frequency.

The output of the phase demodulator is supplied to the 30-kilocycle and 70-kilocycle subcarrier demodulators. The subcarriers are amplified, filtered and limited in their respective demodulators. The resulting signals are fed to discriminators, which are of the pulse averaging type. The outputs of the discriminators are the desired up-data and up-voice.

SIGNAL DATA DEMODULATORS

by
G. Hondros
Goddard Space Flight Center

ABSTRACT

This paper presents an overall view of the function and capabilities of the signal data demodulator which is an integral part of the heart of the Unified S-Band system. The text includes a general discussion of the different types of demodulators. Then, more specific discussion follows explaining in detail the dynamic behavior of each demodulator, tabulation of parameters, operational procedures and integration of the demodulators with the rest of the Unified S-Band system.

INTRODUCTION

Before we begin the discussion of the signal data demodulator system (SDDS), it is necessary to acquaint the reader with the type of signals transmitted from the spacecraft. The spacecraft has the capability of transmitting two carriers simultaneously at different frequencies, one of these is phase-modulated by the information and the other, is reserved for frequency modulation. An examination of typical spectra of the phase- and frequency-modulated carriers, shown in Figures 1 and 2, reveals the necessity for simultaneous ground demodulation of both carriers and recovery of all data. For this reason, two demodulator channels, which will be discussed in subsequent pages, have been designed for the ground stations.

THE SIGNAL DATA DEMODULATOR SYSTEM

The signal data demodulator system is an integral part of the heart of the Unified S-Band system. As Figure 3 indicates, the SDDS is fed by the receiver and in turn, feeds a multichannel tape recorder, provides the inputs to the various data display systems,

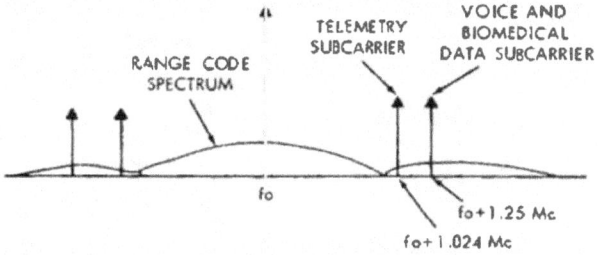
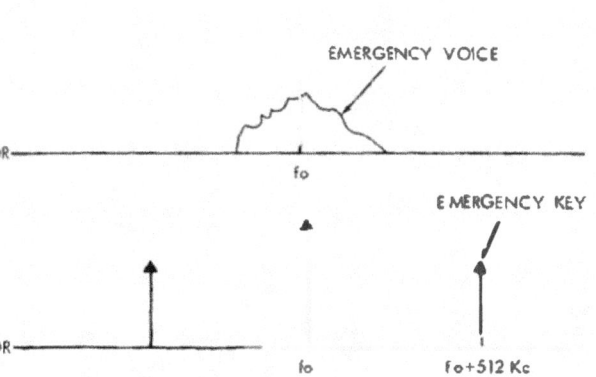

Figure 1—PM spectra of frequency-modulated carriers.

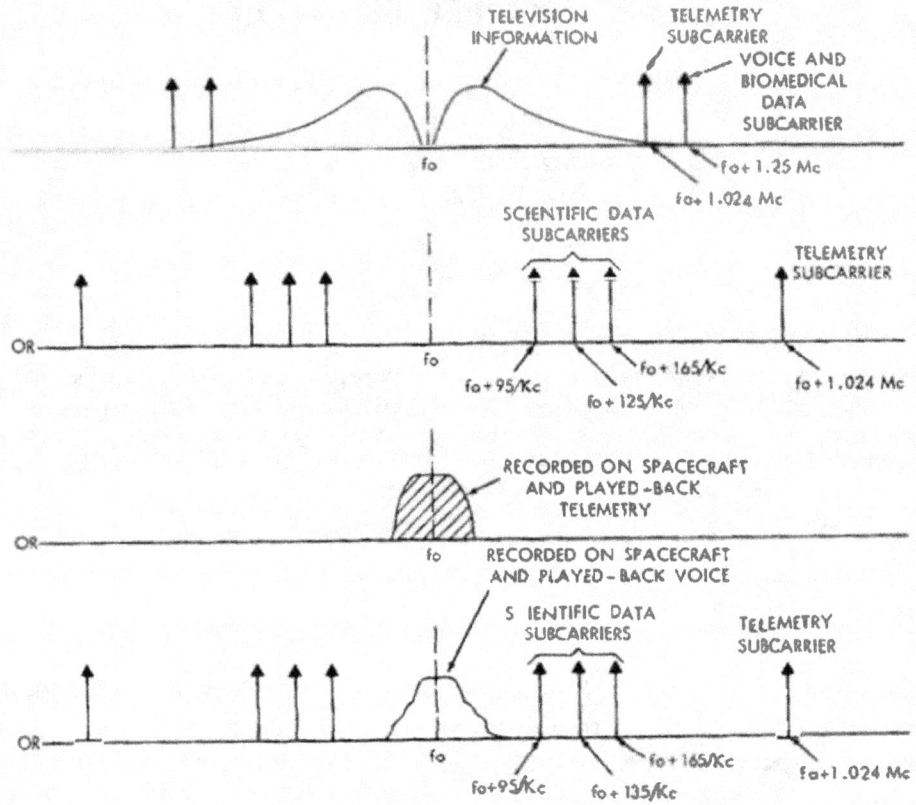

Figure 2—FM spectra of frequency modulated carriers.

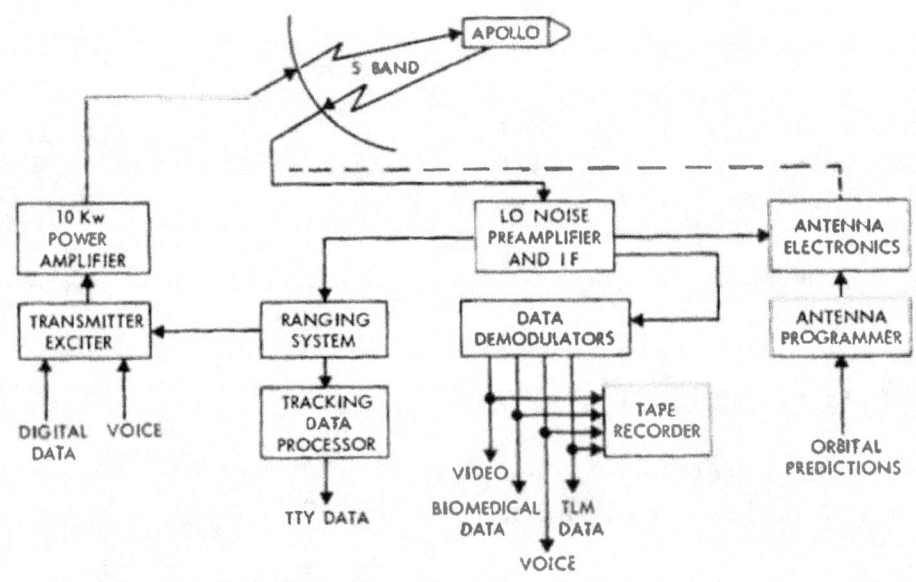

Figure 3—Simplified Apollo Unified S-Band system.

and feeds the data processing equipment such as the PCM system. Figure 4 is a simplified diagram of the SDDS. As the figure indicates, the receiver feeds the demodulators with two signals. One is a 50-megacycle IF which carries the frequency modulation. The other input

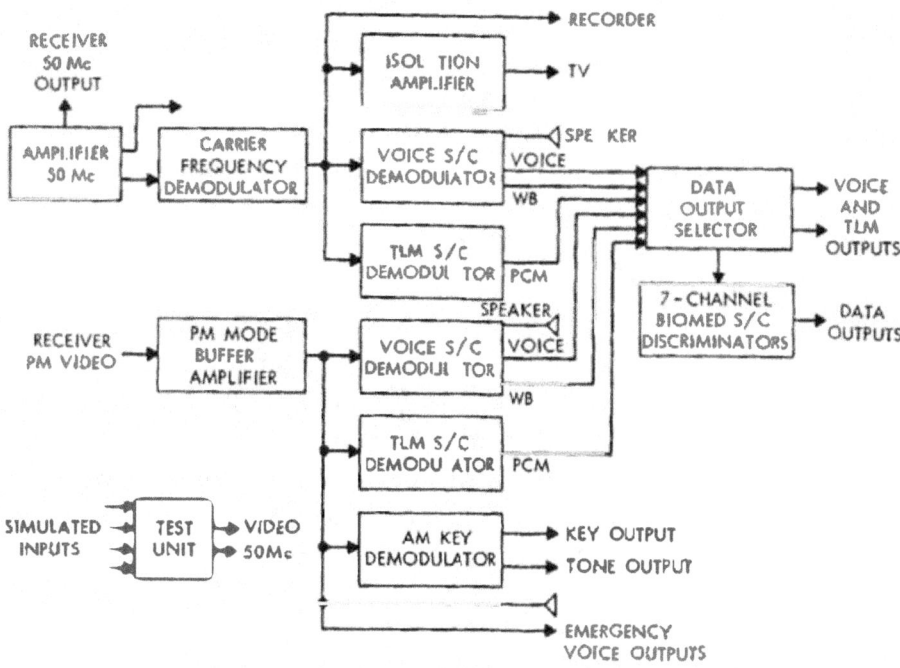

Figure 4—Signal data demodulator system.

from the receiver is at video and contains the phase modulation. Thus the SDDS consists of two channels which may be referred to as the FM and PM channels.

The 50-megacycle IF is routed to the carrier frequency demodulator which reduces the signal to video and feeds a recorder, an isolation amplifier and filter (television channel), the voice and biomedical data demodulator, and the telemetry demodulator. The PM video input from the receiver supplies the inputs to the voice and biomedical data subcarrier demodulator, the telemetry demodulator, and the emergency key demodulator. Also obtained from this channel is the emergency voice information. It should be noted at this point that the telemetry subcarrier demodulators and the voice and biomedical data subcarrier demodulators of the PM and FM channels are identical.

As Figure 4 indicates, the outputs of the voice and biomedical data subcarrier demodulators and the telemetry subcarrier demodulators are routed to a data output selector, which is simply a switch. This allows the ground operators to route the voice, telemetry, and biomedical data to the proper data-processing equipment regardless of whether these data are recovered from the FM or PM channels of the SDDS. In addition, the data selector provides the inputs to seven biomedical subcarrier demodulators for the recovery of the biomedical information.

DEMODULATOR EQUIPMENT

At this point, let us discuss in some detail the various demodulators.

Carrier Frequency Demodulator

The carrier frequency demodulator is shown in detail in Figure 5. The 50-megacycle IF enters the demodulator through an attenuator and is routed to a bandpass filter of either 1- or 4-megacycle bandwidth. However, as shown in the figure, if the switch is in the horizontal position, the input bandwidth is determined by the receiver and it is about 9.3 megacycles. The reason for the use of variable bandwidth is to optimize the performance of the demodulator for the various signals which are transmitted from the spacecraft. The output of the filter is amplified, limited, and converted to 120 megacycles using a local oscillator and doubler. Subsequently, the signal is reduced to video, using a modulation tracking phase-lock loop. Again for optimization purposes, the loop has a 4- and an 11-megacycle closed loop noise bandwidth. The loop bandwidth is selectable through a front panel control. The output of the loop is routed to an output amplifier via a buffer where the video outputs are used to feed the various subcarrier demodulators, television monitors, and tape recorders. The demodulator also contains a threshold detector and a loop lock indicator. A loop disable switch is also available.

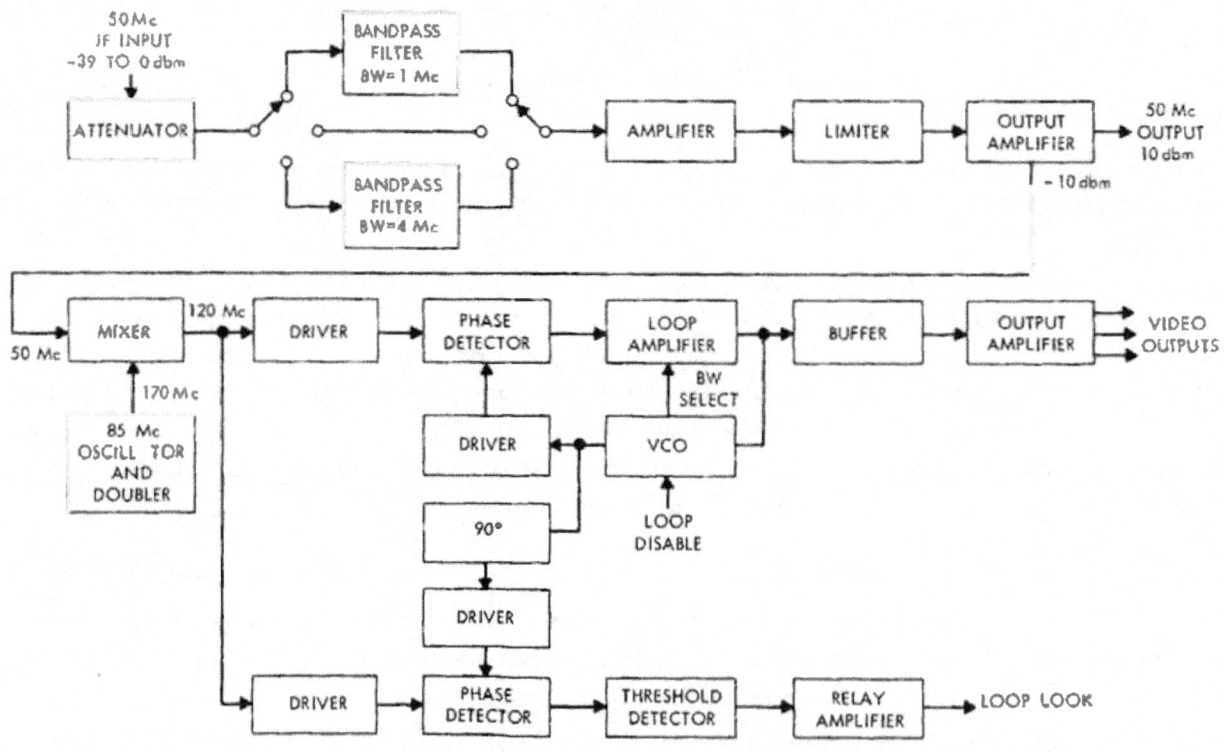

Figure 5—Carrier frequency demodulator.

The PCM Telemetry Subcarrier Demodulator

As previously pointed out, the telemetry demodulators of the PM and FM channels are identical. Therefore, only one will be described here.

The telemetry demodulator is shown in Figure 6. Since the spacecraft has the capability of transmitting 200, 51.2, or 1.6 kilobits per second bi-phase modulated on a 1.024-megacycle subcarrier, it was necessary to provide the demodulator with three different predetection bandwidths. These bandwidths are obtained using a variable bandwidth filter, shown in Figure 6, and their values are 600-, 150-, and 6-kilocycles, respectively. These bandwidths are equal to three times the bit rate and are selected by a front panel control. When the proper bandwidth is selected for

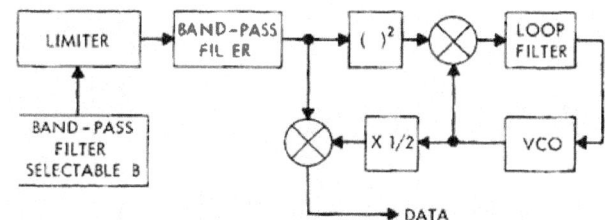

Figure 6—Telemetry subcarrier demodulator.

the particular bit rate transmitted from the spacecraft, the 1.024-megacycle bi-phase modulated subcarrier is routed to a limiter, a filter and a phase detector. The other output to the phase detector is obtained by squaring the 1.024-megacycle bi-phase modulated subcarrier, thus eliminating the modulation and obtaining a 2.048-megacycle stable component which is locked on using a modulation restrictive phase-lock loop. The output of the loop is then routed to the phase detector via an X1/2 multiplier. The output of the multiplier is the PCM information which is routed to the bit synchronizer and PCM data processing equipment.

The Voice and Biomedical Data Subcarrier Demodulator

As previously pointed out, the voice and seven biomedical data subcarriers are frequency-multiplexed and the composite is modulated onto the 1.25-megacycle subcarrier. This is done only when EVA is performed. At any other time the biomedical data are transmitted via the PCM telemetry system and the voice is transmitted alone on the 1.2-megacycle subcarrier.

The voice and biomedical data demodulator is shown in Figure 7. The 1.25-megacycle subcarrier enters the demodulator via an attenuator. If the subcarrier is modulated with the voice information, the 20-kilocycle filter is used prior to detection. If, however, the subcarrier is modulated by voice and biomedical data, then the 35-kilocycle filter is used. When the proper filter is selected, the output is limited and detected using a modulation tracing phase-lock loop. At the output of the loop a low-pass filter is used to recover the voice information and a wideband output is obtained which feeds the biomedical data demodulators via the data output selector shown in Figure 4. The demodulator also contains a threshold detector and in-lock indicator.

The AM Key Demodulator

The emergency key demodulator is shown in Figure 8. A 512-kilocycle band-pass filter is used to recover the subcarrier from the output of the PM buffer amplifier shown in Figure 4. Subsequently the signal is amplified and converted to 1 kilocycle using a 513-kilocycle crystal oscillator. The 1-kilocycle signal is filtered and detected using an amplitude detector. Thus dc or keyed outputs are provided. Also incorporated is an audio output amplifier which provides the required audio outputs.

The Test Unit

For field and laboratory performance evaluation of the various demodulators, the SDDS is provided with a test unit. This unit is shown in Figure 9. It can be seen that this test unit has

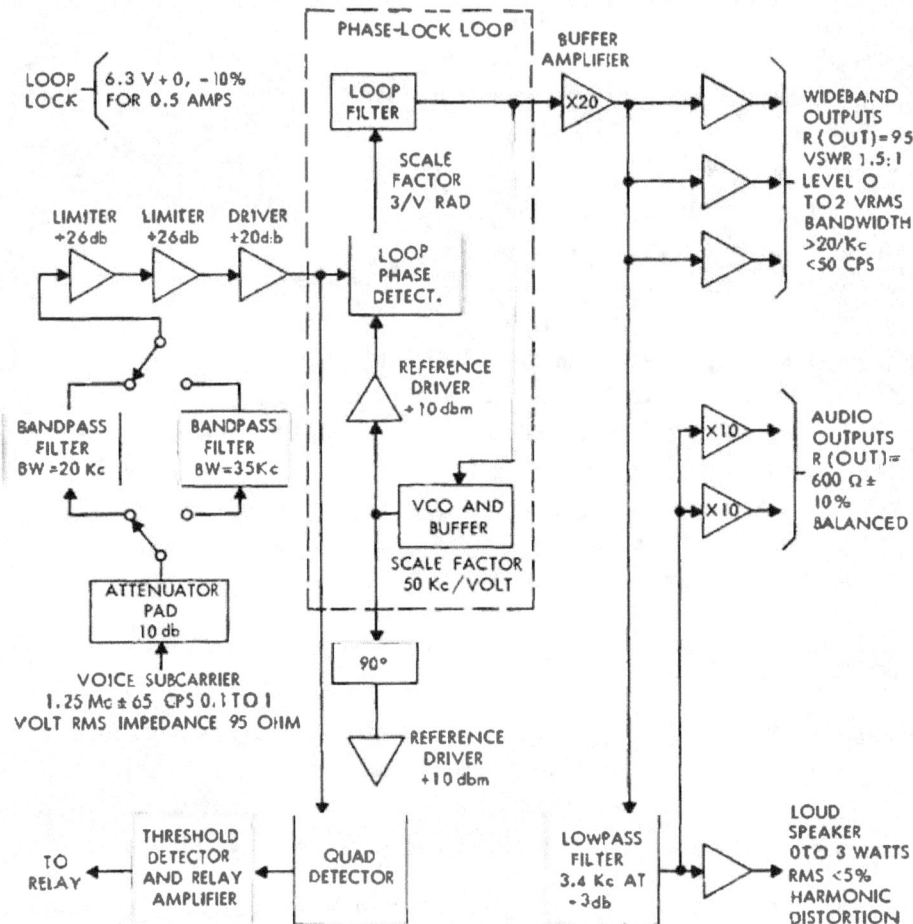

Figure 7—Voice subcarrier demodulator.

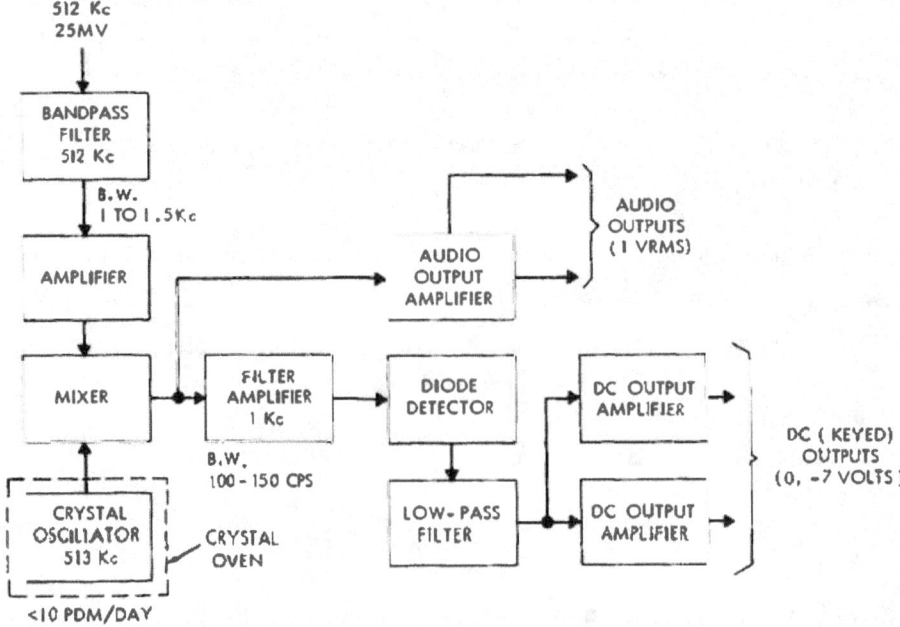

Figure 8—AM key demodulator.

SIGNAL DATA DEMODULATORS

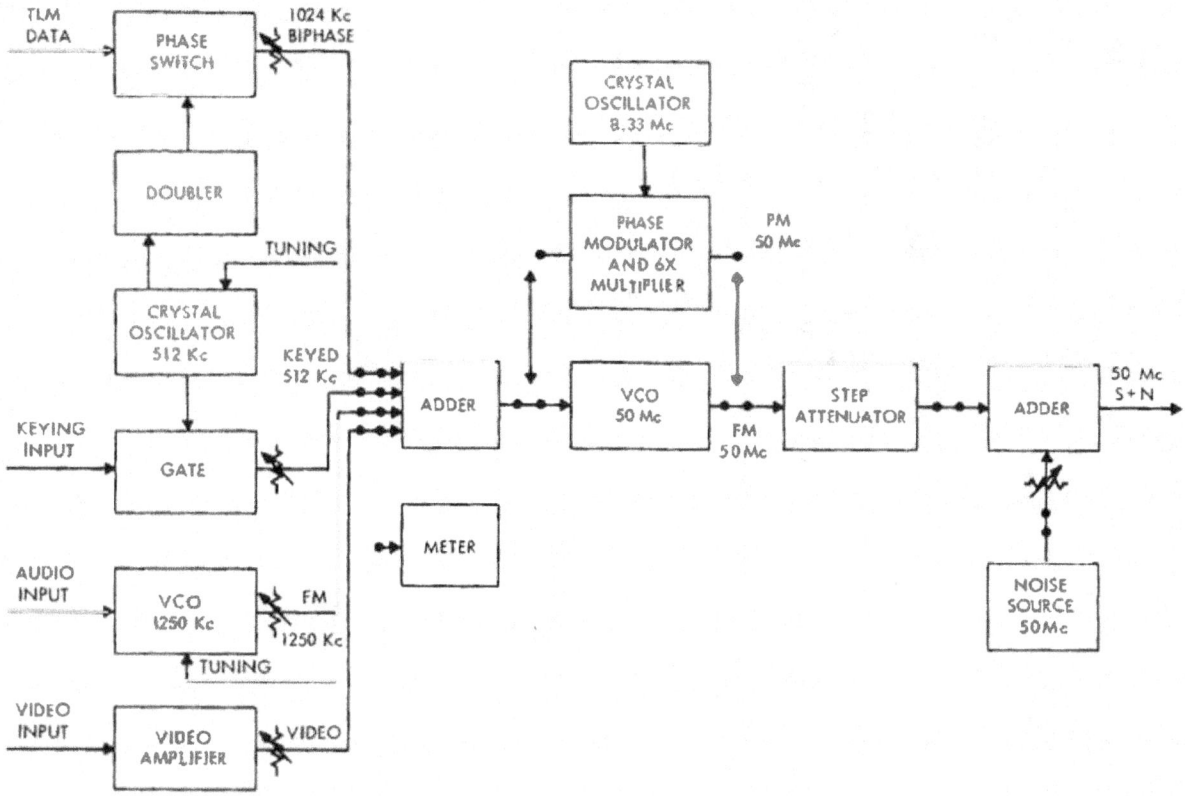

Figure 9—SDDS test system.

the capability of simulating the telemetry subcarrier, the voice subcarrier, the emergency key subcarrier, and the video information. These signals may be summed and routed to a 50-megacycle VCO from which the 50-megacycle IF is derived or to a phase modulator and multiplier where the 50-megacycle PM signal is obtained.

Depending upon which SDDS channel is to be tested, the PM or FM 50-megacycle signal is obtained from the test unit, using a switch, and noise is added to it from a 50-megacycle noise source which is built into the test unit. Thus a signal plus noise at 50 megacycles containing the desired subcarrier and/or video information is available to the SDDS for testing.

Two signal data demodulator systems have been completed and tested. Figure 10 shows the first system. On the top draw there are three loud speakers used for PM voice, FM voice and emergency voice with the various volume controls also shown. In addition, this draw contains the various in-lock indicators and a patch panel for routing the various signals to the various demodulators. The second draw contains the carrier frequency demodulator on the left and the two voice demodulators on the right with their various front panel controls. This draw also contains the emergency key demodulator which has no front panel controls since they are not necessary. The third draw contains the two telemetry demodulators with their bandwidth switching controls and in-lock indicators. The fourth draw contains the various power units. In the fifth draw there are seven phase-lock biomedical subcarrier demodulators with their power supplies.

From the data obtained from testing the first two systems, GSFC is convinced that the SDDS design is very good and there is every reason to believe that this system will operate very well in the field.

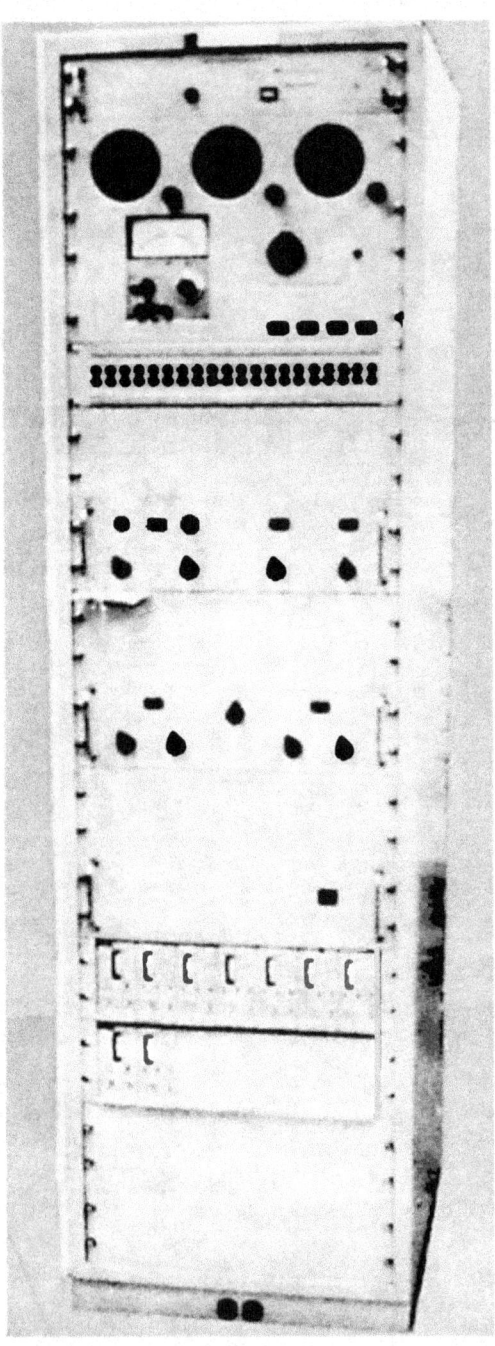

Figure 10—Photograph of signal data demodulator system.

THE UNIFIED S-BAND POWER AMPLIFIER

by
T. E. McGunigal
Goddard Space Flight Center

ABSTRACT

This paper reviews the salient specifications of the 10-kilowatt power amplifier including its interface characteristics and actual performance characteristics determined during the acceptance tests. Special emphasis is given to the RF performance and the more difficult RF tests are discussed. Metering and control circuit operation is covered in the light of the Apollo mission operational requirements. An overall block diagram of the amplifier is included.

INTRODUCTION

The unified S-band (USB) power amplifier will provide uplink data, voice communications, and ranging transmissions to either the Apollo Command and Service Module (CSM) or the Lunar Excursion Module (LEM). In an emergency situation, it would be possible to simultaneously provide two uplinks at 2 kilowatts each to both the CSM and the LEM. Essentially the same power amplifier will be used by all the Apollo ground stations; that is, the 85-foot dish stations, the 30-foot stations, and the shipboard installations. The power amplifiers were designed and manufactured by Energy Systems, Inc. at Palo Alto under subcontract from Collins Radio in Dallas.

SALIENT CHARACTERISTICS OF USB POWER AMPLIFIER

To begin, we will describe the power amplifier by reviewing the salient specifications. It should be noted that in all but a couple of cases which will be mentioned as we go along, these specifications represent demonstrated performance determined during type-testing of the first two units at the manufacturer's plant. The output power of the power amplifier is continuously variable from 1 to 20 kilowatts, cw. The tunable frequency range of the transmitter is 30 megacycles from 2090 megacycles to 2120 megacycles. The bandwidth of the power amplifier is 10 megacycles minimum to the 1db point at all power levels. As a matter of fact, it was determined during type testing that at all but the very lowest power levels the bandwidths are typically 16 to 18 megacycles at the 1db point.

The bandwidth is, of course, wider than the single uplink spectrum to the CSM or the LEM so that it can simultaneously accommodate both spectrums or, on the other hand, provide rapid switching between the CSM and the LEM by switching the exciter.

Tuning time, however, if it is desirable to tune across the band, is less than 10 minutes between any two frequencies in the specified band.

The required drive power to produce the full 20 kilowatts of output is 500 milliwatts. The linearity of the amplifier is such that when driven with two tones, each producing 2 kilowatts of output power and separated in frequency from 1.5 to 8.5 megacycles, the third-order intermodulation products are down at least 30db.

The output power stability of the transmitter was specified to be less than 0.5db of variation for a 24-hour period. Again during type-testing, it was determined that typical variations for a daily period were on the order of 0.1db rather than 0.5 as required. The phase stability and the phase-transient characteristics of the amplifier have not been measured as yet due to the unavailability of a phase-coherent receiver at the manufacturer's plant; however, these parameters will soon be tested by Collins at their Dallas installation. The specifications are that the power amplifier shall contribute less than 1 degree rms residual phase noise when measured with a phase-coherent receiver having a double-sided loop bandwidth of 50 cycles per second. The phase transients shall not exceed 4 degrees peak for power line variations of ±5%.

The wideband noise output of the power amplifier in the receive band, which is from 2270 to 2300 megacycles, will be less than -80dbm per cycle. The in-band noise, that is from 2 kilocycles to 5 megacycles away from the carrier on either side, will be at least 130db per cycle below the carrier level. In order to keep radio frequency interference problems to a minimum, the conducted and radiated interference in the power amplifier is reduced to comply with MIL-I-26600 for Class III equipment.

The interface specifications of the power amplifier are as follows: The input impedance is 50 ohms with VSWR of not more than 1.25:1 over the entire amplifier bandwidth. The output characteristics are such that the amplifier will perform to specifications when terminated with a load having a VSWR of 1.5:1. The input primary power required by the amplifier is 440 volts and 208 volts. The 440 volt-input can be ±10 percent 60 cycle, 3-phase, 3-wire and requires 85 kilovolt-amperes. The 208 volt can also be ±10 percent and is also 60 cycle, 3-phase, 4-wire and requires 6 kilovolt-amperes. It was anticipated that the 85 kilovolt figure would be required for operation in either the single 20-kilowatt mode or the dual 2-kilowatt, linear mode. However, the tube seems to be more efficient than we expected and during type-testing it was determined that a typical value for power consumption at 440 volts is on the order of 66 kilovolt-amperes.

POWER AMPLIFIER SYSTEM

Figure 1 is a power amplifier system block diagram. The exciter-supplied signal comes into the RF input control and monitor circuitry, which consists of an input-isolator, waveguide coaxial switch which permits rapid removal of the drive from the klystron, a directional coupler, and a continuously variable 20db attenuator which permits smooth variations of the input drive to the klystron. From there the signal goes into the klystron and electromagnet assembly. The klystron is an Eimac tube, a 5KM70SJ, which has been modified to actually reduce its standard tuning range, thereby giving greater precision in tuning to preset counter readings and also

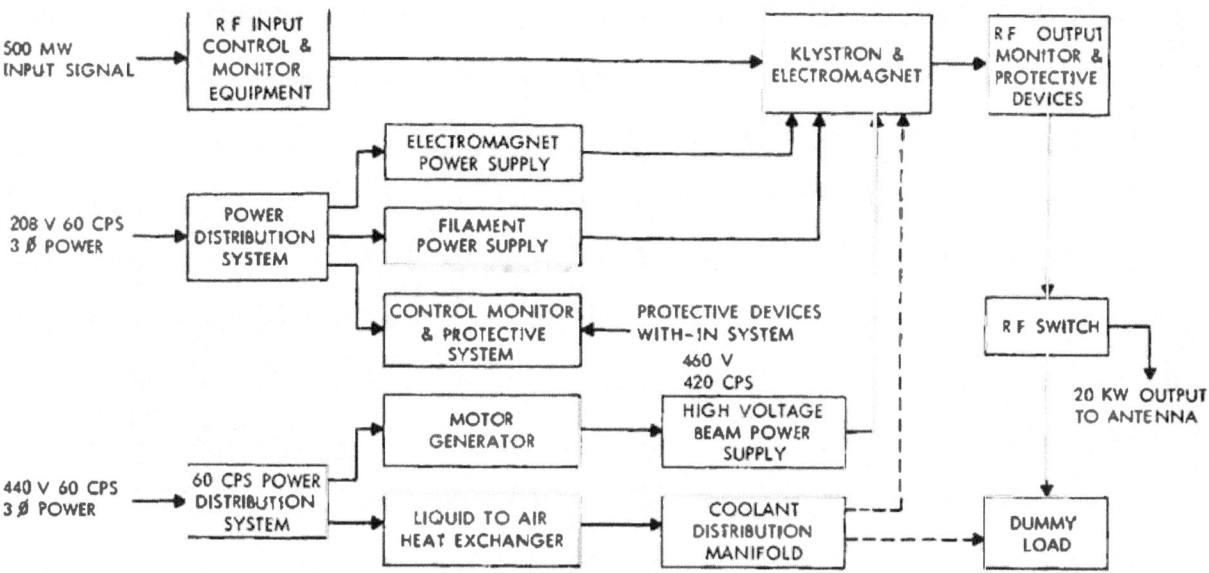

Figure 1—Power amplifier system block diagram.

giving greater transmitter tuning stability. From there, the signal anywhere between the 1- and 20-kilowatt level is fed into the RF output monitor and protective devices which include a 10db output isolator and several directional couplers, one of which feeds the verification receiver, an arc-detector circuit, which detects the presence of an arc in the output waveguide and immediately cuts the drive to the transmitter, and a reflectometer, which senses high output VSWR and again cuts off the transmitter. From there the signal goes into the RF switch which, at the discretion of the operator, controls the flow of the signal either to the 20-kilowatt feed or into the dummy load.

At the middle of the block diagram we see the 208-volt power distribution system which powers the electromagnet power supply, the filament power supply, and the control monitor and protective system. The other input into this circuitry is provided by the various protective devices throughout the power amplifier.

The primary power for the amplifier is 440 volt, 60 cycles, which powers the motor-generator set and the liquid-to-air heat exchanger. The motor-generator converts the 60 cycle per second power to 420 cycle per second power which is then, in turn, rectified by the high-voltage beam power supply (22 kilovolts, 30 amperes). The advantages of using the motor-generator in this system are two-fold. By converting the frequency of the primary power to the high voltage beam power supply, the filtering job can be done better and in less space. Secondly, the motor-generator provides a desired degree of isolation from line voltage variations and transients.

The liquid-to-air heat exchanger is of conventional design. The coolant flows from the heat exchanger into the distribution manifold and then to both the dummy load and the klystron/electromagnet. The flow is regulated in the liquid-to-air heat exchanger so that the temperature of the coolant at the klystron is maintained to within ±5 degrees Fahrenheit of nominal value.

Figure 2 is a unit-identification diagram of the power amplifier system. As will be noted, it is made up of four main enclosures: the power supply enclosure, the RF enclosure, the

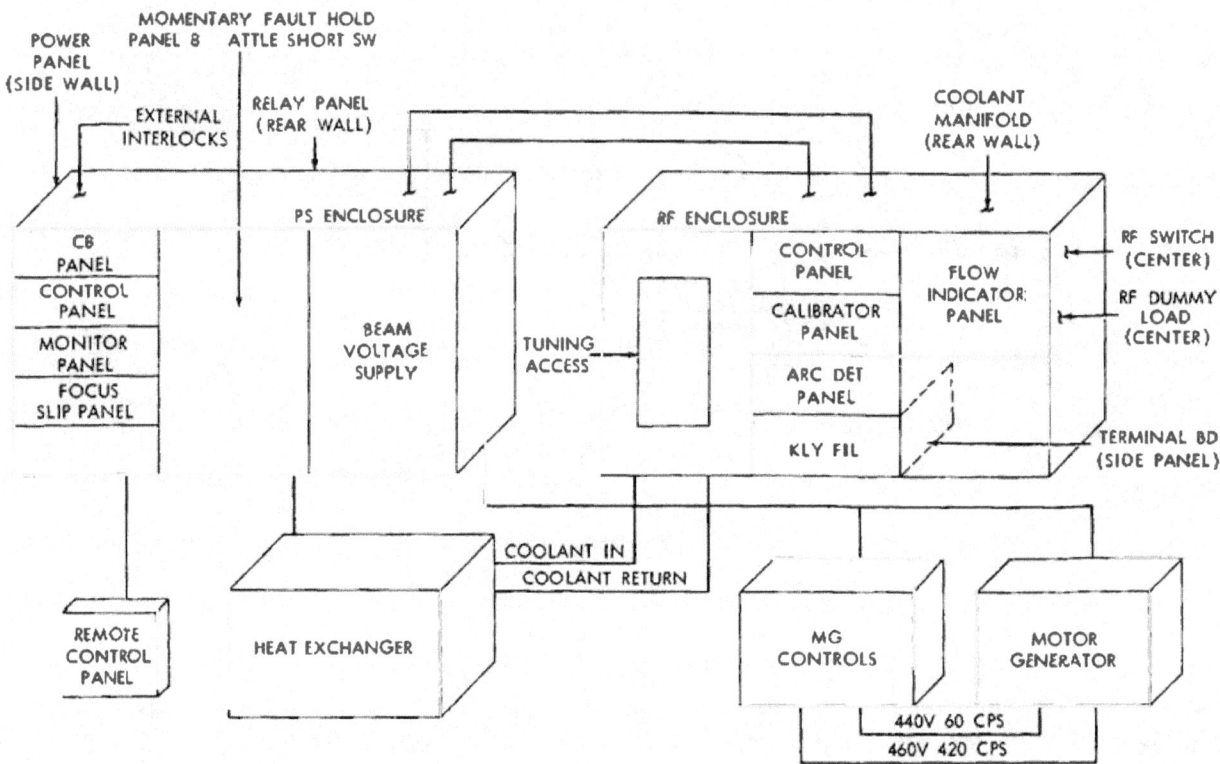

Figure 2—Power amplifier unit identification.

motor-generator and its associated controls, and the heat exchanger. In the power supply enclosure in the far right-hand cubicle, we have the high-voltage beam supply itself. The next cubicle to the left houses most of the control circuitry and the battle-short switch, about which a little more will be said later. On the left-hand portion of the power supply enclosure, we see the circuit-breaker panel which contains the circuit breakers for the whole power amplifier system, the control panel, the monitor panel, and a focus supply panel.

In the RF enclosure in the left-hand cubicle the klystron and the input RF circuitry is housed. A small door is provided in the enclosure so that the transmitter can be tuned without having to open the cubicle door. On the next panel over we have the control and monitor panel which is very much like the one in the power supply enclosure. A calibrator panel which employs a bolometer-type RF power measurement system so that the meters in the various control panels can be calibrated from time to time.

The next panel down contains the arc detector panel and the klystron filament metering and control. On the right-hand panel of the RF enclosure is mounted the coolant flow, pressure and temperature gauges required for monitoring the status of the cooling system. In back of this monitoring panel is mounted the RF switch and the RF dummy load.

In the case of the 30-foot and the shipboard installations, the RF enclosure will be mounted on the ground and power will be fed through a waveguide system and rotary joints to the feed point. In the case of the 85-foot stations, the RF enclosure will be mounted on the steerable portion of the antenna eliminating the waveguide run and the rotary joints.

On the lower left-hand portion of Figure 2 is shown a remote control panel which is again essentially identical to the other two control panels. This remote panel is mounted within the operations building at each site so that the transmitter can be operated from the central control area during a mission.

Figure 3 is a closeup view of the remote control panel which will serve to demonstrate the protective circuit and monitor circuit philosophy employed in the power amplifier. Across the top of the panel are two rows of lights which indicate that a failure has occurred causing the transmitter to cycle down, either by removal of the beam voltage or by cutting back the RF drive, or both.

The particular faults which will cause the power amplifier to cycle down are:

1. Loss of cabinet air flow.
2. Open cabinet door somewhere in the system.

Figure 3—Remote control panel

3. Failure of the heat exchanger.

4. Excessive temperature of:
   a. Magnet.
   b. Body.
   c. Collector.
   d. RF load.
   e. Isolator.

5. Klystron filament undercurrent.

6. Klystron filament air flow failure.

7. Phase failure of AC line.

8. AC overcurrent.

9. Excessive beam voltage.

10. Excessive beam current.

11. Magnet undercurrent.

12. Body overcurrent.

13. Excessive forward or reflected output power.

14. Occurrence of an arc in output wave guide.

15. Loss of waveguide pressurization.

In the case of the Apollo system external interlocks are used to provide protection of the RF horizon. If the power amplifier is illuminating the feed and is directed to a point on the horizon which would be hazardous either to personnel or equipment, RF drive is removed from the transmitter, leaving the beam voltage up so that as soon as the antenna comes above the hazardous point, the transmissions are immediately resumed. If a failure has occurred which causes the beam voltage to cycle down, it takes about 20 or 30 seconds for this to occur, and while this is happening, the beam-voltage lowering light is illuminated. The large light in the middle of the panel indicates that the battle-short switch is in the battle-short position, which is an extreme emergency measure because it wipes out the protective features just described.

The metered quantities on the control panel are the RF output power in either the forward or the reflected direction, the body current, the beam current, the beam voltage, the status of the input circuit, the forward driving power, the reflected power in the input circuit, and the position in db of the input-variable attenuator which is controlled by a switch below it, allowing the operator to manually raise or lower the drive to the power amplifier.

An interlock light test is included which should light all of the fault-indicator lights if the bulbs are in satisfactory condition. A pushbutton also allows the operator to flash a small light in the output wave guide which simulates an arc and should stop both the RF drive and lower the beam voltage. A beam-voltage safety switch is key-operated and when placed in a safe position on any one of the three control panels, precludes the operation of the high voltage beam power supply.

A control under the beam-voltage meter allows the operator to raise or lower the beam manually. If it should become desirable to operate it automatically, there is a switch in the control circuitry which permits him to have the beam-voltage cycle up automatically to a preset level simply by pushing the beam voltage on light.

The first two switches at the bottom of the control panel are the main system on/off. Next is the dummy load antenna-selector switch followed by the ready light, and the two beam-voltage switches. On most of these meters, a second needle is found (on the power supply enclosure control panel only) whose function is to indicate the particular setting of the overvoltage or undercurrent, or whatever it is that is going to represent a fault which will cause the transmitter to cycle down.

Figures 4-8 are photographs of the actual equipment. Figure 4 is a picture of the RF enclosure including the control panel, RF calibration panel, arc detector, reflectometer panel and coolant flow, pressure and temperature gauges. The waveguide can be seen but the klystron itself cannot. Also not visible in Figure 4 are the dummy load and RF switch which are right in back of the coolant monitoring panel.

Figure 5 is a view of the klystron itself. In the upper left-hand corner of the cubicle is mounted all of the RF input circuitry. The tuning controls and the associated mechanical

Figure 4—RF enclosure.

Figure 5—Klystron.

counters which indicate the position of the plunger in the cavity are lined up — five of them since it is a five-cavity klystron — on the body of the electromagnet. To facilitate removal of the klystron from the power amplifier system the rather large chassis slides are mounted so that the whole power amplifier — the klystron and the electromagnet — can be slid out of the cubicle and removed by using a hoist. The cooling connections are the quick-disconnect, leak-proof type.

Figure 6 is a view of the liquid-to-air heat exchanger which will be used at the ground stations. It is approximately six feet tall to the top of the fan.

Figure 7 is a view of the liquid-to-liquid heat exchanger which will be used in the ships' systems. Its use was necessitated in spite of the fact that the ships have a built-in liquid-to-liquid heat exchanger, because the regulation of the temperature of the coolant is not adequate to permit the power amplifier to maintain the specifications; thus, this one is used to regulate the temperature of the coolant at the klystron to ±5 degrees F.

Figure 6—Liquid-to-air heat exchanger.

Figure 8 is a view of the motor-generator and its associated controls. There are two because at the Dallas installation, the contractor is operating two power amplifiers; but, of course, one power amplifier requires only one motor-generator.

In general, the transmitter tunes smoothly and accurately to the preset counter readings and meets most of the requirements by a rather comfortable margin.

Figure 7—Liquid-to-air heat exchanger for use in ships' systems.

Figure 8—Motor-generator.

RANGING SUBSYSTEM – MARK I

by
P. Lindley
Jet Propulsion Laboratory

ABSTRACT

This presentation covers the functional characteristics of the Mark I ranging subsystem including its general description, parameters, constraints, and interfaces with other subsystems of the ranging complex.

The following main functions of the Mark I are discussed: Code generation and synthesis, code synchronization, code shifting, doppler detection, range tallying, output, ranging code acquisition and tracking, and range monitoring.

INTRODUCTION

The JPL ranging system measures the round-trip propagation time of a signal from a ground transmitter to a spacecraft transponder and back to a ground receiver. The accuracy and resolution are independent of the velocity of the spacecraft relative to either the ground transmitter or the ground receiver.

The measurement is made continuously and can be sampled on demand. The unit of measurement is called the range unit (RU) which has the dimension of time. The RU is defined and determined by the frequency of the transmitter S-band carrier and is otherwise invariant. Specifically, the RU is independent of any doppler shift on the signal received from the spacecraft.

The JPL ranging system transmits an S-band carrier, phase modulated by a particular type of pseudo-random binary code (called a ranging code), to a transponder in a spacecraft. The code modulation is detected in the transponder and used to remodulate a down-link S-band carrier (shifted in frequency), which is then received by a ground receiver using the same antenna as is used for transmitting. The ground receiver is a type of phase-locked receiver which tracks both the S-band carrier and the ranging code.

The subsystems directly involved in the determination and readout of range data are the S-band exciter and transmitter, the S-band receiver, the tracking data processor, and the Ranging Subsystem Mark I.

BASICS OF PSEUDO-RANDOM-CODE RANGING

The basic nature of pseudo-random-code ranging is probably best explained by starting with a basic, though inadequate, concept and increasing its complexity as shortcomings become apparent. To this end a series of what in German are called "gedanken-experimente" or *thought experiments* are conducted in this presentation.

Range Measurement On a Stationary Reflecting Target

By assuming a reflecting target rather than a transponder, and by stipulating that it be anchored in space, as shown in Figure 1, its range may be determined in the most straightforward manner.

This target constitutes a standard frequency source which serves to modulate an S-band carrier with periodic single pulses. The reflected modulation signal is detected

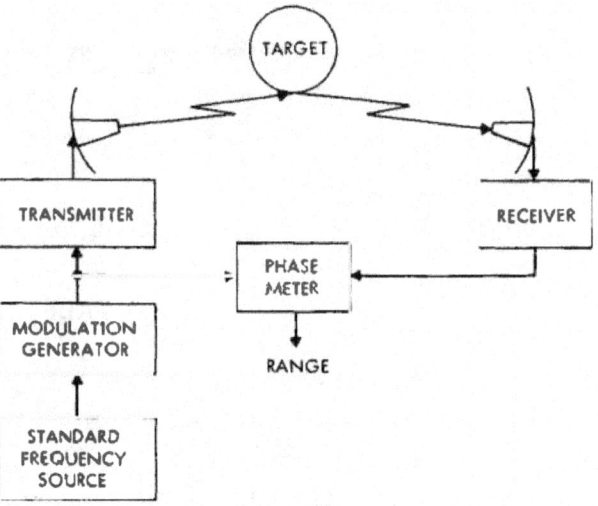

Figure 1—CW radar ranging system.

in a receiver and, by means of a phase meter of some sort, the phase difference between modulation transmission and reception is determined. It will be found that the period of the pulse modulation (i.e., the interval between single-pulse transmissions) must be greater than the round-trip transit time. Otherwise, there will be ambiguities of integral pulse periods. Conventional pulse radar works in this way.

Range Change Measurement on a Moving Reflecting Target

Assuming the reflecting target is permitted to move, our concern is to detect the resultant changes in range. As the target moves, the phase meter reading changes, *increasing* if the target moves *away*.

Resolution of Range Measurements

The resolution of the range increment detection and the initial range determination depend on the precision of the phase meter. By designing the phase meter as a digital device as shown in Figure 2, it is possible to attain almost any desired resolution, which will then be invariant.

The transmitter is shown to be modulated at a much higher frequency which is, in turn, continuously compared with the received frequency in a doppler detector consisting of a mixing device and a counting

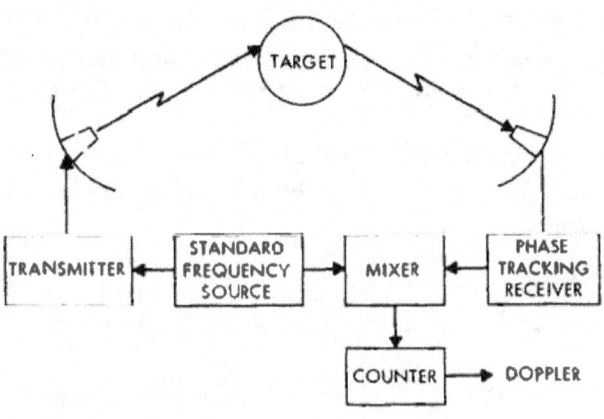

Figure 2—Doppler measurement by coherent CW radar.

device; the shorter the period of the modulating pulses, the finer the resolution of measurement.

The General Ranging Principle

In general, ranging consists of filling the up-link and down-link path with uniformly transmitted cycles of known period, determining the number of cycles in space at the start of ranging acquisition, and subsequently adding or subtracting cycles in accordance with motion of the target.

DETERMINATION OF FRACTIONAL CYCLE OF INITIAL RANGE

Again, considering the target anchored in space, by subdividing the transmitter local oscillator frequency, a transmitter clock signal is derived which serves as one input to a clock doppler detector and also drives a transmitter coder which generates a continuous code (101010...) two bits in length, referred to as transmitter clock code. This then modulates the transmitter coherently with the carrier, as shown in Figure 3.

A receiver clock signal is derived from the received modulation and fed to the other input of the clock doppler detector. In the absence of doppler (since the target is stationary), the received clock code will be delayed with respect to the transmitted clock code by some unknown integral number (n) of clock code periods (τ), plus a delay (d) equal to some unknown fraction of τ. In other words, total round-trip delay = $n\tau + d$.

A clock transfer loop is then provided to help determine the value of d and concern about the number n is postponed until later.

A range tally is provided in the form of a digital accumulating register, in which range numbers, in range units, are tallied in accordance with outputs from the clock doppler detector.

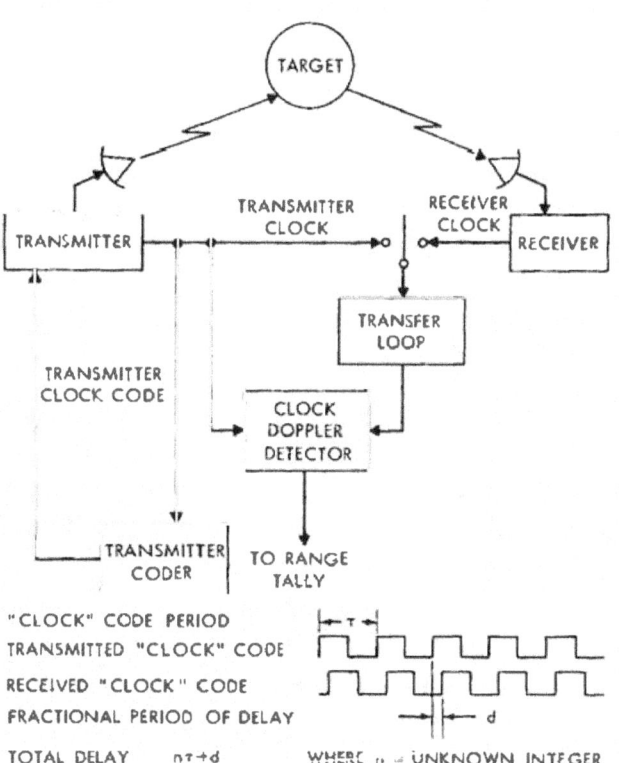

Figure 3—Determination of fractional cycle.

At the start of range acquisition, the input to the transfer loop is switch-connected to the transmitter. The inputs to the clock doppler detector are then identical and there is no output. The range tally is set to zero range units.

The transfer loop is now switched to the receiver. As the transfer loop tracks into the phase without loss of lock, the doppler detector keeps track and causes tallying of range numbers in accordance with what *appears* to be a slight spacecraft motion. This then corrects what would otherwise have been an error in range corresponding to the fractional clock-cycle delay, d.

DETERMINATION OF INCREMENTAL CYCLES OF RANGE

Assuming again that the target is moving, the resultant increments in range will be detected, clock cycle by clock cycle, in the clock doppler detector and will be continually tallied in the range tally.

THE COMPLETE RANGING EQUATION

The determination of total range at time t is based on the relation

$$R_t = R_0 + \int_0^t \dot{R}\, dt$$

where R_0 is the range at some reference time t_0 and the integral is the sum of range increments since that time. The mechanization of the ranging system is quite analogous to solving this integral equation:

First the integration is performed by determining the incremental range throughout the time required for acquisition and the subsequent time of tracking. This is accomplished by continual tallying of range units corresponding to doppler cycles which, in turn, are derived from comparison of received carrier submultiple with transmitted carrier submultiple.

Secondly the constant of integration R_0 is determined by determining the fixed range at the start of ranging acquisition. This is accomplished by tallying range units corresponding to the time offset (or delay) between transmission and reception of a given point in the ranging code at the start of range acquisition. This, in turn, comprises the determination of the fractional clock-cycle delay d (already accomplished) and the determination of the integral number of clock cycles n (next step): $R_0 = d + n_0 \tau$. The operations required to determine R_0 are referred to as range acquisition and are the *only* operations requiring the use of the pseudo-random codes.

MODULATION PATTERN DESIDERATA

For the purpose of precisely determining the number of clock cycles n, a modulation pattern is desired having the following four characteristics:

1. A detectable overall periodicity greater than the maximum anticipated round-trip time. This is required to prevent ambiguous results, and means in effect that the measuring tape should be longer than the distance to be measured.

2. A detectable, fixed, high-frequency periodicity within the overall modulation pattern. This is required for the sake of high resolution or precision of measurement. The clock code period of slightly over 2 microseconds, which we have previously discussed, will serve this requirement.

3. The characteristic of two-level autocorrelation. This means that the overall pattern is required to be such that if the pattern is compared with the same pattern displaced by integral numbers of bits, the two patterns will match exactly in one relative position, and they will fail to match to the same degree in all other relative positions. The *firm* requirement here is that there be only one relative position that yields maximum correlation. If it is possible to have all other relative positions yield uniformly low correlation, the correlation detection is, of course, greatly simplified because it becomes a binary (or true-false) problem, rather than one of precise measurement.

4. The characteristic of being essentially balanced, i.e., of having as many 1's as 0's in it. While this is not an absolute requirement, balanced use of power in the carrier sidebands makes for higher efficiency and better system design.

RANGING CODES

The problem is solved by the use of a pseudo-random binary sequence continually generated in the form of 1's and 0's in digital equipment.

Figure 4 shows two cycles of such a sequence having fifteen binary digits per cycle. Also shown is the rectangular waveform of a ranging code derived from the sequence where 1 is represented by a low level and 0 by a high level.

To see whether and how this code satisfies the requirement for two-level autocorrelation, consider it matched against a second code, identical to that shown, but displaced by any number of digits other than 0, 15 or a multiple of 15. It will be found that the measure of correlation, i.e., of digit-by-digit matching, is uniformly low. It is high when the two codes are in phase, which occurs every 15 displacements in this example.

The resolution obtainable from a code as such is inversely proportional to the digit period. The maximum round-trip time which can be determined without ambiguity corresponds to the total length of the code (here, 15-digit periods).

In Figure 5 a transmitter coder has been provided to generate the repetitive

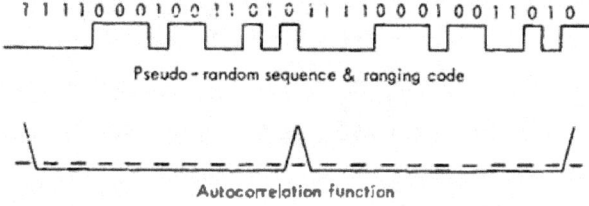

Figure 4—Pseudo-random binary sequence and ranging code waveform.

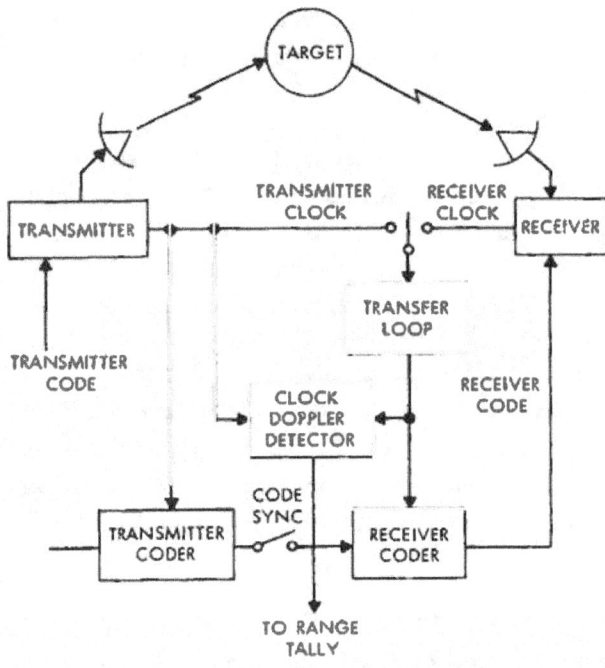

Figure 5—Phase modulation of S-band carrier.

pseudo-random ranging code to be used to bi-phase modulate the transmitted S-band carrier.

A receiver coder has been provided to generate the same code as the transmitter coder, with additional features whereby this code can be matched to the received code in the receiver. It must therefore be time-movable by bits with respect to the received code or, in a way, with respect to the transmitted code. A reference must, of course, be provided for this receiver code shifting. Thus, when the transfer loop is still connected to the transmitter and the range tally reset to zero at the start of acquisition, the receiver coder is code-synchronized to the transmitter coder, as shown schematically by a switch.

THE OVERALL CODE AND CODE COMPONENTS

With respect to the overall code to be used, a bit period of 1/992,000 second or slightly more than 1 microsecond has been chosen for Manned Space Flight Network (MSFN) use. This corresponds roughly to 300 meters of round-trip distance or to 150 meters of one-way range. It was intended that the Mark I reach to 800 million meters, requiring then a code of no less than 800/150 or 5-1/3 million bits. Such a code can be generated directly, but its acquisition would require 5-1/3 million correlation readings to determine the proper match.

It is possible on the other hand to generate such a long code by combining, bit by bit, several repetitive shorter subcodes or code components cleverly chosen. These components must meet the same requirements as the total code. We have chosen five code components whose designations and lengths in bits are:

CL code component of length 2 bits
 X code component of length 11 bits
 A code component of length 31 bits
 B code component of length 63 bits
 C code component of length 127 bits

Provided their lengths in bits have no common factors, the length in bits of the total code is the product of the lengths in bits of the individual components, or 5,456,682 bits.

Further, it is possible to acquire the total code by acquiring the components individually in turn. This reduces the number of correlation readings required from the previously suggested 5-1/3 million to 232. It must be noted that the 2-bit CL component is not acquired by digital means in the Mark I, but rather by the process of locking up the clock loop in the ranging receiver.

Therefore, the transmitter code contains the five components CL, X, A, B, and C, combined bit by bit in accordance with a certain Boolean logical relationship. The receiver code as generated by the Mark I itself contains only the components X, A, B, and C.

THE DOUBLE-LOOP RANGING RECEIVER

Figure 6 shows a schematic diagram of a part of the ranging receiver. Here the CL component is designated as *clock*, the components X, A, B, and C in combination are designated as *code*, and the combination of all five components as *code X clock*. The code generator shown is the receiver coder of the Mark I. Its code output, matched against the received code X clock in a balanced detector, will provide a clock output whose average amplitude is a measure of the degree of correlation between the received code and the receiver code.

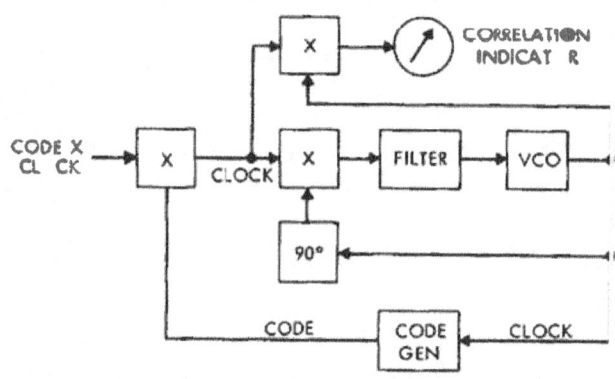

Figure 6—Double-loop code tracking system.

The inner phase-locked loop, or clock loop, is initially locked up to the incoming clock component which it subsequently tracks, whether or not there is any code present.

The outer, or code loop is held in gear by the locked state of the clock loop. It serves no other purpose than to match the received code to the receiver code.

CODE CORRELATION: DETERMINATION OF INTEGRAL CYCLES OF INITIAL RANGE

This matching is accomplished by digitally shifting the components of the receiver code and measuring the correlation indication at each relative shift position until a maximum is obtained.

The total ultimate shift of the receiver code from its initial phase is a measure of the initial range at the start of acquisition or, more correctly, a measure of $R_0 - d$ (both R_0 and d being in units of time).

Each shift of each component in the process of acquisition is noted by adding the appropriate number of range units into the range tally whenever such a shift is made. This in no way interferes with the adding (or subtracting) of the previously mentioned clock doppler tallies, as required by target motion, which can occur simultaneously.

RESOLUTION OF MEASUREMENT IN THE MARK I

The resolution of measurement was indicated earlier as being ±1 clock doppler cycle, for ease of presentation. Since this represents 2-bit periods of about 1 microsecond each, it corresponds roughly to ±2 microseconds or ±600 meters of round-trip distance or ±300 meters of range. Actually clock doppler tallies are made every *quarter* cycle, for a resolution of about ±0.5 microsecond or ±75 meters of range.

Once acquisition has been accomplished, the Mark I automatically switches from tallying every 1/4 clock doppler cycle to tallying every 16th S-band doppler cycle. This improves the resolution by a factor of 72 to ±1 RU or approximately ±1 meter.

MODULATION CHANGE FROM CODE TO CLOCK

At the same time, or any time thereafter, it is possible to disable the full code modulation and modulate the carrier instead, with the 2-bit clock component only. There is, as previously indicated, no further need for the code, the clock component being alone responsible for keeping the clock loop in lock. The advantage of changing from full code to clock code lies in the fact that this not only cuts down on the required sideband power, but also limits the spectral distribution of ranging frequencies to two single spectral lines — 496 kilocycles above and below the carrier frequency.

THE MARK I RANGING SUBSYSTEM

Many of the statements and illustrations in this paper have been purposely simplified to present the basic principles of digital precision ranging as developed at JPL, and as employed in the ranging subsystem Mark I (Figure 7a).

The Mark I is a special-purpose binary digital computer with special input and output interface devices. As part of the receiver-exciter-ranging system, it makes on demand range determinations without *a priori* knowledge of approximate range. Its construction is almost completely modular. Monitor and display equipment, power supplies, and controls are located in the upper half of the single cabinet. Some 300 pitch-wired, solid-state, digital logic modules are mounted in the lower half on movable frames as illustrated in Figure 7b. This subsystem is not really complicated, but is definitely complex.

The principles of digital ranging are essentially straightforward, consisting mainly of counting integral cycles, a fractional cycle, and incremental cycles due to motion, on the two-way radio link between ground and a spacecraft. It has been shown why and how pseudo-random code components are combined and thus used in a ranging code for the purpose of a *fix*. Once the code has been acquired, it is possible and desirable to shut it off and continue to track doppler.

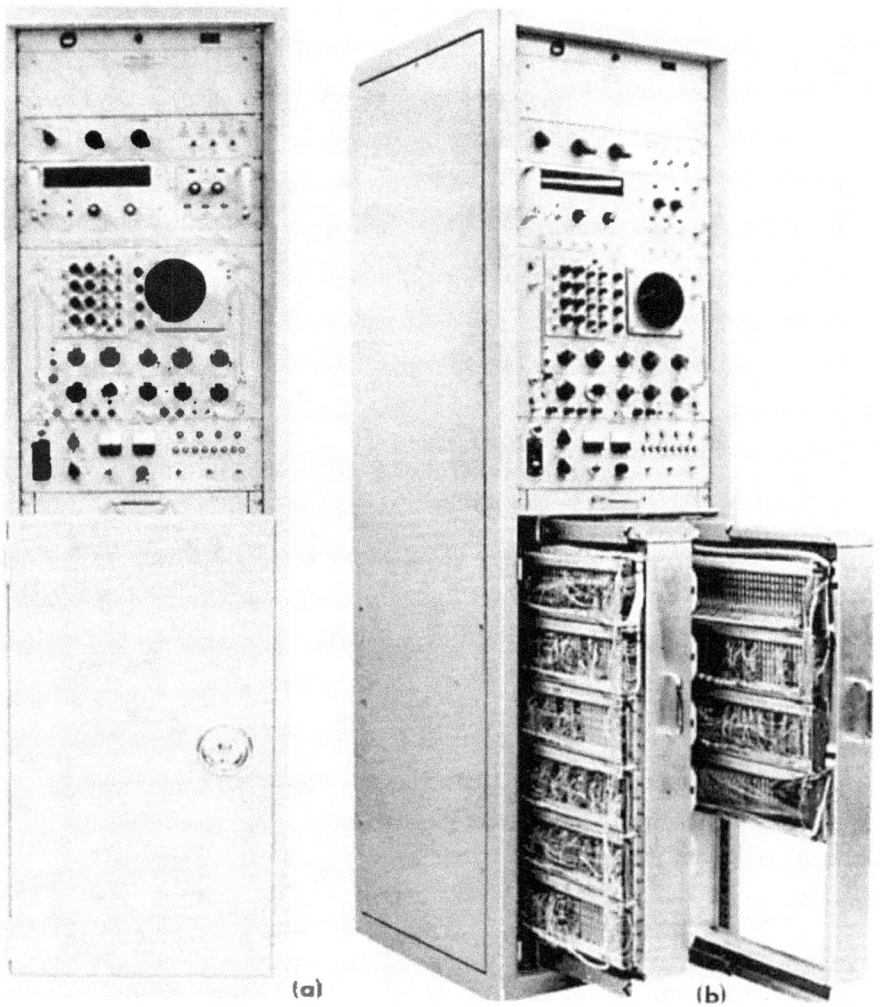

Figure 7—(a) Ranging subsystem Mark I (closed).
(b) Ranging subsystem Mark I (open).

PERFORMANCE CHARACTERISTICS OF THE MARK I

To summarize the performance parameters of the Mark I:

Its maximum unambiguous range of 800,000 kilometers is twice the distance to the moon.

Its resolution is ±1 range unit (RU), which is defined as

$$\frac{2^{21} \text{ light-seconds}}{30 \times \text{transmitted frequency}}$$

and is of the order of ±1 meter.

Overall system inaccuracies of no more than ±15 meters are attributable to drifts and instabilities in ground and space loops.

Minimum range acquisition time is 1.6 seconds at strong signal levels and may possibly go as high as 30 seconds at lunar distances in the MSFN configuration.

Range data output is in binary range units and can be effected once per second.

SHIPBOARD DOPPLER COUNTER, ANTENNA PROGRAMMER, AND TRACKING DATA PROCESSOR

by
W. Hocking
Goddard Space Flight Center

ABSTRACT

This paper describes three subsystems within the Unified S-Band System: the antenna position programmer (APP), the tracking data processor (TDP), and the shipboard doppler counter (SDC). The discussion includes the relationship of these subsystems to all interfacing subsystems of the overall S-band system, and detailed description of their functions and mode of operation. Range and range-rate data problems and characteristics are also treated.

INTRODUCTION

Two important tracking modes within the Unified S-Band System are autotrack (prime) and program (acquisition and backup). The antenna position programmer subsystem provides the backup or program mode. The programmer accepts real X and Y angular data from the angle encoders mounted on the antenna. Command X and Y angle data (predictional data) are entered into the programmer via punched paper tape. The spacecraft prediction data are processed by computer into a five-level punched paper tape with X, Y, and time (command) information existing in Baudot code.

The tracking data processor subsystem collects and formats the Apollo tracking data on-site, and prepares these data for communication to centralized Apollo computers. The tracking data parameters included in the format are antenna X and Y angular information, spacecraft range and range-rate data, and Greenwich mean time (GMT). Apollo S-band transmitter frequency information is also inserted into the data format as required. The tracking data processor provides the Apollo USB system with both a high-speed data rate (up to 2400 bits per second) and low-speed (teletype) data rates.

The shipboard doppler counter subsystem accepts a 1 megacycle biased doppler signal from the JPL range and range-rate subsystem, and operates in two modes: non-destruct and destruct. In the non-destruct mode the doppler signal is counted directly, and is read out upon operator decision. A dual shipboard doppler counter accepts and processes simultaneously two doppler signals: Lunar Excursion Module (LEM) and Command and Service Module (CSM).

The Antenna Position Programmer (APP) is found in all 85 and 30-foot dish sites along with either a "single" or "dual" Tracking Data Processor (TDP); "single" or "dual" pertains to the capability of the USB tracking site to track one spacecraft only or two simultaneously. The TDP system contains a doppler counter for measuring the doppler information prior to insertion into the TDP data format. This doppler counter is packaged separately (SDC) with augmentation to make the unit independent in operation and environmentally acceptable for shipboard use. The SDC is included in the five USB Ships (2 single and 3 dual).

ANTENNA POSITION PROGRAMMER

The relationship of the APP with all interfacing USB subsystems is shown in Figure 1.

There are two antenna tracking modes, autotrack and program. In autotrack mode (prime), the spacecraft RF signals are received and processed by the tracking receiver. The tracking receiver sends to the antenna servo system angular error signals (ϵ_x and ϵ_y) for both the X and Y antenna axes. The servo system converts these signals into X and Y axes drive signals which ultimately move the antenna. A backup mode exists, called program mode, in the event that spacecraft autotracking fails. In this program mode, the APP generates the angular error signals for both X and Y and supplies these signals to the servo system. The autotrack mode may fail if, for instance, the tracking receiver fails or becomes intermittent, if the spacecraft antenna or transmitting system becomes erratic or fails, and if the spacecraft attitude is such as to cause the received RF signal to drop below the autotracking threshold.

The APP receives accurate X and Y axes positional information from angle encoders mounted on the two axes of the antenna. These X and Y angular data are called true or real data since they are the true or actual antenna pointing information. True or real time information

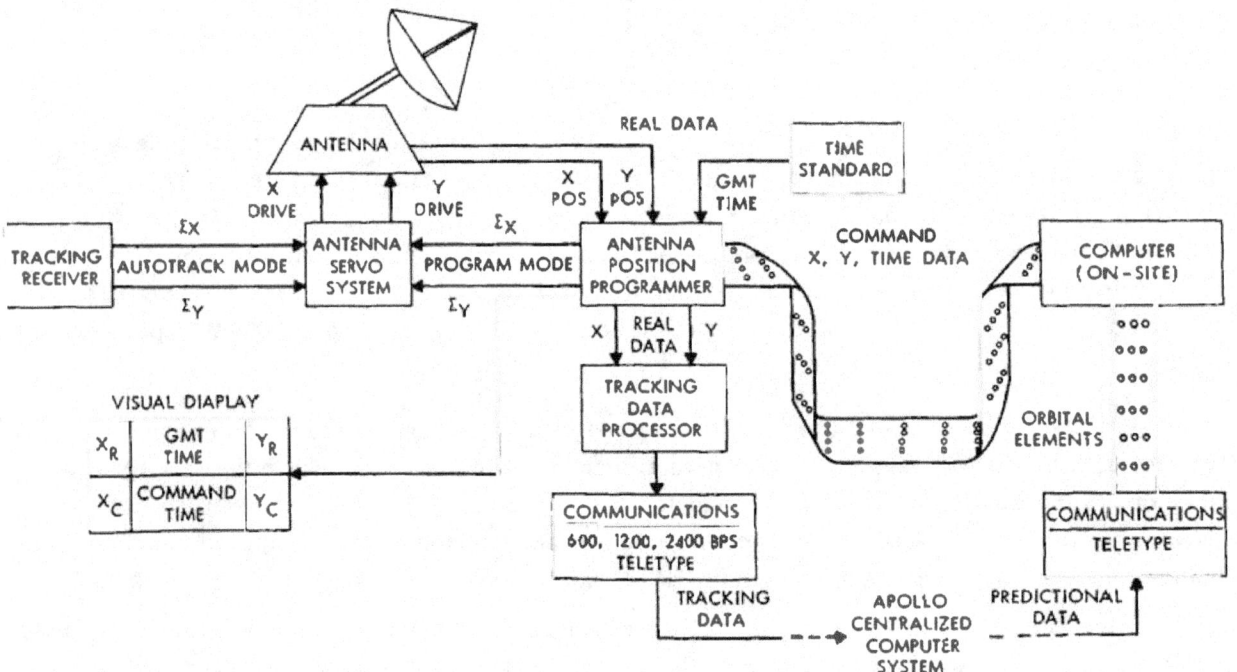

Figure 1—USB system antenna position programmer.

is entered into the APP from the USB time standard. The APP now knows where the antenna is pointing as a function of time. The APP is ready to accept X, Y, and time command data, which are spacecraft predictional information. These command data represent the best estimate of where the spacecraft should be as a function of time.

The APP then compares where the antenna is pointing at a given time (X and Y real data) to where the spacecraft should be at that time (X and Y command) and takes differences. If the angular differences are zero meaning the antenna is pointing to the best estimate of spacecraft position, then the APP feeds ϵ_x and $\epsilon_y = 0$ signals to the servo system. If, however, the antenna is not pointing to the estimated spacecraft position, non-zero ϵ_x and ϵ_y signals are given to the servo system which attempts to null out ϵ_x and ϵ_y. Thus, in program mode, spacecraft tracking is achieved by comparing existing X and Y angles with predicted angles. The accuracy of spacecraft tracking is therefore a function of how well the real antenna angles are measured, as well as a function of the accuracy of the predictional data. Under similar antenna tracking conditions experienced within the GSFC space tracking and data acquisition network, program mode tracking has been a 0.1 degree or better for both 85- and 40-foot antennas.

The predicted or command X, Y, and time data are generated in the following manner: The APP sends to the tracking data processor real X and Y angular data. The tracking data processor accepts these data as well as range, range rate, and time information; formats and transmits these tracking data through communication circuits to the Apollo centralized computer system (ACCS). The ACCS digests this tracking message as well as the tracking data messages from other Apollo facilities and generates an orbit. From this orbit, predictional data are generated and transmitted back to the tracking facility to an on-site computer system. The predictional data are transmitted to the tracking site in an abbreviated form to conserve communication transmission time. The on-site computer accepts the transmitted predictional data and converts these data to antenna drive tapes containing command X, Y, and time data words once each second. The antenna drive tapes (the familiar 5-level teletype paper tape) are produced in advance of the actual tracking operations.

The antenna system is placed in the program mode for spacecraft acquisition purposes. The antenna slews to that portion of the horizon and awaits the spacecraft horizon ascent to occur at the predicted time on the drive tape.

Figure 2 is a simplified block diagram of the APP. Real X and Y angular information from antenna mounted encoders is supplied to the APP in straight binary form (these binary signals are also sent to the tracking data processor). Translation from binary to binary coded decimal (8-4-2-1 BCD) takes place before the X_R and Y_R angles are inserted into the "difference" arithmetic unit. The APP operator has the option of selecting either the actual angles from the antenna or from an antenna simulator. The use of this "select" function is to input to the APP X_R and Y_R angles for test and/or maintenance purposes (simulator mode) without actually requiring the use of the antenna.

The command angles (X_c and Y_c) may be inserted into the APP in one of three ways: from drive tape, on-site computer, or manually inserted by means of digit switches. After command data input selection has taken place, the X_c and Y_c angles may be updated by adding or subtracting bias X and Y angles. The APP provides a visual means of determining the quality of

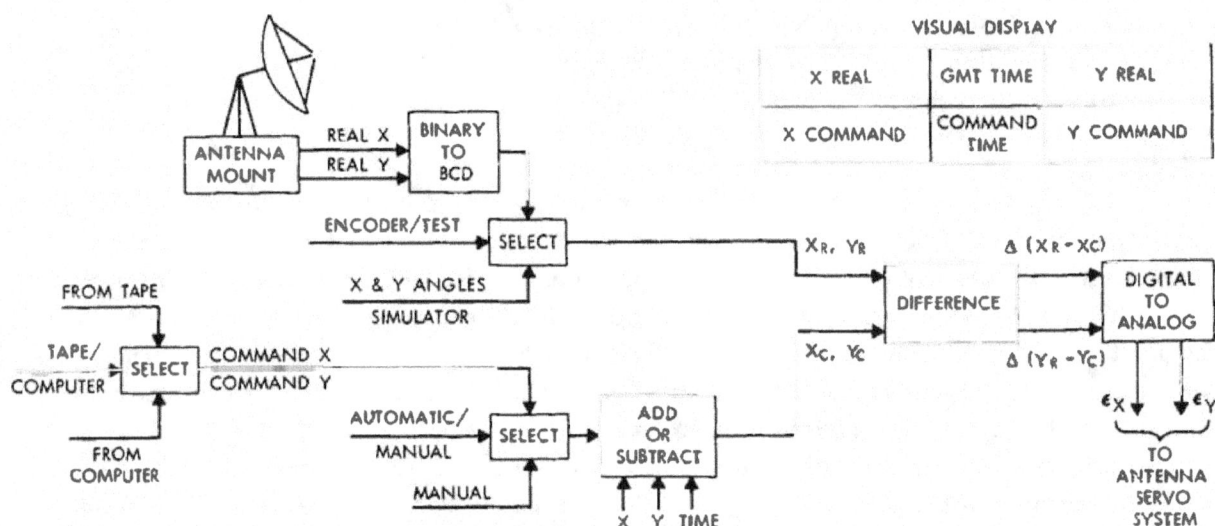

Figure 2—Antenna position programmer.

the command information while the system is in autotrack mode. Any bias error that may exist in the predictional data may be minimized by using this add/subtract bias function. The updated command angles are then passed to the arithmetic difference unit where $(X_R - X_C)$ and $(Y_R - Y_C)$ signals are generated. These difference signals are in digital form and are therefore passed through a digital to analog converted unit which outputs the angular error signals (ϵ_X and ϵ_Y) to the antenna servo system.

The APP also provides visual displays as shown in Figure 2.

Figures 3 and 4 show the local and remote control panels respectively. The antenna simulated angles, as can be seen from Figure 3, are inserted into the APP by means of digit switches.

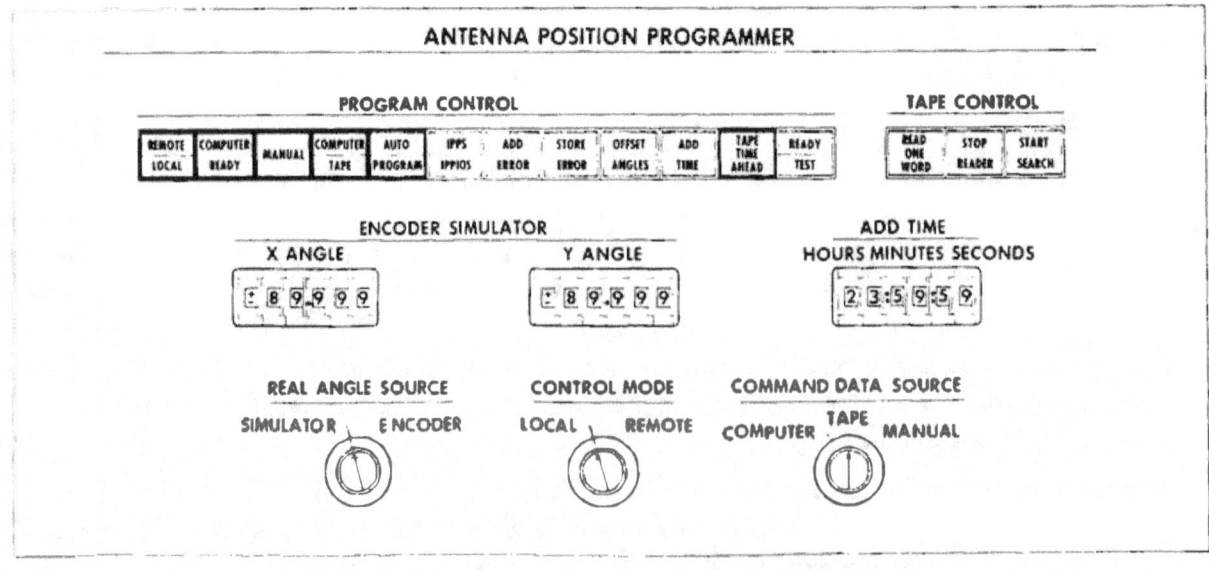

Figure 3—Antenna position programmer control panel (local).

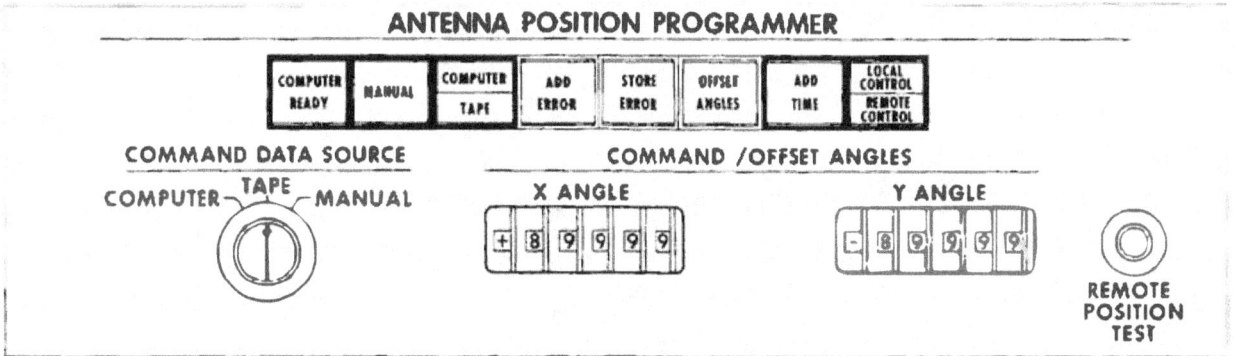

Figure 4—Antenna position programmer control panel (remote).

Changing the command time may be accomplished with the bank of digit switches on the extreme right. Note that the time bias may also be *subtracted* from command time by adding the complement of the time to be subtracted (to subtract 1 hour and 10 minutes add 22 hours and 50 minutes). The dark edged boxes indicate visual displays only, while the undarkened boxes represent both visual display and switch function.

The APP has the capability (while in autotrack mode) of storing any error that may exist between the real X and Y antenna angles and the command X and Y angles. The operator need only actuate the STORE ERROR button during the autotrack mode. If the USB system is forced into the program mode, then this stored error may or may not be added to the command X and Y angles at the discretion of the operator (ADD ERROR). The purpose of the ADD ERROR, STORE ERROR and biasing X and Y digit switches is to provide the means to most effectively update command or predictional data during a mission.

As seen in Figure 4, the operator at the remote location (servo control console) may select the input command source (computer, tape, or manual), provided the local (APP) operator has relinquished control. If the manual input is selected (usually for test and maintenance purposes), the operator can dial in the command data by means of the two banks of digit switches shown. The "remote position test" button allows the remote operator to know the status of his control panel when the APP control is in "local". This information is necessary before transfer of control can safely be made.

TRACKING DATA PROCESSOR

The tracking data processor (TDP) is interfaced with other sub-systems within the unified S-band system (Figure 5). The two receiver systems (representing a dual USB station) each output a thirty binary bit range word and a range rate signal. Certain data identification information is also fed to the TDP, e.g., one-way or two-way doppler, good/bad data determination. Time of year information from the Apollo time standard system is supplied in straight binary form for the high speed section of the TDP and in BCD form for the low speed section of the TDP. Several timing control signals are also used by the TDP.

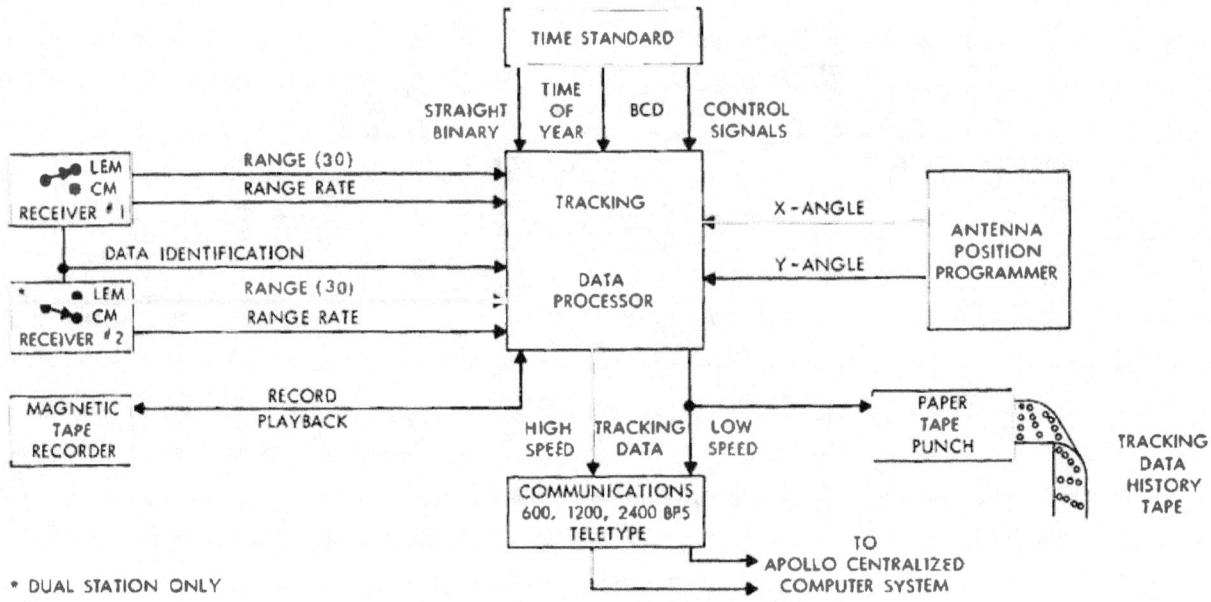

Figure 5—USB system tracking data processor.

As mentioned previously, the straight binary real X and Y antenna angles are received by the TDP from the antenna position programmer. The function of the TDP therefore is to accept range, range rate, X and Y angles, and time information; format these data into a 240-bit frame and prepare these tracking data for communications to the Apollo centralized computer system. The TDP records on a magnetic tape recorder the output serial bit stream in the event that communications circuits are down during a mission. This emergency makes it necessary to playback the data in "post" time through the TDP to the ACCS.

The same precaution is applied to the low speed data. As the low speed data is transmitted to the ACCS, it is recorded on teletype 5-level punched paper tape. This "history data" tape (not to be confused with the antenna drive tape) may be fed into a tape reader and recorded on a page printer in readable form. The TDP must be capable of interfacing the tracking data to a family of communication circuits, namely, 600, 1200, and 2400 bits per second and 6 and 10 characters per second teletype.

Figure 6 is a simplified block diagram of the TDP. The data input unit accepts the input data and distributes them to the high and low speed data sections. The range rate signal from the Jet Propulsion Laboratory (JPL) range and range rate system is a one-volt rms sinusoid of frequency 1 ± 0.18 megacycle. This signal is passed through a doppler counter for measurement prior to entering the high or low speed sections of the TDP. The doppler counter will be described later with a discussion of the shipboard doppler counter. A measurement of the 22 megacycle voltage control oscillator (VCO) transmitter frequency is inserted into the TDP frequency counter which counts this VCO signal directly. This measurement is made automatically and inserted into the data formats (both high and low speed) when a new range measurement is made. The frequency measurement may also be performed at operator discretion by controlling the "manual frequency measurement" button. The frequency data are inserted into the range word (bits 93 through 122 of the high speed format) after measurement takes place.

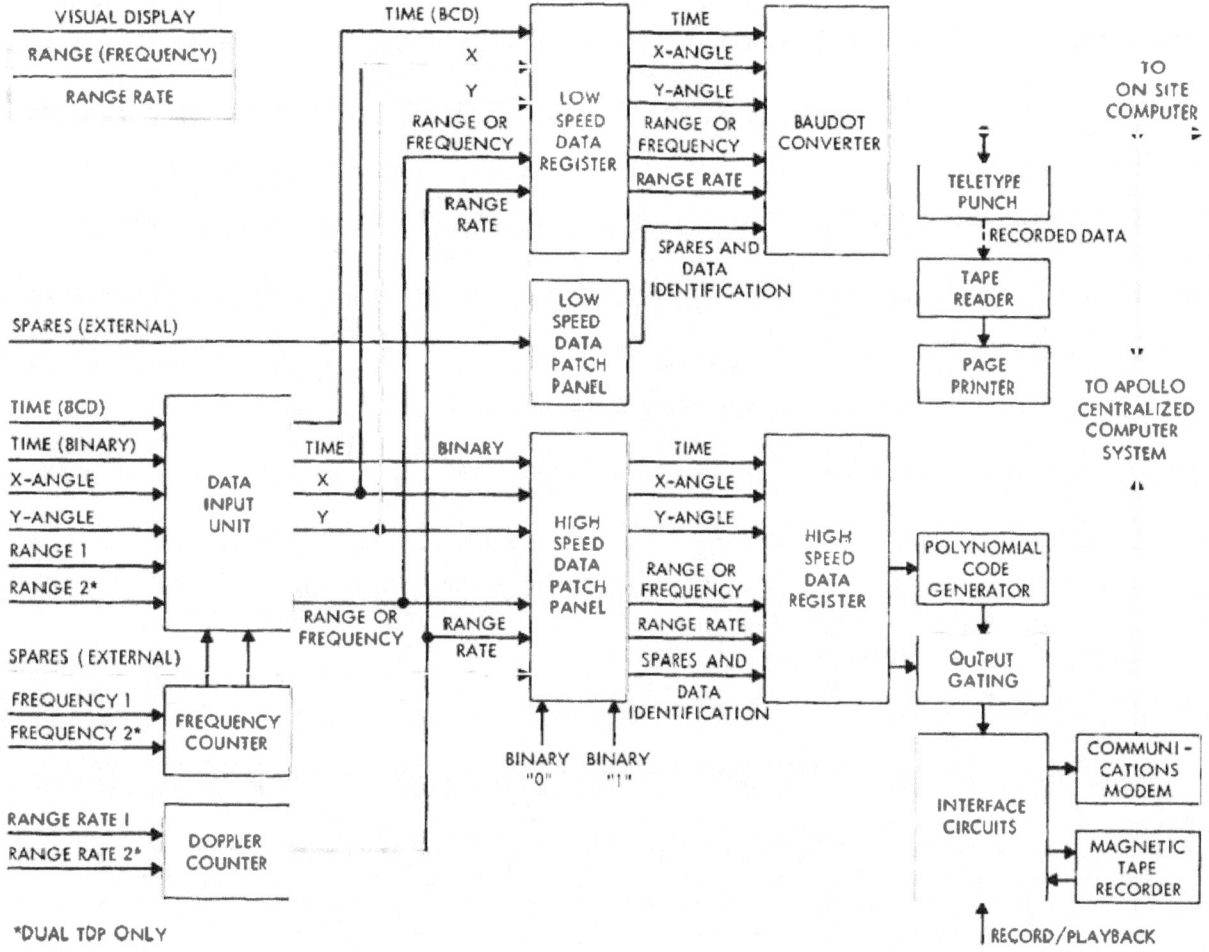

Figure 6—Tracking data processor.

It should be noted here that the TDP high and low speed data formats are transmitted in complete blocks; no partial blocks are gated out as a result of switching, as an example from "receiver number one" to "dual", or from 10 characters per second to one character per second frame rate.

The low speed data are passed through a Baudot converter (converts to familiar teletype code), where the data are permanently stored on punch paper tape. The low speed data are also sent to the on-site computer system where some data smoothing or processing is being considered. The teletype data are transmitted to the ACCS in real time if the facility has available 10-character per second teletype communications.

The high speed data are passed through the output gating circuit and to the interface circuits.

A polynomial code generator accepts the high speed bit stream and generates a powerful 33-bit error detection word which is inserted at the end of each 240-bit frame. The 33-bit error detection word travels with the block of data through all communication circuits and input and output buffers until the data are received at the ACCS. At this time the ACCS generates a second 33-bit error detection word by knowing the original polynomial, assuming no errors due to communication circuits have occurred in the data bit stream itself. The transmitted 33-bit word and the ACCS generated 33-bit word must be identical; otherwise an error has been detected, thus enabling the ACCS to reject that particular frame of data. The 33-bit code working with the seven "start of frame" bits provide a powerful format or data synchronization to be applied to the tracking data.

The TDP system has the capability of inserting a slightly less powerful 22-bit error detection code word. The advantage of the smaller code is that 11 more data bits may be inserted into the data format, should this prove necessary in the future; under good communication transmission conditions, the power of the 22-bit code may be all that is required for protection of the USB tracking data. The normal mode of operation is to transmit to the ACCS the data in real time. The magnetic tape recorder provides a backup to the communication circuit so that the data are stored and may be fed back through the TDP to the ACCS in "post" time.

The high speed format is shown in Figure 7. The Apollo shipboard tracking data format is identical to that in Figure 7. It is hoped that, where possible, any future systems handling Apollo tracking data will use this format. The first seven bits, start of frame (SOF), and the succeeding five station identification bits (SID) are inserted by toggle switches. All data format bits from bit number 13 to bit 205 (when utilizing the 33-bit error detection word) are controlled

1	2	3	4	5	6	7	8	9	10	11	12	13	14	15	16	17	18	19	20	21	22	23	24
SOF	→					→	SOF	SID	→			→	SID	DID	→							→	DID
25	26	27	28	29	30	31	32	33	34	35	36	37	38	39	40	41	42	43	44	45	46	47	48
DID	TOY	→																					TOY
49	50	51	52	53	54	55	56	57	58	59	60	61	62	63	64	65	66	67	68	69	70	71	72
TOY	→				→	TOY	SIGN X	X^{17}	→													→	X^1
73	74	75	76	77	78	79	80	81	82	83	84	85	86	87	88	89	90	91	92	93	94	95	96
SIGN Y	Y^{17}	→														→	Y^1	RDQ	R_A	R_{30}	→	→	R_{27}
97	98	99	100	101	102	103	104	105	106	107	108	109	110	111	112	113	114	115	116	117	118	119	120
R_{26}	→																					→	R_3
121	122	123	124	125	126	127	128	129	130	131	132	133	134	135	136	137	138	139	140	141	142	143	144
R_2	R_1	SP	SP	SP	SP	SP	SP	JPL ID	RDQ	$\dot{R}_{35}$	→											→	$\dot{R}_{22}$
145	146	147	148	149	150	151	152	153	154	155	156	157	158	159	160	161	162	163	164	165	166	167	168
$\dot{R}_{21}$	→																	→	$\dot{R}_1$	SP	SP	SP
169	170	171	172	173	174	175	176	177	178	179	180	181	182	183	184	185	186	187	188	189	190	191	192
SP	→																					→	SP
193	194	195	196	197	198	199	200	201	202	203	204	205	206	207	208	209	210	211	212	213	214	215	216
SP	→										→	SP	EC_{33}	→								→	EC_{23}
217	218	219	220	221	222	223	224	225	226	227	228	229	230	231	232	233	234	235	236	237	238	239	240
EC_{22}	→																			→	EC_1	CSS	CSS

Figure 7—Tracking data processor high speed format.

by a patch panel which enables a binary "1" or "0" or *external* input to be inserted into the format. Therefore all tracking data may be shifted in any sequence desired or deleted and replaced by ones and zeros; the 33- or 22-bit error detection code words are always at the end of the format, followed by two "fixed wired", communication synchronous signal (CSS) bits.

Data identification (DID) bits describe the sample data rates, frequency or range information, and in general any pertinent information the ACCS needs to consider in processing the tracking data. Time is 29 bits of straight binary which is required to give time of year (TOY) information to 0.1-second resolution. X and Y angles are next, followed by the range data quality (RDQ) bit, which is a binary "1" when the range word is good, and binary "0" when bad. Range acquisition (RA) is a binary "1" for one range reading only, when range acquisition has occurred and "0" all other times. RA is succeeded by thirty bits of range data which have a resolution of 1.5 meters; thirty bits therefore represent an unambiguous range measurement of 10^6 miles. The 22-megacycle VCO frequency measurement bits are inserted in the range word when frequency measurement takes place. Spare (SP) bits 123 to 128 are unique spares in that any data placed into these bit locations are also placed into the low speed data format (characters 58 and 59). Bit location number 129 (JPL ID) is a binary "0" when the station operator considers that for any reason the range and/or range rate data are not good.

This information is inserted manually and may be used effectively to override the RDQ bits. $\dot{R}$DQ ($\dot{R}$ represents range rate) serves the same function as the RDQ bit. Thirty-five bits of range rate enables the TDP to count maximum doppler frequency without ambiguity for a period of approximately seven hours. Bits 166 to 205 are at present designated as spare bits. The 33 error control or error detection bits (EC) and the communication synchronization signal bits (CSS) terminate each 240-bit frame of data.

Figures 8a and 8b show when the TDP multiplexes range and range-rate data in the high-speed format for both single and dual TDP systems.

The low-speed data format is shown in Figure 9. This format consists of 60 teletype characters and is transmitted to the ACCS in real time over a 10-character per second teletype circuit. The characters 1, 2, and 60 (line feed, figures, and carriage return) offer the low speed data "hard copy" readability when the tracking data (history tape) are played through a page printer. Station ID and spare characters are inserted into the format by means of toggle switches. The time of year, unlike time in the high-speed format, is binary coded decimal (BCD). All other data in the format are the same as the data contained in the high speed format with the exception of the spare, EC, and CSS bits. As an example, consider the X angle information in the high-speed format (17 bits plus one sign bit, straight binary). The first three least significant bits of the 18-bit X angle sequence are grouped together to form the octad, 8^0 (character 26). The next three bits are grouped to form 8^1. All 18 X angle bits are grouped into 6 octads with the most significant bit of the 8^5 octad (character 21) being the X-angle sign bit. Approximately four hours of low speed data can be recorded on one roll of paper tape when the recording rate is maximum (one frame each six seconds).

Figures 10a and 10b show when the TDP multiplexes range and range rate data in the low speed format for both single and dual TDP systems.

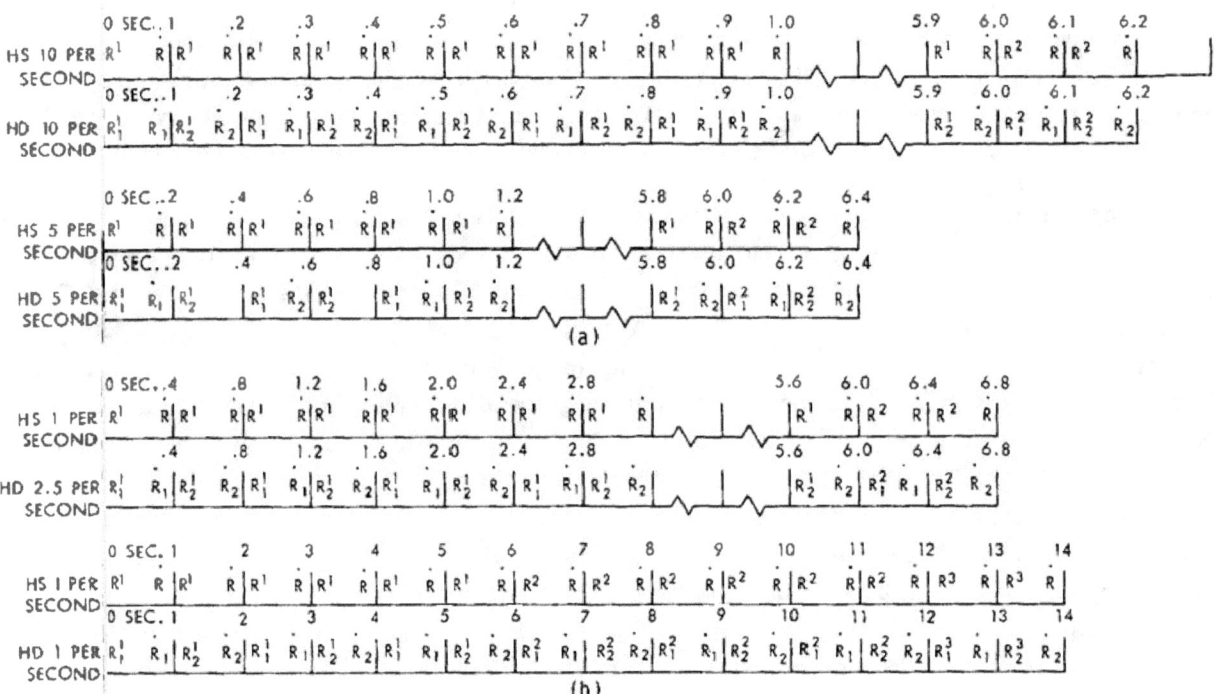

Figure 8—Range and range rate recording - high speed.

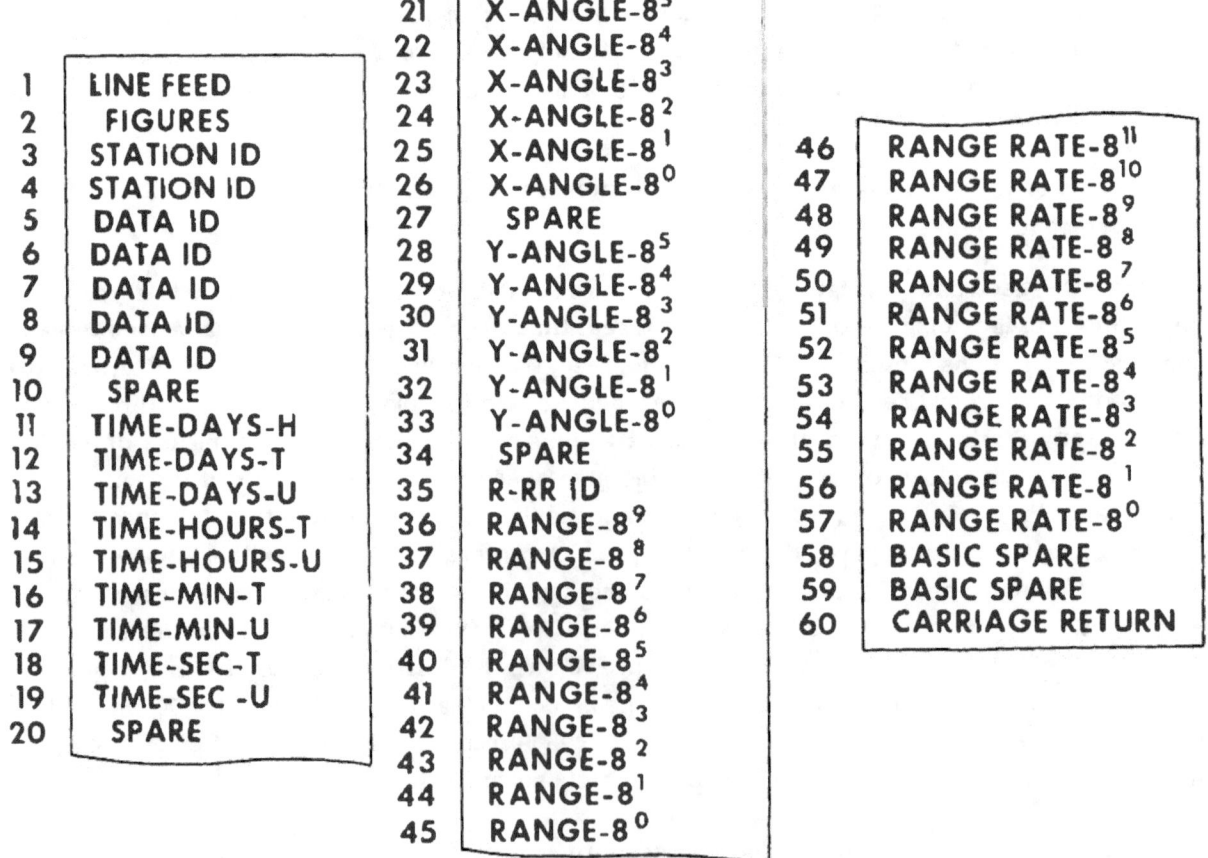

Figure 9—Tracking data processor low speed data format.

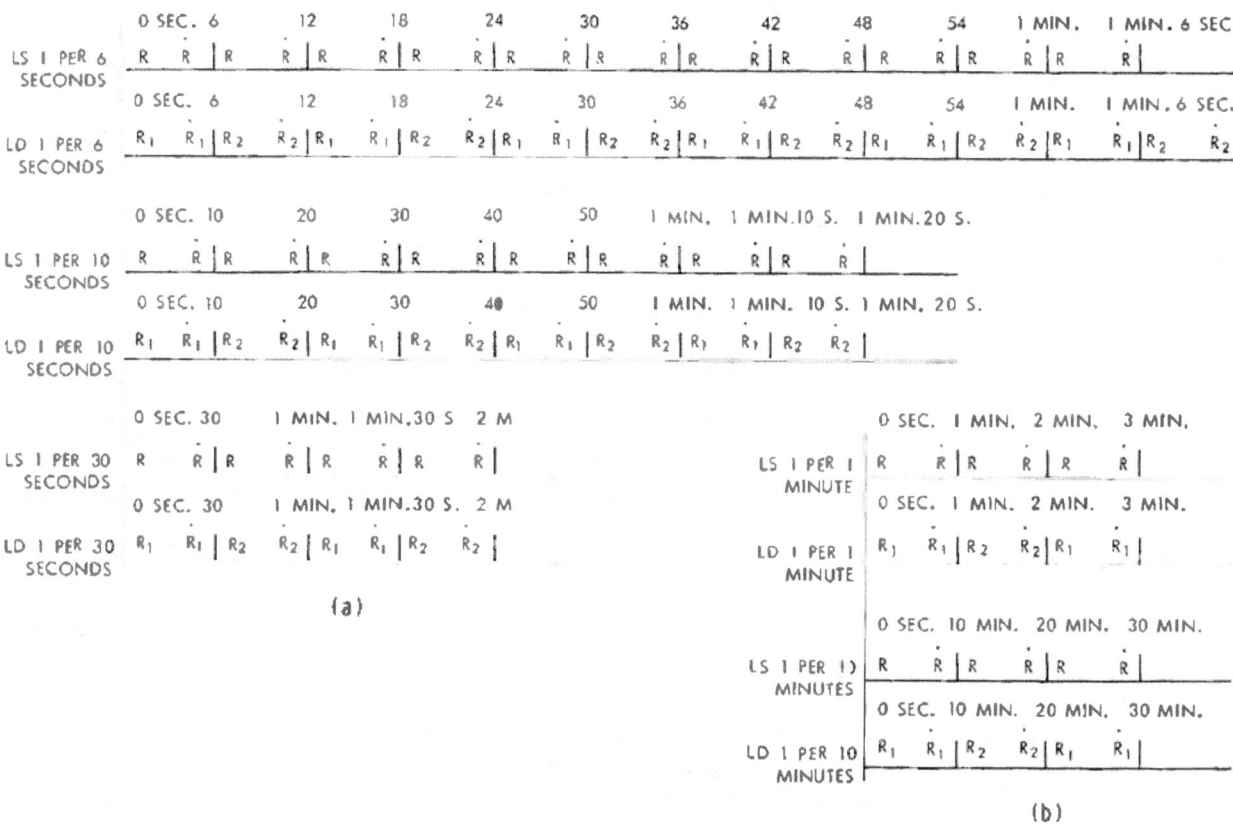

Figure 10—Range and range rate recording - low speed.

SHIPBOARD DOPPLER COUNTER

Figure 11 is a simplified block diagram of the shipboard doppler counter (SDC). The doppler counter existing within the TDP is identical in function to the SDC; for shipboard use, the doppler counter was repackaged and designed for shipboard environment and to be independent from a control and maintenance consideration. The SDC has two modes, destruct and non-destruct.

In the destruct mode the 100-megacycle counter performs a high resolution time measurement of a predetermined number (N) of doppler cycles. After the measurement is made (N_1 for 10 times a second measurement rate and N_2 for 1 time per second rate), the contents of the 100-megacycle counter is transferred to the shipboard centralized computer system.

In the non-destruct mode, the N_∞ (N very large) switch position is selected which enables the N-counter to count continuously without reset. Extreme care is exercised in transferring the N-counter contents to storage each 0.1 second or once per second. During the non-destruct mode the measurement of doppler is such that a pulse is never gained or lost throughout the entire spacecraft tracking time. The input doppler signal is S-band doppler (the SDC is designed for plus or minus 180-kilocycles biased about a standard 1Mc frequency). The input signal is a one-volt rms sinusoid.

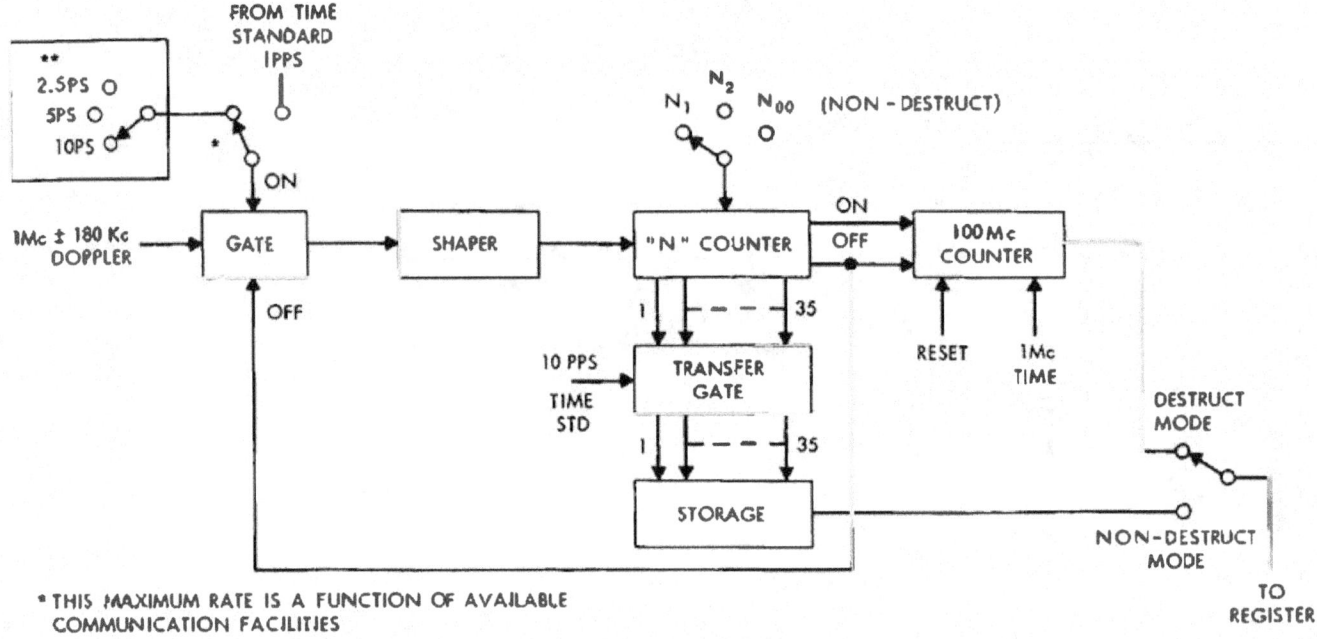

Figure 11—Shipboard doppler counter.

The destruct measurement is made in the following manner. The on pulse to the gate comes from the time standard system. The 10 characters per second is selected when the communication circuit available is 2400 bits per second (10 per second times 240 bit frames). When the gate opens, the first positive zero crossing of the first gated doppler sinusoid is shaped and advances the N-counter to state one. This transition turns the 100-megacycle counter on, which starts immediately (within 10 nanoseconds) counting 100-megacycle pulses (synthesized 1 megacycle input from the Time Standard System). The gating of the approximate 1-megacycle doppler signal continues until the count of "N" (N_1 or N_2) has been reached in the N-counter. At the positive going zero crossing of the Nth pulse, an off pulse is generated which turns off the gate and the 100-megacycle counter. The result is a 100-megacycle counting operation for precisely the period of time between positive going zero crossings of the first and Nth doppler sinusoid. All gating and counting operations are performed with Apollo time standard coherent pulses. The advantage of N-counter techniques is briefly that high resolution measurements (10 nanoseconds) are available with short measurement periods (100 milliseconds).

In the non-destruct mode the same operation takes place, except that "N" is very large and the off pulse never occurs to turn off the gate. The counting operation continues without disturbance or reset throughout the spacecraft tracking period. The advantage of the non-destruct technique is that a continuous, undisturbed doppler measurement is made with no data "gaps" such as exist in the destruct mode.

APOLLO USB HIGH SPEED DATA FORMAT

Bit	Function	Description
1 - 7	Start of Frame	Selectable by switch
8-12	Station or Site Identification	Selectable by switch
13	Data Identification	Range Rate Destruct Mode is binary "one"* Range Rate Non-Destruct Mode is binary "zero"*
14	Data Identification	Range Rate N1 Mode "1" Range Rate N2 Mode "0"
15	Data Identification	High Speed Format 10, 5 or 2.5 per second "1" (The rate 10, 5 or 2.5 depends on communication circuits available at site. This is identified by the SID bits 8 - 12) High Speed Format 1 per second "0"
16	Data Identification	Real data "1" Test data "0"
17	Data Identification	Object number - not defined
18	Data Identification	Object number - not defined
19	Data Identification	Object number: LEM "1" CSM "0"
20	Data Identification	Auto Track Mode "1" Other "0"
21	Data Identification	Time, X and Y angle data (manual) Good "1" Bad "0"
22-23	Data Identification	Doppler Mode Bit 22 23 0 0 one way doppler 0 1 two way doppler 1 0 multiple (non-coherent) 1 1 multiple (coherent)
24	Data Identification	Frequency Standard Identification Rubiduim "1" Crystal "0"
25	Data Identification	Range data (bits 93 - 122) "1" Frequency data (bits 93 - 122) "0"

*Binary one is designated by "1" and binary zero by "0".

Bit	Function	Description
26-54	Time of Year	
55-72	X-Angle	
73-90	Y-Angle	
91	Range Data Quality	
92	Range Acquisition	
93-122	Range	
123-128	Spare	Information inserted into these bit locations also go into the Low Speed Format (characters 58 and 59)
129	Manual Good/Bad Data Information	Range and Range data (manual) Good "1" Bad "0"
130	Range Rate Data Quality	
131-165	Range Rate	
166-205*	Spare	Information inserted into these bit locations is not inserted into the Low Speed Format.
206-238	Error Control Bits	These bits provide <u>error detection</u> to the data.
239-240	Communication Sync Signal	

*When 22 bit error control is utilized bits 166-216 are spare.

APOLLO USB LOW SPEED DATA FORMAT

TTY Character		Function	Description
1	Baudot	Line Feed (LF)	Fixed for hard copy, computer, and communication purposes.
2	Baudot	Figures (FIGS)	Fixed for hard copy, computer, and communication purposes.
3	Decimal	Station ID-Tens	Variable
4	Decimal	Station ID-Units	
5	Octal	Data Identification	Data ID bits 13, 14 and 25
6	Octal	Data Identification	Slow Speed System data rate 0 = 1P6S 1 = 1P10S 2 = 1P30S 3 = 1PM 4 = 1P10M 5 = Manual 6 = Spare 7 = Spare
7	Octal	Data Identification	Data ID bits 16, 20 and 21

TTY Character	Function		Description
8	Octal	Data Identification	Data ID bits 22, 23 and 24
9	Octal	Data Identification	Data ID bits 17, 18 and 19
10	Baudot*	Spare	Variable (patch panel)*
11	Decimal	Time-Days-Hundreds	Time of day and year
12	Decimal	Time-Days-Tens	
13	Decimal	Time-Days-Units	
14	Decimal	Time-Hours-Tens	
15	Decimal	Time-Hours-Units	
16	Decimal	Time-Minutes-Tens	
17	Decimal	Time-Minutes-Units	
18	Decimal	Time-Seconds-Tens	
19	Decimal	Time-Seconds-Units	
20	Baudot*	Spare	Variable (patch panel)*
21	Octal	X Angle - 8^5	High Speed Data Bits 55, 56 and 57**
22	Octal	X Angle - 8^4	High Speed Data Bits 58, 59 and 60
23	Octal	X Angle - 8^3	High Speed Data Bits 61, 62 and 63
24	Octal	X Angle - 8^2	High Speed Data Bits 64, 65 and 66
25	Octal	X Angle - 8^1	High Speed Data Bits 67, 68 and 69
26	Octal	X Angle - 8^0	High Speed Data Bits 70, 71 and 72
27	Baudot*	Spare	Variable (patch panel)*
28	Octal	Y Angle - 8^5	High Speed Data Bits 73, 74 and 75
29	Octal	Y Angle - 8^4	High Speed Data Bits 76, 77 and 78
30	Octal	Y Angle - 8^3	High Speed Data Bits 79, 80 and 81
31	Octal	Y Angle - 8^2	High Speed Data Bits 82, 83 and 84
32	Octal	Y Angle - 8^1	High Speed Data Bits 85, 86 and 87
33	Octal	Y Angle - 8^0	High Speed Data Bits 88, 89 and 90
34	Baudot*	Spare	Variable (patch panel)*
35	Octal	R-R ID	High Speed Data Bits 91, 92 and 130
36	Octal	Range - 8^9	High Speed Data Bits 93, 94 and 95
37	Octal	Range - 8^8	High Speed Data Bits 96, 97 and 98
38	Octal	Range - 8^7	High Speed Data Bits 99, 100 and 101
39	Octal	Range - 8^6	High Speed Data Bits 102, 103 and 104
40	Octal	Range - 8^5	High Speed Data Bits 105, 106 and 107
41	Octal	Range - 8^4	High Speed Data Bits 108, 109 and 110
42	Octal	Range - 8^3	High Speed Data Bits 111, 112 and 113
43	Octal	Range - 8^2	High Speed Data Bits 114, 115 and 116
44	Octal	Range - 8^1	High Speed Data Bits 117, 118 and 119
45	Octal	Range - 8^0	High Speed Data Bits 120, 121 and 122
46	Octal	Range Rate - 8^{11}	High Speed Data Bits 129, 131 and 132

*A TTY "space" will be patched into the format until character is needed for data

**All X, Y, R and Ṙ characters, the first bit listed (i.e., 55 in character 21) is the most significant bit of that octad. The least significant bit of the octad is the last bit listed (57 in character 21).

TTY Character	Function		Description
47	Octal	Range Rate - 8^{10}	High Speed Data Bits 133, 134 and 135
48	Octal	Range Rate - 8^9	High Speed Data Bits 136, 137 and 138
49	Octal	Range Rate - 8^8	High Speed Data Bits 139, 140 and 141
50	Octal	Range Rate - 8^7	High Speed Data Bits 142, 143 and 144
51	Octal	Range Rate - 8^6	High Speed Data Bits 145, 146 and 147
52	Octal	Range Rate - 8^5	High Speed Data Bits 148, 149 and 150
53	Octal	Range Rate - 8^4	High Speed Data Bits 151, 152 and 153
54	Octal	Range Rate - 8^3	High Speed Data Bits 154, 155 and 156
55	Octal	Range Rate - 8^2	High Speed Data Bits 157, 158 and 159
56	Octal	Range Rate - 8^1	High Speed Data Bits 160, 161 and 162
57	Octal	Range Rate - 8^0	High Speed Data Bits 163, 164 and 165
58	Octal	Basic Spare	High Speed Data Bits 123, 124 and 125
59	Octal	Basic Spare	High Speed Data Bits 126, 127 and 128
60	Baudot	Carriage Return (CR)	Fixed for hard copy, computer, and communication purposes

NOTE: All octal and decimal characters are Baudot encoded for transmission over teletype communications circuits.

APOLLO PRECISION FREQUENCY SOURCE AND TIME STANDARD

by
R. L. Granata
Goddard Space Flight Center

ABSTRACT

A brief description is given of the Apollo precision frequency source and time standard.

The precision frequency source will be the source of accurate frequency reference for the Apollo time standard, S-band ranging system, tracking data equipment, and other site functions. This unit contains four frequency references, two rubidium resonators and two crystal resonators. These four sources are processed through combiner circuitry and distributed throughout the S-band system. A brief description of the operation of this equipment is discussed.

The Apollo time standard will be utilized to generate station time and to correlate station and spacecraft events. The Apollo time standard consists of redundant clocks and time-code generators to enhance reliability. Various time codes and standard frequencies are generated by this equipment for use by the S-band system. Peripheral instrumentation is also included with this system to aid in synchronizing the network to a common source. Time synchronization is established with the National Bureau of Standards Station, WWV, and frequency correlation is maintained to one part in 10^{10} with VLF techniques.

INTRODUCTION

The basic or primary function of the Apollo precision frequency source and the time standard is to provide a reliable and accurate tag or reference scaler for tracking and telemetry data. These systems are utilized within the S-band system to supply precision frequencies and pulse repetition rates to various subsystems, such as the tracking data processor and modem, the antenna position programmer, the ranging subsystem, and the digital command subsystem.

SYSTEM DESCRIPTION

Precision Frequency Source

The precision frequency source contains two rubidium frequency standards and two crystal frequency standards. The operator has the option to select any one of the four standards as the primary standard and to also select the order of preference for the other three. Normal operating procedure is to select the more accurate of the two rubidium standards as the primary unit.

Detection of failure, which is performed in the control logic, can be noted in several ways: excessive drift rate, amplitude variations, and power supply failures. Upon detection of a failure in an oscillator, that unit will be switched off line. If this happens to be the operating standard, the second preference unit is made operational. If a failure occurs in a secondary unit, the ones lower in preference are moved into a higher position. Detection of failure is also indicated on the control panel to inform the operator of the present equipment status.

The control logic also performs another important function, that of frequency control of the secondary frequency standards. The secondary standards are compared to the primary standard in a phase detector, and the resulting phase error signal drives a proportional and an integral control loop. The proportional loop controls phase variations of the output signal by changing the control voltage on the varicap frequency control in the oscillator. The integral loop drives a servo and nulls out the oscillator drift or long term errors. The servo also drives an indicator dial which is calibrated for frequency correction. The rubidium control loop is similar in nature except that current variations are made to adjust the magnetic field around the gas cell. The maximum tracking rate of this servo will follow an error of 2×10^{-9}. A stepping motor is employed in the integral control loop to eliminate the need for the generation of 60 or 400 cycles per second power.

The output of the combiner (5 Mc), which is the selected standard, is then synthesized by redundant paths into two additional output frequencies, 1 megacycle and 100 kilocycles. These three frequencies are then expanded in the distribution amplifier to furnish the required output configuration. This unit interfaces with the time standard at this point. All users of these frequencies in the S-band system obtain their outputs directly from the time standard system.

In order to obtain a highly reliable unit, each frequency standard contains its individual power supply and battery pack. The control logic and distribution system have redundant power supplies and battery packs, each capable of supplying the required power. The power units for the two rubidium standards, the control logic, and the distribution amplifier are identical and can be interchanged in case of a major failure.

Time Standards

The time standard contains redundant clocks and time code generators. Logic gating within the signal distribution area of all output functions allows the operator to select the operational time standard. Switching is performed manually and at the discretion of the operator. All other functions, binary coded decimal (BCD) to binary conversion of time, status clock control signals, special frequencies, and clock synchronization signals are generated after the time standard selection circuits. The easiest method of describing the functions of this system is to logically follow the signals through each unit.

The frequency divider portion of the digital clock receives two redundant one megacycle signals from the precision frequency source. These signals are added through a resistive network. The combined signal is then limited and squared by means of a zero crossing detector developing pulses for driving the digital circuits. The two inputs signals are generated in the precision frequency source and thus are always phase coherent. Loss of one input signal results in a phase change of less than 20 nanoseconds. The 1 megacycle square wave is then divided by means of 8-4-2-1 BCD decades to a 1 pulse per second rate. At the 100-kilopulses per second point in the divider chain, two methods for obtaining time synchronization are employed, an analog and a digital technique. The analog method employs a continuous phase shifter utilizing a sine-cosine potentiometer. This method is used for daily time corrections to compensate for oscillator drift as measured by the VLF equipment. The digital method adds or subtracts pulses from the 100-kilopulses per second bit stream. Several rates of correction are manually selectable by the operator. This technique is utilized to aid in synchronizing the frequency divider to WWV or to the redundant time standard.

The digital clock accepts the 1 pulse per second from the frequency divider and divides by appropriate factors to obtain time of day and day of year information. For synchronization with WWV, time of year information is inserted into the digital clock by means of a bank of switches. A visual display is also present above this unit to indicate time. Outputs from the frequency divider and digital clock are employed in the time code generator to develop the four NASA time codes.

The outputs of the flip-flops in the frequency divider, time of year information from the digital clock, and serial time codes are presented to the logic switch in the signal distribution unit for selection, as a group, to develop the output signals.

SIGNAL DISTRIBUTION

As well as performing the time standard switching function, this unit develops all the output signals; contains expansion for the AC and DC line drivers; modulates the time codes with their appropriate carrier frequencies; develops the special frequencies for the tracking data processor; such as 2400, 1200, and 600 pulses per second; and develops the necessary signals for driving the time synchronization equipment.

The BCD to binary converter generates time of year information into straight binary data to be compatible with the tracking data processor requirements. This information is available in three different granularities, one second, one tenth of a second, and one millisecond. One complete conversion of the time of year information is made within 100 microseconds.

The STATUS clock develops count down and elapsed time information to display the mission status. Outputs from this unit are provided to drive multiple displays for use within the ground station.

The VLF and WWV equipment are provided to aid in the time synchronization and maintenance of time synchronization of the station clock. The WWV signal is displayed upon an oscilloscope and used to synchronize the station one pulse per second to the received WWV signal.

The VLF equipment phase locks a 100-kilocycle output from the precision frequency source with the received VLF signal. A phase plot is obtained which defines the frequency error of the precision frequency source and the accumulated time error since the previous VLF measurement.

The patch panel provides a convenient method of connecting signals to the other station subsystems.

ACCURACY & RELIABILITY

System accuracy can be broken down into two basic categories: frequency and time.

Frequency

The frequency accuracy is determined by the inherent stability of the rubidium gas cell, VLF tracking capabilities, and operator ability.

The specification placed upon the Apollo precision frequency source is to maintain a frequency setting within 5×10^{-11} for a one year period. This value is placed upon the system for worst case operation. These units could deviate to the maximum on a daily basis and still meet the specification, but past data have shown that the daily mean frequency does not exhibit deviations greater than 2×10^{-11}.

VLF tracking is now widely utilized to compare frequencies of a house standard to that of a stable reference transmitted in the 10 to 30-kilocycle band. In order to achieve the system accuracy required by the S-band, VLF signals must be monitored on a continuous basis. The VLF equipment in this system has a resolution of one microsecond which gives approximately a frequency resolution of 1×10^{-11} over 24 hours. Due to diurnal shifts and ionospheric noise, the system at best can resolve five microseconds or 5×10^{-11} over 24 hours.

If this data is closely monitored and tabulated for a period of one week or more, an accuracy of 2×10^{-11} is achievable. These results have been repeatedly obtained in our laboratory and we feel that with properly trained site personnel, the same results can be achieved. This then leads into the question of the site personnel effects on the system frequency accuracy. The author has already stated that the best achievable results are 2×10^{-11} and that on a daily basis 5×10^{-11} can be maintained. With reasonable performance on the part of the site personnel, the author feels that our site will maintain its frequency standard between these two values.

Time

Time accuracy is determined by the synchronization technique employed and the frequency accuracy of the precision frequency source. The method of obtaining time synchronization is to employ the WWV, HF time signals. These transmissions are received and displayed upon an oscilloscope. The station pulse is then compared with the WWV tick, a one pulse per second signal, and aligned to be coincident with it. The jitter on the WWV pulse allows setting of the station clock to no better than ±0.5 millisecond. Propagation times from the transmitting station have been calculated previously. The station pulse which is coincident with the WWV signal is a delayed ($\Delta 1$ pps) pulse. The amount of delay being that calculated for the particular site. By synchronizing the $\Delta 1$ pulse per second with the received WWV signal, the stations undelayed or normal 1 pulse per second output is synchronized to the WWV transmitted signal.

In this way, the S-band network will be synchronized to a common base. The propagation delay times are then the second source of station time error. The calculated values for each station are based upon both experimental data and mathematical calculation. The estimated error for these values is in the order of one millisecond. This is partly due to the uncertainty in the mathematical model used, seasonal ionospheric variations, and ionospheric disturbances.

The third source of time error is that derived from a frequency error. As stated earlier, the maximum expected frequency error is 5×10^{-11}. This error would contribute approximately 5 microseconds a day to the time error. The author mentions this source of error to show its magnitude, and because if left unchecked, it accumulates and contributes to the other errors.

VLF monitoring of frequency is done by displaying phase error. The display is exhibited directly as time error. The operator can then adjust his clock on a daily basis to eliminate the time error due to frequency offset. The total system error based upon these facts is in the order of one millisecond. Other techniques are now being developed which show promise for improving this error by an order of magnitude. These may be incorporated into this system at a later date if mission requirements show the need for improvement.

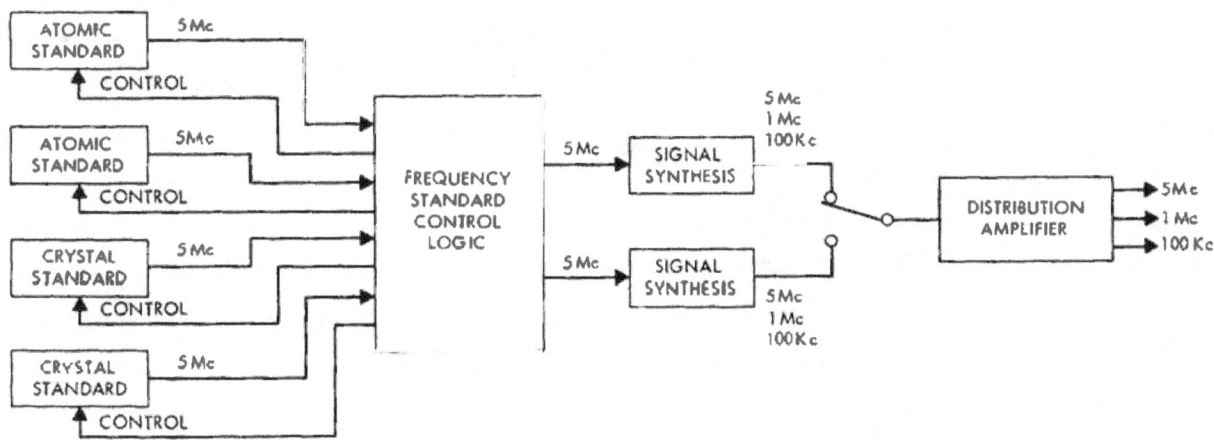

Figure 1—Apollo precision frequency source.

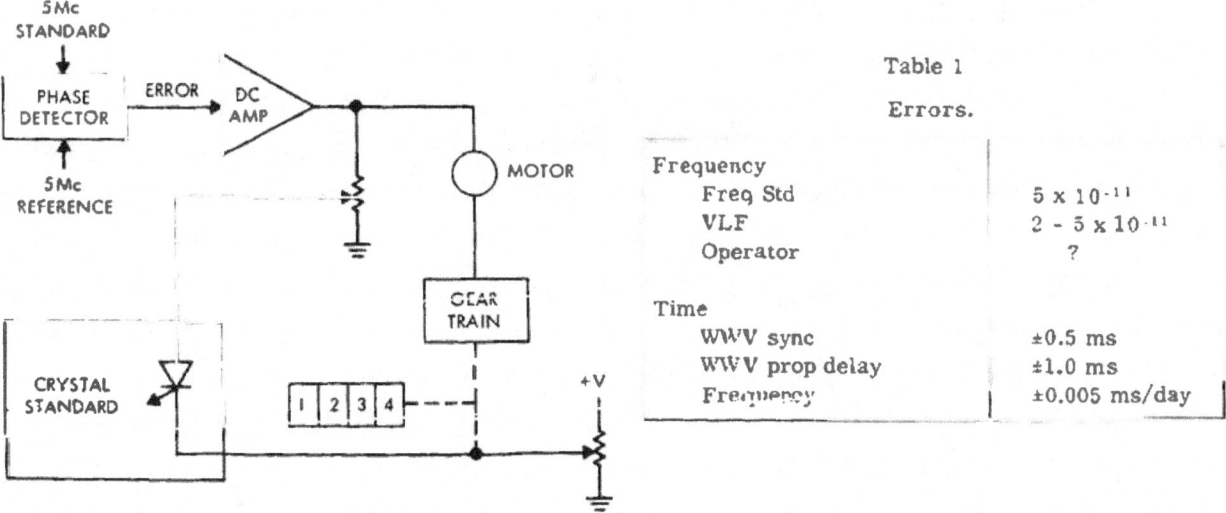

Figure 2—Frequency control loop.

Table 1

Errors.

Frequency	
Freq Std	5×10^{-11}
VLF	$2 - 5 \times 10^{-11}$
Operator	?
Time	
WWV sync	±0.5 ms
WWV prop delay	±1.0 ms
Frequency	±0.005 ms/day

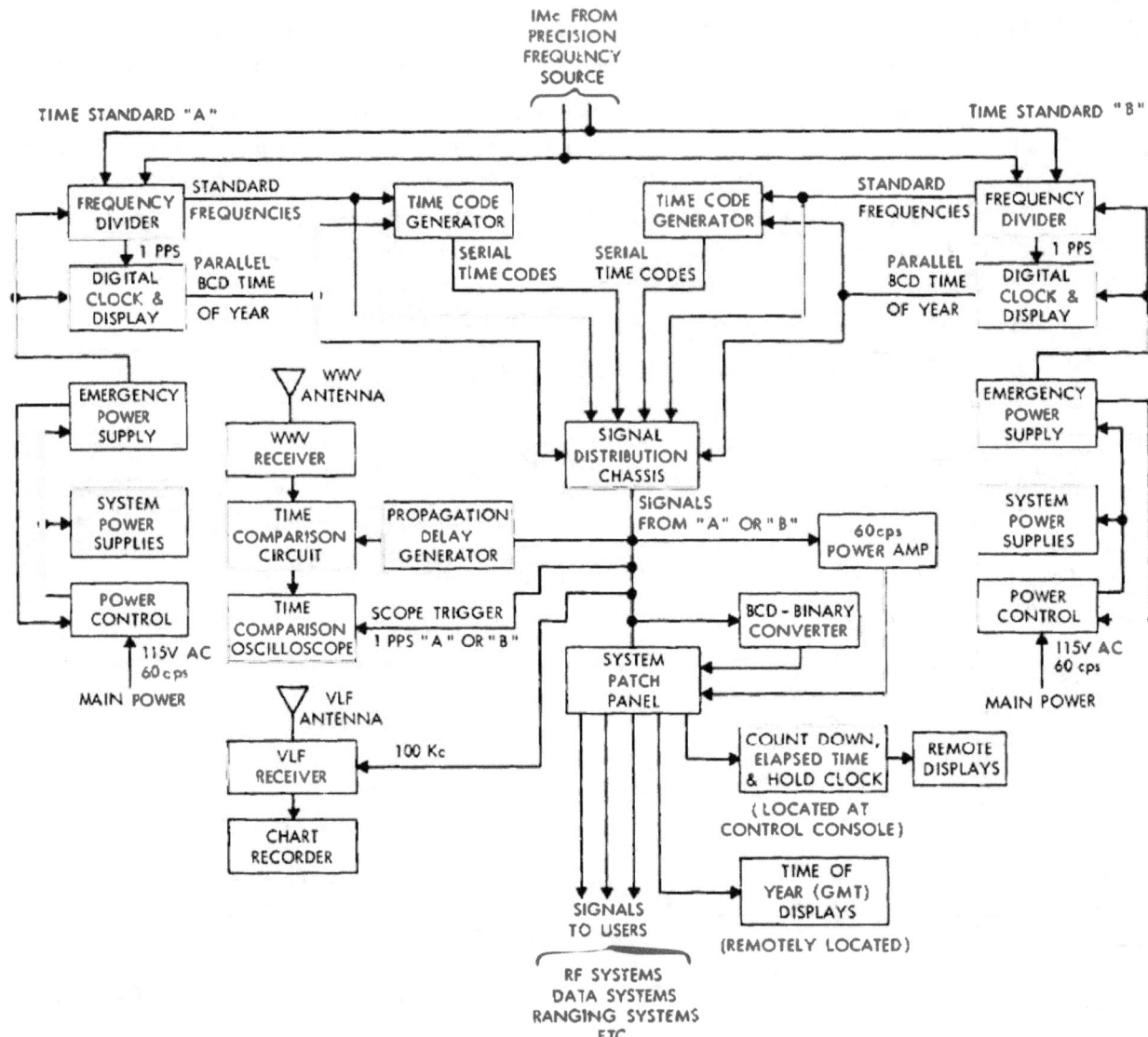

Figure 3—Time standard block diagram.

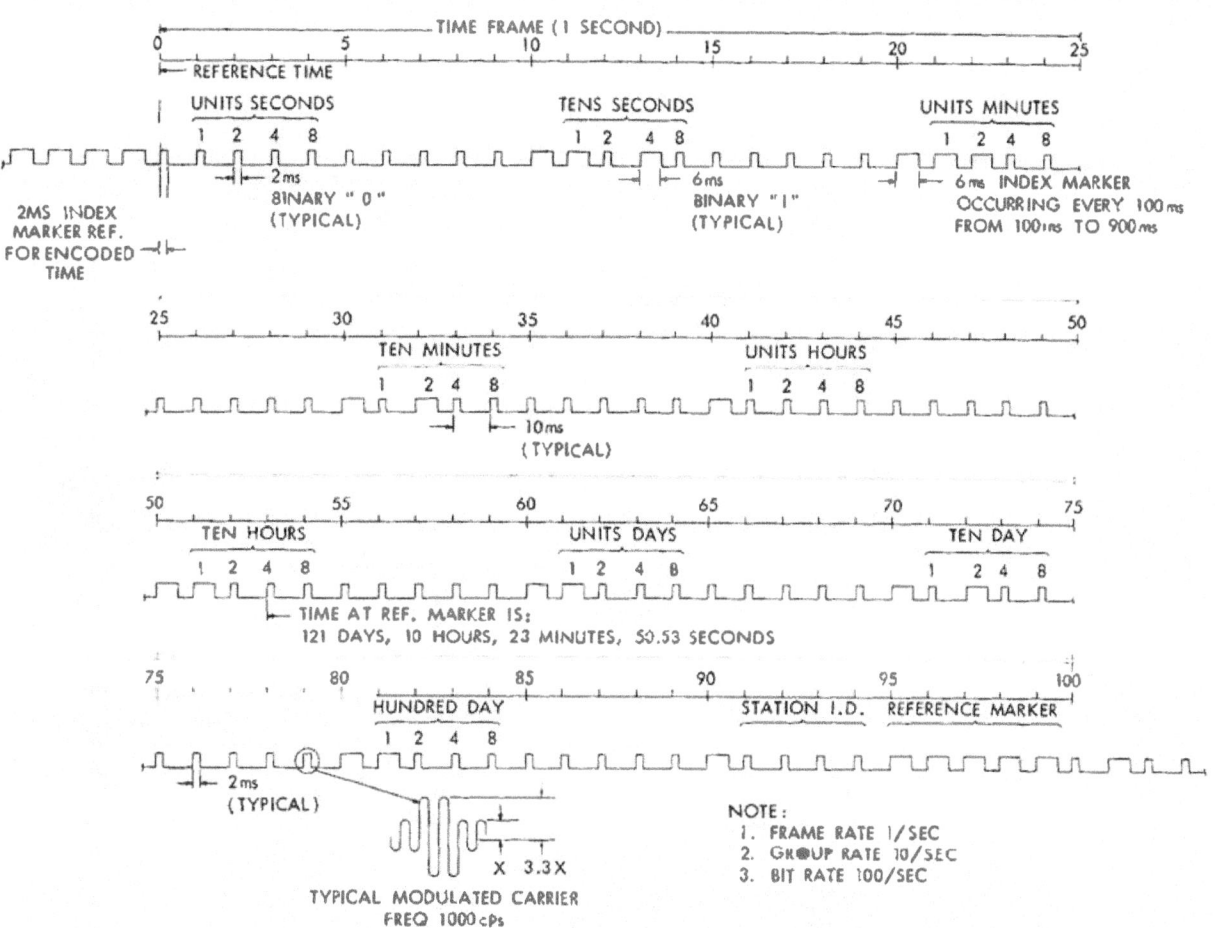

Figure 4—NASA 1/second binary time code.

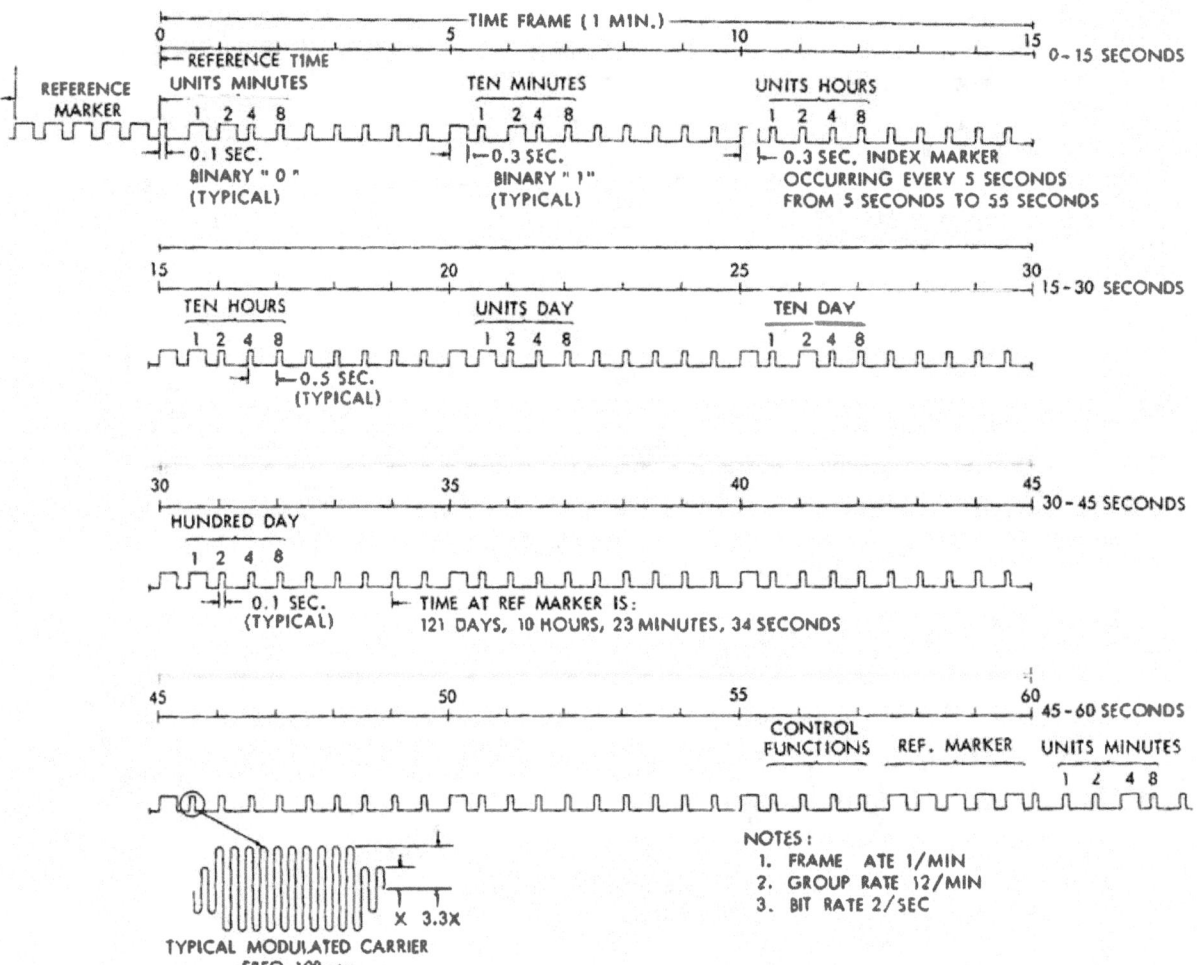

Figure 5—NASA 1/minute binary time code.

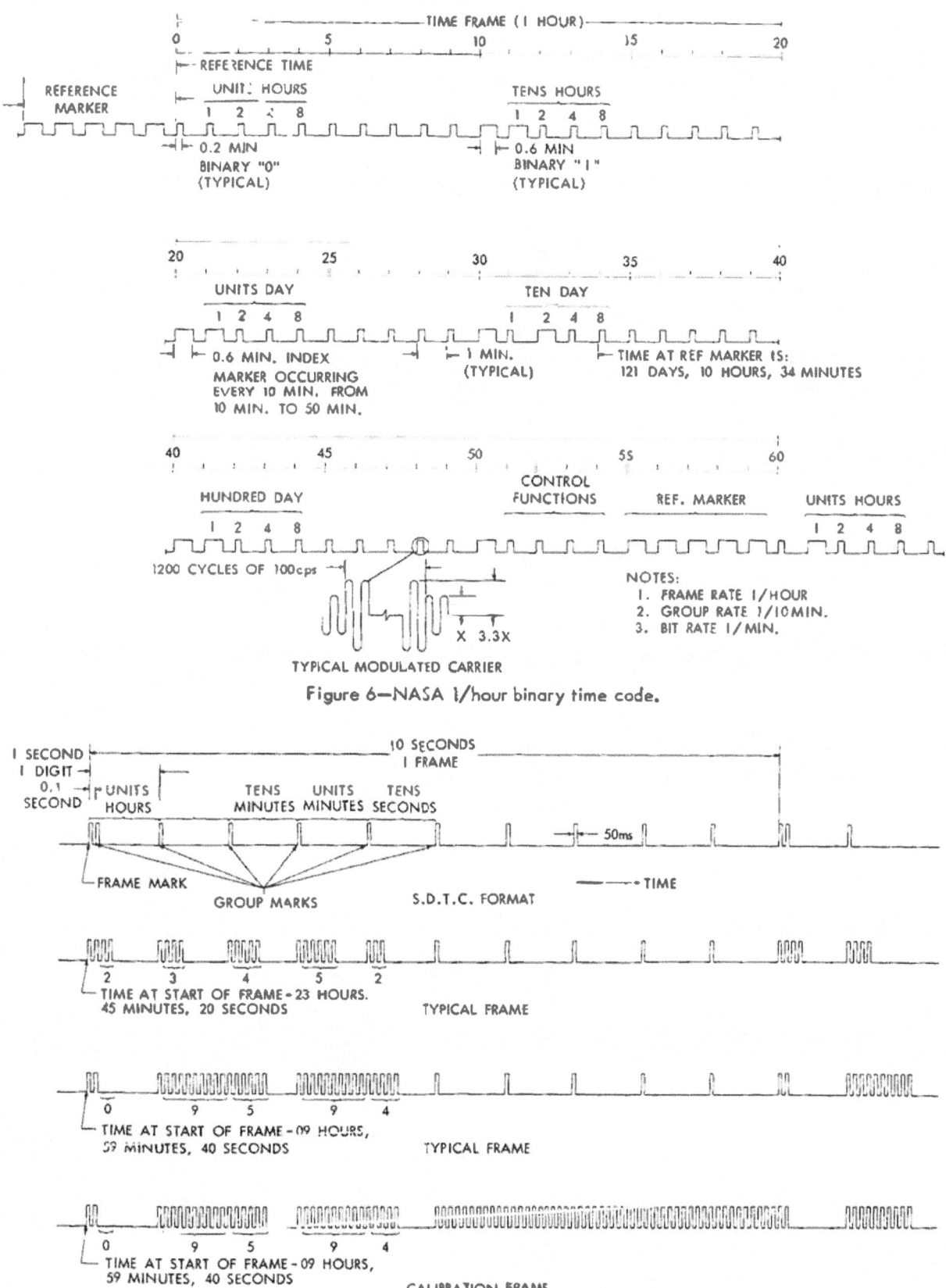

Figure 6—NASA 1/hour binary time code.

Figure 7—Format serial decimal time code.

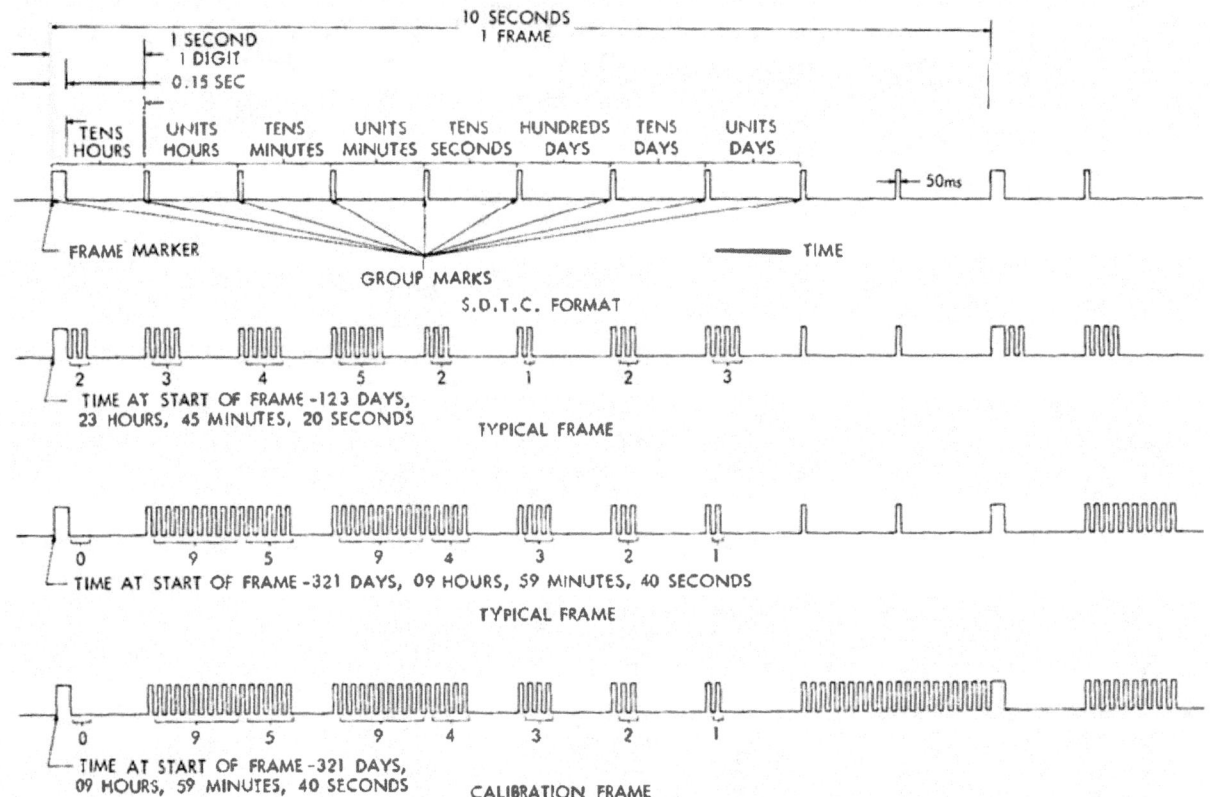

Figure 8—Apollo format serial decimal time code.

APOLLO MISSION PROFILE

by
J. J. Donegan
Goddard Space Flight Center

ABSTRACT

A typical Apollo mission profile is presented to provide an understanding of the tracking requirements placed on the Unified S-Band System in support of the Apollo missions. Characteristics of the Apollo spacecraft are included, as well as the timing sequence of the Apollo spacecraft events from initial liftoff through lunar touchdown, lunar liftoff, reentry, and earth touchdown. Orbital parameters, flight constraints, and navigational problems are also discussed.

The Unified S-Band System is implemented to provide ground instrumentation support for Apollo missions. In this role it will provide monitor and realtime control capability to flight controllers on the ground from liftoff to landing. In order to understand the tracking requirements on the Unified S-Band System, it is necessary to know the mission profile the system will be required to support. This presentation describes a typical Apollo mission profile.

Figure 1—Saturn V vehicle.

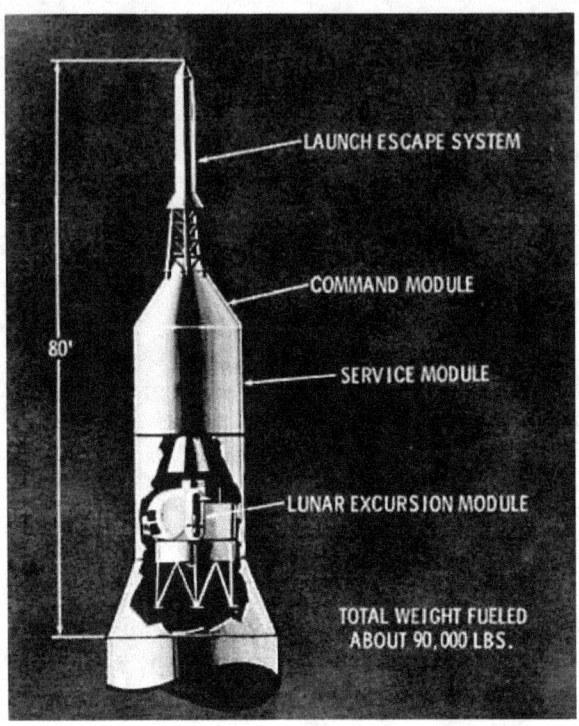

Figure 2—Apollo spacecraft.

The Saturn V vehicle is 360 feet tall as compared with 109 feet for the Gemini Titan II vehicle. It delivers 7-1/2 million pounds of thrust as compared with 430,000 pounds delivered by Titan II. It is a three stage launch vehicle. The first stage is designated S-1C, the second S-11, and the third S-IVB. The third stage or S-IVB is a restartable engine.

Elements of the Apollo spacecraft are the launch escape system, the Command and Service Module (CSM), and the Lunar Excursion Module (LEM). Totally fueled, this configuration weighs about 90,000 pounds at liftoff.

The operational control of the Apollo mission will reside on the ground in the Houston Control Center, even though the spacecraft will be designed with the capability of executing the mission and all abort options without use of ground information.

Figure 3—Apollo operational control center.

For a typical mission the timing sequence of events is given in the following table:

TIME Δt ELAPSED TIME AFTER LIFTOFF	EVENT
2.8 hours	Lunar injection
3.3/3.5 hours	Transposition (turn around) discard third stage
72.8 hours	Arrive at moon (3 days)
74.5 hours	Lunar touchdown 24 hours stay on moon
98.5 hours	Lunar liftoff 24 hours stay on moon
99.9 hours	Rendezvous at moon
103.5 hours	Leave lunar orbit for earth
196 hours	Start reentry
196.6 hours	Earth touchdown (8 days)

At liftoff the Saturn V weighs approximately 3000 tons and has the capability of transporting 45 tons to the moon any day of the month. The first stage (S-1C) burns for about 2-1/2 minutes

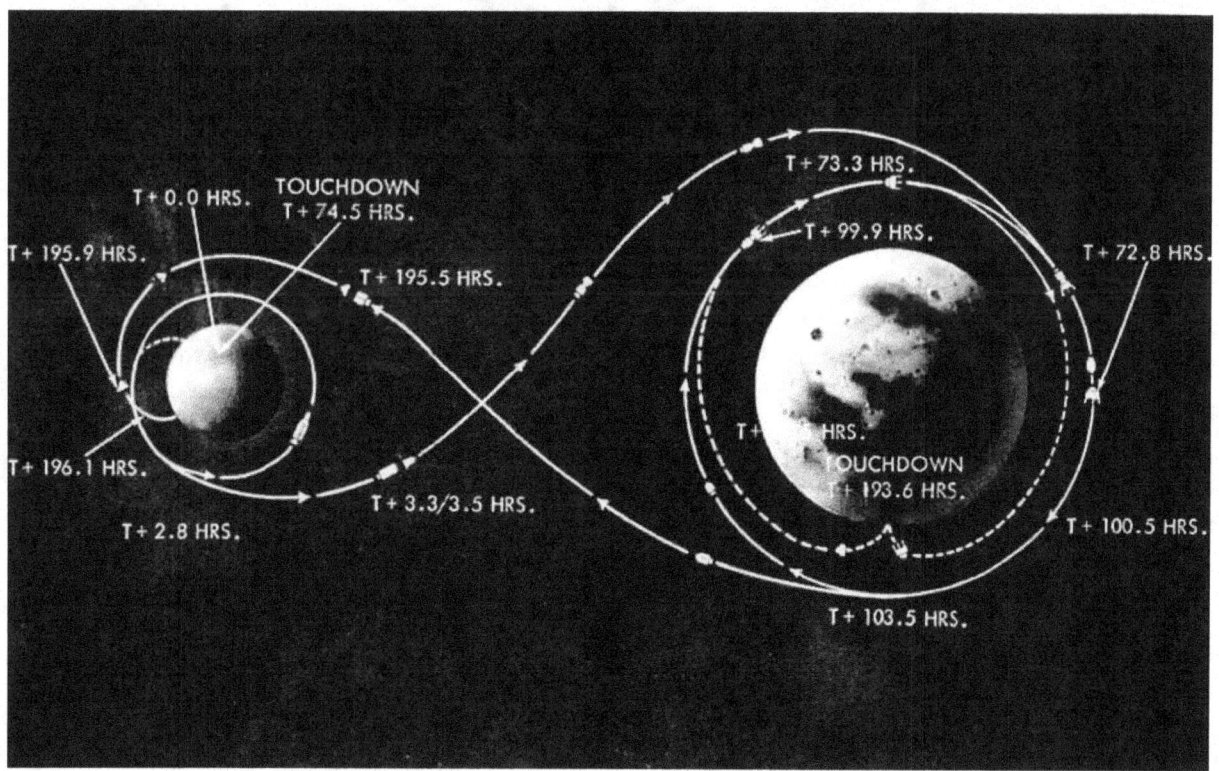

Figure 4—Timing sequence of events for typical mission.

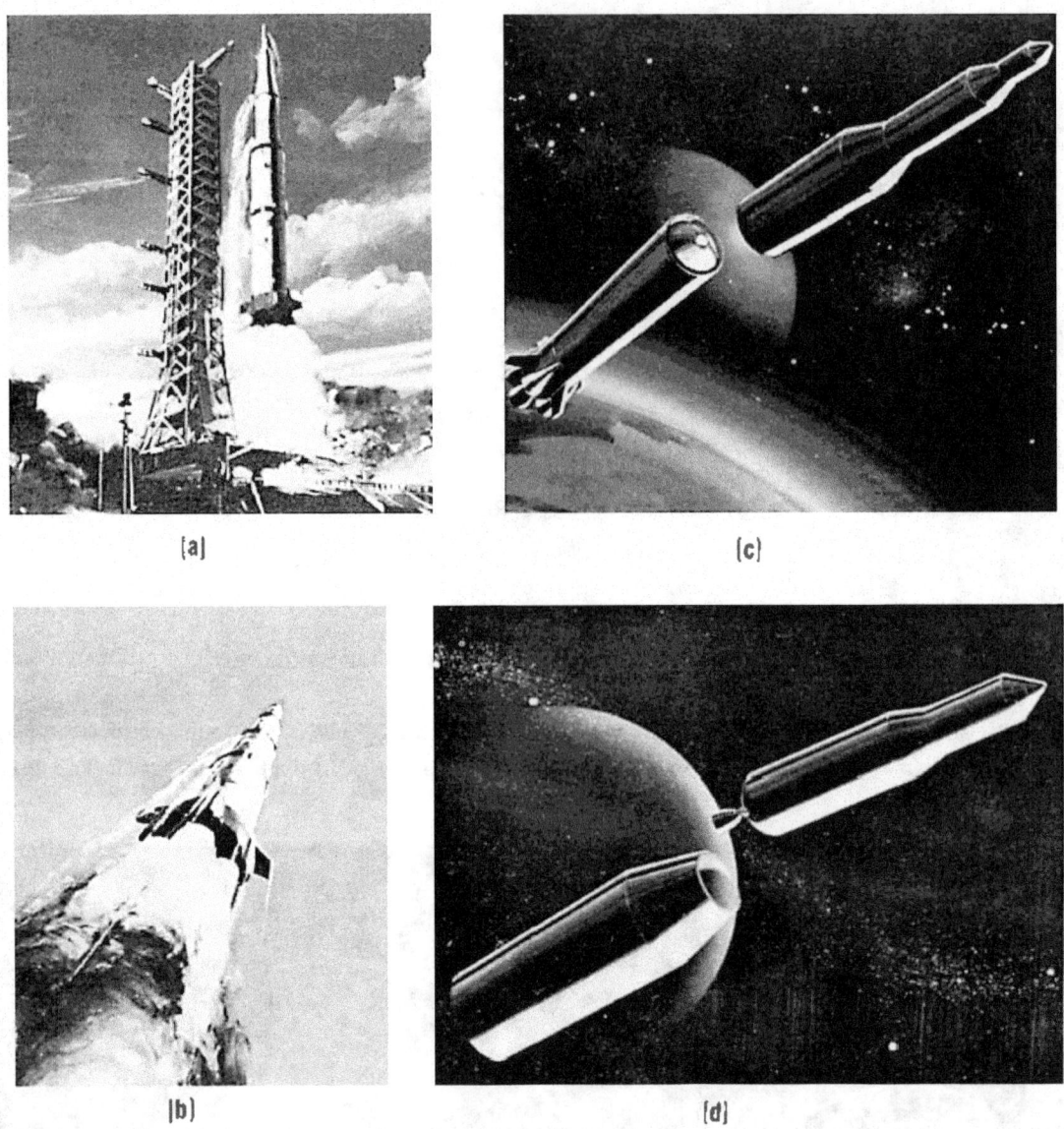

Figure 5—Saturn liftoff, first stage burning, first stage separation, and second stage separation.

to approximately 200,000 feet. After first stage separation, the second stage (S-11) ignites, producing a thrust capability of approximately 1 million pounds and burns for about 3-1/2 minutes to an attitude of 600,000 feet. At this point the second stage separates. The launch window will be about 2-1/2 hours based on the restraint of a variable launch azimuth limited to 26 degrees and on the basis of one tracking ship covering the insertion phase.

During the second stage burn the tower launch escape system is jettisoned.

The third stage or S-IVB which is a restartable engine is then fired briefly to attain a velocity of 25,520 feet per second and places the spacecraft in a 100-nautical mile parking orbit. During this phase crew and equipment will be checked out to see if they qualify to

Figure 6—Earth orbital checkout.

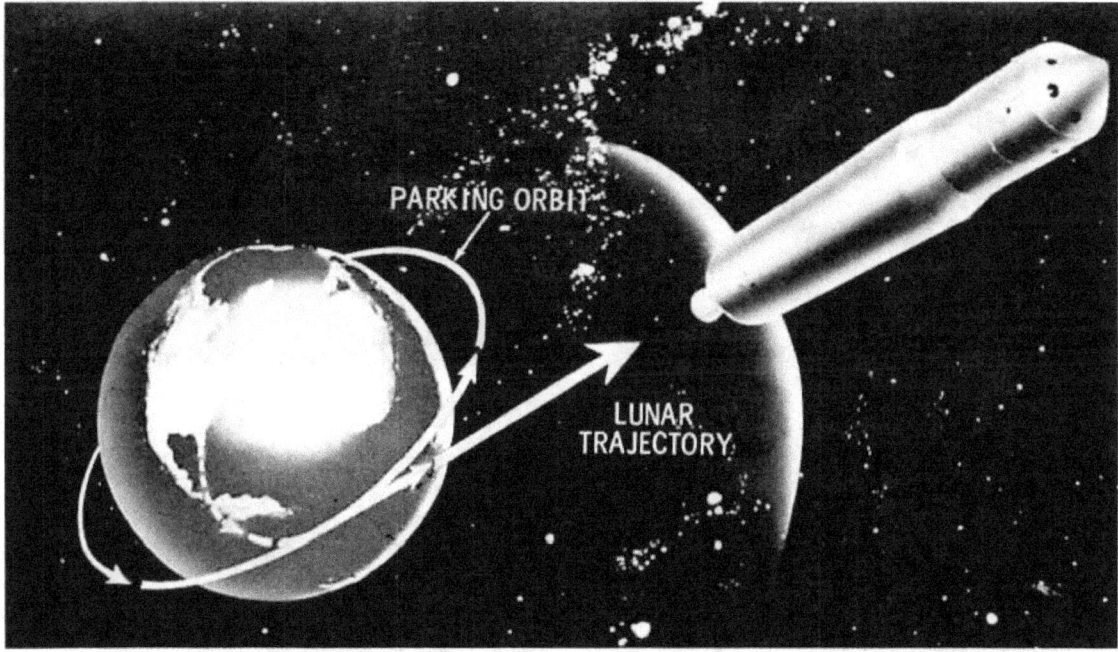

Figure 7—Lunar trajectory insertion.

perform the complete mission. The plane of the parking orbit should include the target or anticipated lunar landing point to avoid costly out of plane maneuvers.

At approximately 2.8 hours after liftoff the S-IVB engine re-ignites, propelling the spacecraft to a velocity of approximately 35,640 feet per second and injecting it into the translunar trajectory. The spacecraft then goes into a translunar coast and during this phase it is necessary to determine the orbit quickly to make a "go/no go" decision on the translunar phase of the mission prior to transposition. This will require about 10 minutes of tracking.

Apollo will introduce new and complex operations. One of these is the transposition or turnaround maneuver. During this maneuver the CSM will be separated from the S-IVB/LEM configuration, turned around, and coupled up again, freeing the engine of the Service Module (SM) for use. Figure 8 shows the explosive separation of the forward section of the spacecraft/LEM adapter, and the turn around maneuver. It is presently estimated that this phase will take about 30 minutes.

Figure 8—Command service module separation, turnaround maneuver, docking and coupling, separation of S-IVB stage, midcourse correction, and breaking maneuver.

Also shown in Figure 8 are the docking and coupling up of the CSM to the LEM/S-IVB, and the separation of the S-IVB stage which is now discarded. If required, midcourse corrections are then performed by the astronauts using the service engine to establish the proper course. This will occur about 5 to 8 hours after injection. It will take about 72.8 hours to reach the moon. Using the SM propulsion system, the astronauts will perform a braking maneuver to achieve the proper lunar orbit. This will be approximately a 100-nautical mile circular orbit above the moon's surface, at an injection speed of approximately 7500 feet per second.

Sometime later two astronauts will transfer from the Command Module (CM) to the LEM, and one astronaut will remain in the CSM in lunar orbit.

When all is ready, the astronauts will separate the LEM from the CSM and turn around the the LEM to descent attitude. First they will make a reconnaissance pass coming to a Pericynthian of 50,000 feet above the anticipated landing point. If all looks good, they will start the actual landing approach. The rate of descent will be carefully controlled. The LEM will reach a hover point 300 feet above the lunar surface before final landing. Lunar touchdown then occurs. Immediately upon landing, the LEM will be prepared for relaunch before either astronaut sets foot on the moon. Lunar landing occurs at 74.5 hours elapsed time.

While on the moon the astronauts will perform scientific experiments, gather geological samples, take photographs, and do some exploration. They will also leave some scientific instruments behind for transmitting scientific data back to earth.

After a 24-hour stay on the moon, the astronauts will fire the liftoff engines using the four-legged adapter as launch pad and leaving it behind. Lunar liftoff occurs approximately at 98.5 hours.

The lunar launch must be timed to permit rendezvous with the CSM. This is a critical maneuver which imposes severe requirements on ground tracking. Rendezvous will occur at 99.9 hours. Upon docking the two astronauts will return to the CM, detaching the LEM and leaving it in lunar orbit. If everything checks out at approximately 103.5 hours, the astronauts will fire the service module for the return trip to earth.

It is very important to determine the transearth trajectory early. From ground tracking, midcourse corrections will be made to assure that the spacecraft enters the correct reentry corridor about 40 miles thick. A miss can mean up to 350 g's or can mean skipping back into outer space, or can mean encountering exceedingly high temperatures during reentry. The determination of the orbit quickly in this phase is of paramount importance. Adjustments of time enroute to earth will determine where landing takes place on earth. Fuel penalty for trajectory adjustments early in this phase are less than for later in the transearth phase.

Before entering the earth's atmosphere, the astronauts will jettison the SM. It must be separated so that there is not a re-contact problem between the SM and CM, and so that the anticipated impact point of the SM is not in a populated area.

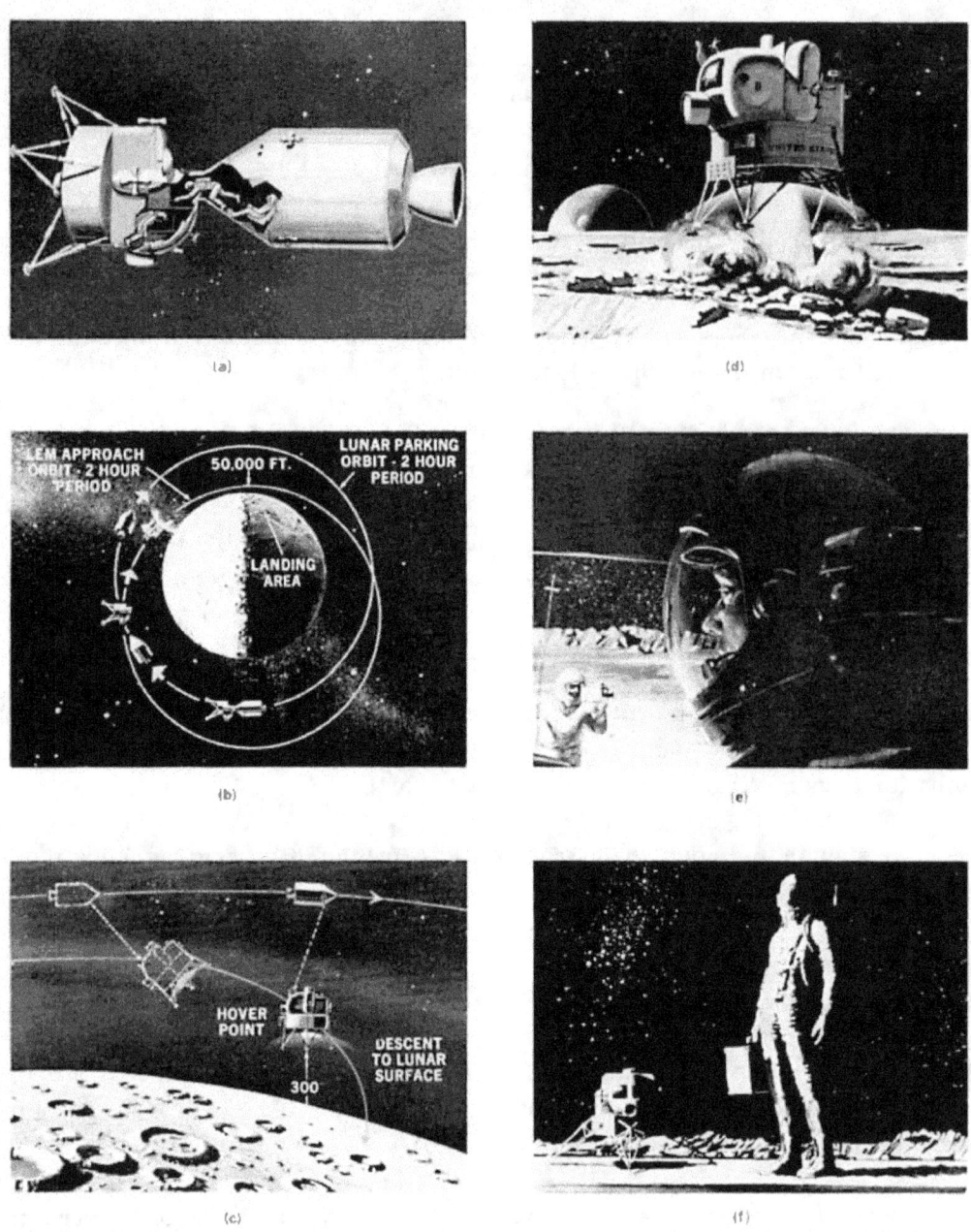

Figure 9—Transfer to LEM, lunar approach orbit, lunar descent, lunar touchdown, photography, and exploration.

The CM is then placed in proper attitude for reentry. The Apollo spacecraft like the Gemini spacecraft is a lifting vehicle. Its landing footprint gives the astronauts some control of their landing point. Apollo reentry is a very critical maneuver. Reentry speed is about 35,787 feet per second and reentry range varies from 2100 and 5000 nautical miles. Ionization phenomena are intense during this phase, creating tracking problems for the ground during the blackout periods. Drogue chute deployment and main chute deployment are shown in Figure 11.

The Apollo mission terminates in a water landing in the Pacific after approximately 196.6 hours elapsed time. Two possible landing areas are contemplated. one in the

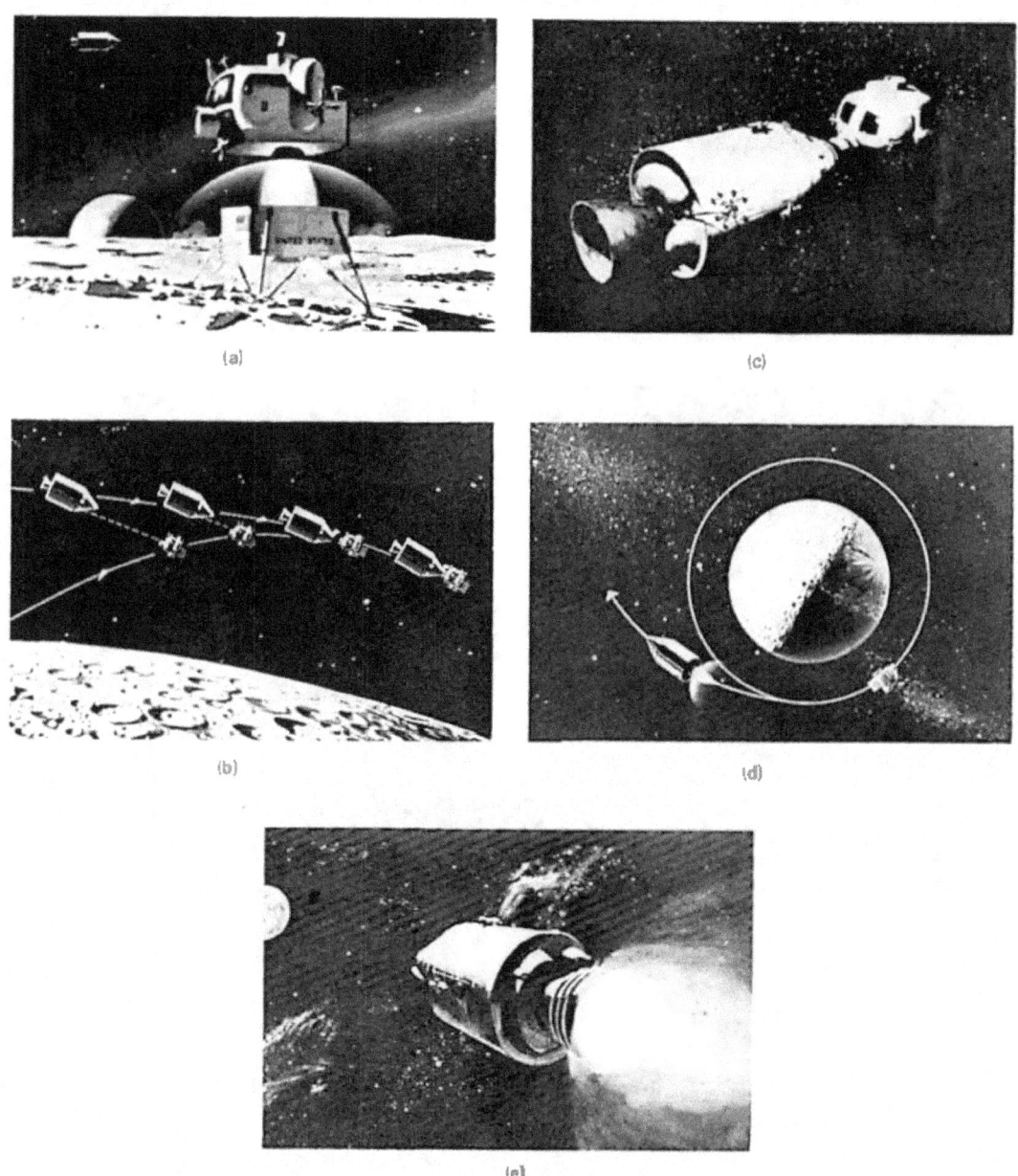

Figure 10—Lunar liftoff, lunar orbit rendezvous, docking, leaving lunar orbit, and return trip to earth.

northern and one in the southern hemisphere. These are near Hawaii and Pago/Pago.

As seen from study of the mission profile, the Apollo project introduces new and complex tracking problems, which must be resolved to provide realtime control of the mission from liftoff to reentry. The Apollo Unified S-Band System is being designed to achieve this result.

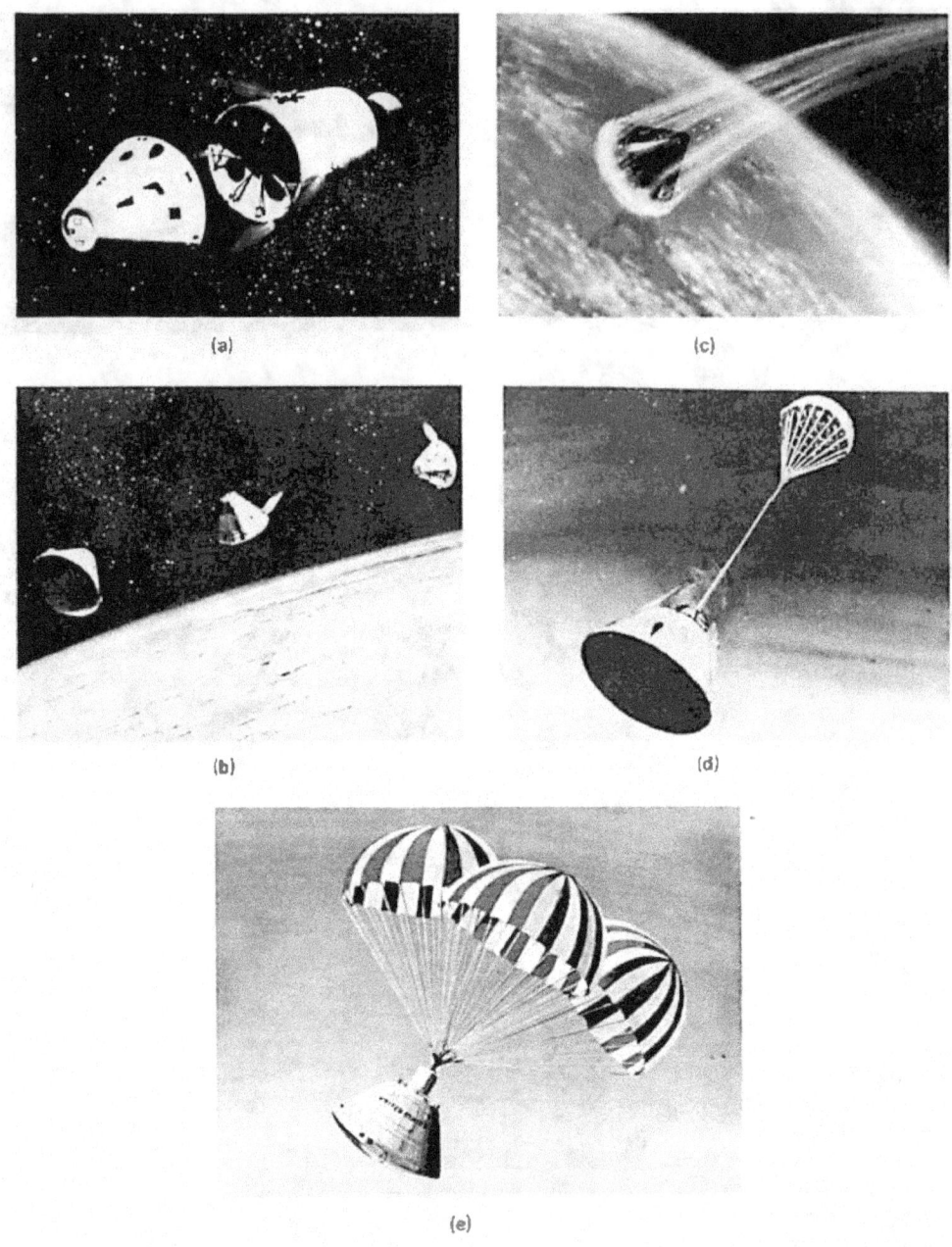

Figure 11—Service module jettisoning, reentry attitude orientation, reentry, drogue chute deployment, and main chute deployment (terminal descent).

COMPUTER TEST PROGRAM TO QUALIFY USB SYSTEM

by
J. Barsky
Goddard Space Flight Center

ABSTRACT

The GSFC computer processor will utilize data from the Unified S-Band (USB) System to document its operating characteristics during actual orbital track. The USB system will be calibrated by the C-band network of FPS-16 and FPQ-6 radars which have been proven in Mercury and Gemini. The basic computer program utilized will be the Gemini program, revised to receive and process the outputs of the two systems both simultaneously and separately to allow for maximum comparison of the results. The results will be measurements of the noise and bias in the USB network and the orbital determination accuracy as a function of this noise and bias.

System testing of data flow to and from a USB site conducted at GSFC will utilize developed tests from the CADFISS program. These tests will be used to determine (1) the degree to which the computer-related portions of the system have fulfilled their system design requirements, (2) those portions of the systems which are not functioning properly, and (3) the operational capability of the system to support a mission.

INTRODUCTION

The computer test program to qualify the Unified S-Band (USB) System will consist of two phases. The first is the CADFISS program which will check each site as it is implemented and check the whole network before a mission, supplementing remote site testing. The second is a check of the system during a mission with an orbit computation program capable of utilizing both USB data and C-band radar data.

CADFISS PROGRAM

The responsibility of the CADFISS program testing for the USB will be to verify the operational readiness of hardware and software configurations which may affect the data content of messages used for computer computations at GSFC and/or MSC. Particular emphasis will be placed on subsystem interfaces which are not checked during unit testing. Where possible, the tests utilize the operational program which will be used for mission support. This enables the CADFISS tests to perform an authentic pre-mission checkout of the applicable systems.

The general equipment areas that are involved are communications, radar tracking, including boresight, antenna programmer and range and range rate, the digital command system (DCS) at high and low speed and the PCM telemetry tests at high and low speed. All of these tests are applicable both to Manned Space Flight Network remote sites and the ships, except the boresight test which cannot be run on the ships. All GSFC tests will have the capability to process variable input rates for both high and low speed tests.

CADFISS utilizes an automatic program concept, all phases of testing being under control of the computer program. The 7094 is the center of the testing system and controls data flow activity between remote sites and the Goddard Space Flight Center.

In discussing the tests necessary to accomplish the testing objectives it is assumed that GSFC will have access to all high- and low-speed communication lines that exist between MSFN sites and MSC, Houston, and that GSFC's realtime systems will have access to high speed command circuits, high-speed tracking circuits, and low-speed teletype circuits required to perform the proposed CADFISS lists. Facilities will be available at GSFC to accept and format data from six high-speed telemetry circuits simultaneously, and error codes required on the outgoing command circuits will be affixed by the GSFC realtime system.

Communications Tests

The communications tests are required to ascertain the condition and continuity of each GSFC remote site circuit. The communications circuits are common to all areas of testing; therefore, a simple end-to-end test is required. The testing will consist of sending data from GSFC to the site, where it is compared against an expected pattern and scored, and the results are transmitted to GSFC. The site, in turn, sends data to GSFC, to be compared with an expected pattern and status of the circuit is established.

Radar Tracking System Tests

The testing of the radar tracking system consists of three aspects; the range and range-rate (doppler) test, the radar boresight test and the antenna position programmer test.

Range and Rate Tests

The purposes of range and rate tests are to: (1) verify the operation of the ranging system for one discrete value of simulated range; (2) insure that the voltage controlled oscillator (VCO) frequency is inserted into the range format position of the first output message following range code acquisition; (3) verify the proper operation of the range/frequency indicator bits to provide a coarse check on X, Y angular data; (4) check time; (5) test the ranging system for both Lunar Excursion Module (LEM) and Command Service Module (CSM), on dual radar sites; (6) check the "n" counter; (7) check operation of the 100-megacycle interval counter and (8) test the doppler readout circuitry operation.

According to the procedure set up for range and rate tests, the site upon cues from GSFC acquires phase lock with the collimation transponder and the transponder acquires phase lock with the ground station, after which ranging is started. When range acquisition is achieved, VCO frequency is transferred to the range output register and then to the tracking data processor. Successive output messages will contain the range units which correspond to the delay inserted at the transponder simulator. In addition to range and frequency, the message identification characters and the time word are checked for proper operation.

During the range test, the doppler counter is tested by using a stable 1 megacycle generator as a simulated doppler source analog signal and is checked for the destruct and nondestruct modes and for N1 and N2 count periods. The approximate length of test is 5 minutes. Tests will be required for both high and low speed testing to satisfy launch area and network radars and will be capable of operating with variable speed input data.

Radar Boresight Test

The purpose of the radar boresight test is to check the angular alignment of the radar system and check on the time and message identification. Upon receipt of a cue from GSFC, the site acquires and locks on the boresight signal. "N" frames of data are transmitted to GSFC over low-speed teletype circuits. The data is compared with the tower survey value to determine angular alignment. A high speed test will also be required to test radars used in the launch area but is not applicable to the ships as they will not have a collimation tower.

Antenna Position Programmer Test

The purpose of the antenna position programmer test is to verify the proper operation of the subsystems and interfaces used to position the antenna. The operational program will be used in the on-site data processor. GSFC will send acquisition points to an on-site computer via low speed teletype. The operational program in the computer uses these points to compute and punch a pass tape which contains command angle data to direct the antenna position programmer (APP). The tape is entered into the tape reader for the APP where it will direct the antenna to the specified orbit search. The radar encoder outputs the angular position of the antenna to the tracking data processor (TDP) where it is formated and transmitted to GSFC.

The acquisition points sent to the on-site computer are used at GSFC to construct the command angles which were used to direct the APP. These angles will be used to construct a simulated orbit. The data received from the site will be compared against this orbit with some small amount of errors allowed. Test time is approximately 5 minutes. A high speed test is also required.

Prior to getting an operational program for the on-site computer, a tape will be prepared at GSFC and used to simulate an orbit pass. This tape may also be programmed for discrete antenna positions and the boresight tower coordinates.

Digital Command System

The digital command system will be tested to verify proper operation of the subsystems and interfaces used operationally as part of the digital command system including an input data check (program functions), an up-link check and a validate and retransmit check. The procedure will be to prepare output command loads in the GSFC's IBM 7094 computer, affix error code to data, and transmit to the appropriate site. The remote site program in the command data processor (CDP) will perform a validity check on the data and store commands or request retransmissions as required.

The second phase of the test up-links the data to the collimation tower or the dummy antenna load. The up-linked data feeds the verification receiver and is then fed to the CDP via the input buffer. The CDP performs a comparison with the up-linked command and indicates those commands which do not compare. The addresses of failing commands will be sent back to GSFC via low-speed teletype as a program function or a manually prepared remote site report. Sequential switching of circuits at Honolulu and London will require that this test be run in three passes to test all sites.

The operational program will be used in the CDP. This test will vary slightly in operational procedures at the site due to the different modes of operating the command system. Mode 1 requires site personnel to up-link the command data. Mode 2 operation up-links the data upon receipt of an execute command from GSFC and Mode 3 up-links the command immediately upon receipt of data and validation. The test will be limited to approximately 10 minutes per site.

PCM Telemetry Tests

The PCM telemetry will be tested to verify the data flow path from the sub-carrier demodulator to the telemetry processor and the output of the telemetry processor via high speed to GSFC. The PCM simulator will be used to input directly to the PCM demodulation distribution panel or modulate the S-band downlink at the collimation tower if the latter is available. Each vehicle format will be checked as well as each decommutation station.

The operational program is used in the telemetry processor during this test. The data transmitted to GSFC are compared against tables of expected values and the test results transmitted back to the site. The high circuit switching at London and Honolulu limits the number of sites which may be tested simultaneously, therefore, three passes will be required to test all sites. The CADFISS program will then check the entire network and allow the orbital computation to be performed with confidence in the equipment.

ORBIT COMPUTATION PROGRAM

The ultimate test of the USB system as a tracking system is its ability to provide data to determine an orbit. Theoretical studies can show what the capabilities of a system should be, but only actual track of an orbiting vehicle can prove its real capabilities.

One problem associated with determining the capability of a tracking system is a good standard of comparison. In Mercury and Gemini we simply used the best tracking systems available, the C-band radars FPS-16 and later the FPQ-6. These proved fully capable and provided excellent orbit determination. This then provides an excellent measuring stick for the USB.

Although the USB has the added capability of measuring doppler or range-rate, the specification for angles and range are not as good as either the FPQ-6 or FPS-16. The results of recent Gemini missions show the rms errors for the C-bands to be roughly 0.1 mil in angles and 5 yards in range as compared to specifications of 0.6 mil in angles and 15 yards in range for the USB.

The USB is primarily designed for tracking to lunar distances but does have definite near-earth functions. Once a vehicle is far from the earth, the angle tracking ceases to be of value and the doppler and range are the prime sources of information. Therefore the USB has to be evaluated in two ways: first, as a complete system with angles; second, as a source of range and range-rate alone.

The comparison then will be made on vehicles carrying both C-band and USB transponders. The central computer will accept data from all sites and perform orbit calculations in three modes: C-band track alone, USB track alone, and combined C-band and USB track.

The residuals and rms errors will be computed for the USB as a function of all three solutions. These errors will be analyzed to determine the biases or systematic errors in the various sites, which may be due to static errors such as station location, X and Y angle boresight and boresight misalignment, frequency standard, and dynamics errors such as antenna lag.

Other errors are always present which complicate an analysis. The mathematical model or equations of motion are never exact, particularly in the case of a satellite relatively close to the earth and subject to all of the earth's harmonics perturbation and especially to atmospheric drag.

The model will contain Cowell equations of motion integrated with an improved 8th order central difference integrator. All necessary perturbations will be accounted for in the equations of motion. The data will be corrected for all known effects indicating local vertical, refraction, light time and delays, and timing errors.

One of the principal problems associated with tracking has been the nature of the satellite itself. If it is unstabilized and unsymmetric, a random tumble area is usually used with a fixed coefficient of drag for drag calculations. If it is tumbling at a high rate, the main source of error is the coefficient of drag, which is difficult to estimate for an odd shaped vehicle. If, in addition, the tumble rate is slow compared to the orbital period and the orbit elliptic, the problem of the precise orientation during the period of maximum drag becomes very significant.

Another associated problem occurs if the vehicle is stabilized by on-board thrusters. This tends to act as a small net thruster which perturbs the orbit greatly in precision orbit determination; therefore a stable orbit is required for performing the tests described before. Ideally for these tests, the orbiting vehicle should be round to minimize errors in surface area and coefficient of drag computations. It should be unstabilized to eliminate effects of thrust and should be in a fairly high orbit to minimize perturbations on the orbit. If uncertainty in the drag and thrusting characteristics are allowed to dominate the solution, no definitive analysis can be made.

A crucial role of any tracking system occurs when the orbit has to be defined or redefined on the basis of one station. This is where the performance in terms of low data noise is very important. As has been pointed out, there are many phases in which a single station will have to determine the orbit in the Apollo mission.

In Mercury program, however, it was found that once a sufficient amount of data had been accumulated (e.g. about one orbit), the solution from a "poor" tracking system and a "good" tracking system did not differ appreciably. The case in point was the Verlort versus the FPS-16. At that time the relative noise of the two systems was 1.0 mil and 40 yards for the Verlort, compared to 0.2 mil and 10 yards for the FPS-16. However, after one orbit, the solutions using Verlort alone or FPS-16 alone did not differ greatly. This condition depends on the two systems having only a difference in noise levels where one of them is much noisier than the other but there are no significant biases present.

Where the superior capability of the FPS-16 appeared was in the ability of one station to determine an orbit. Here the systems differed vastly in their results, the FPS-16 being an order of magnitude better in velocity determination. Therefore, by testing the single station solution of USB against a best combined solution, a real figure of merit will be obtained for one of the most critical roles of the system - the ability of a single station to redefine the orbit.

In summary then, the residuals of the combined solutions should provide a good estimate of the possible biases and errors in the system and the single station solution error should provide a real measure of the capability of the USB system.

NETWORK SYSTEMS

by
C. O. Roberts
Goddard Space Flight Center

ABSTRACT

This presentation outlines the configurations and capabilities of the network equipment to be installed on the remote sites for the Apollo program. Discussion includes site and system design considerations, system parameters, functions, and modes of operation.

The major systems, including PCM decom telemetry and DCS processors and console systems, are described in detail.

The discussion also includes the capabilities to be provided for closed-loop tests of the equipment at the remote stations, and the design of the equipment required to process and distribute the data from the Unified S-Band System.

The data flow from the control center to the remote sites is described, as well as equipment arrangement at a typical remote site.

The quantities and types of new network equipment being procured for the Apollo project are listed to provide an indication of the magnitude of the Manned Space Flight Network (MSFN) implementation program for the Apollo project.

INTRODUCTION

The Unified S-Band System is the major system located on the remote sites of the Apollo Network. This system combines the various up-link data and the down-link data on a single carrier. The system required to instrument the remote sites of the Manned Space Flight Network (MSFN) for the Apollo project is described in this paper.

DESIGN CONSIDERATIONS

Many factors were considered in the development of specifications for the individual systems procured for the Apollo program. This program requires instrumentation for three vehicles as well as the booster. This fact dictated the necessity for increased flexibility in the design of the systems. Each of the three vehicles will transmit PCM telemetry. The network was designed to transmit digital commands to each of the vehicles. Increased data processing capability is required for processing and displaying significantly larger amounts of information.

In addition to these features related to the space vehicles, other factors were considered. Redundancy was considered necessary in all of the major systems. Modularity of design was considered to be an important factor. Maximum flexibility was necessary to provide the capability of instrumenting the network with systems which would not be outdated as vehicle parameters were further defined or modified. To meet required operational dates, major systems were required to be procured prior to detailed definition of the spacecraft equipment parameters. The state-of-the-art digital equipment was employed throughout the network wherever possible. This presentation attempts to show how these factors influenced the design of the network systems.

SITE DESIGN

A typical remote MSFN site for Apollo is shown in simplified block diagram form in Figure 1. More detailed block diagrams of the individual systems are presented and explained later.

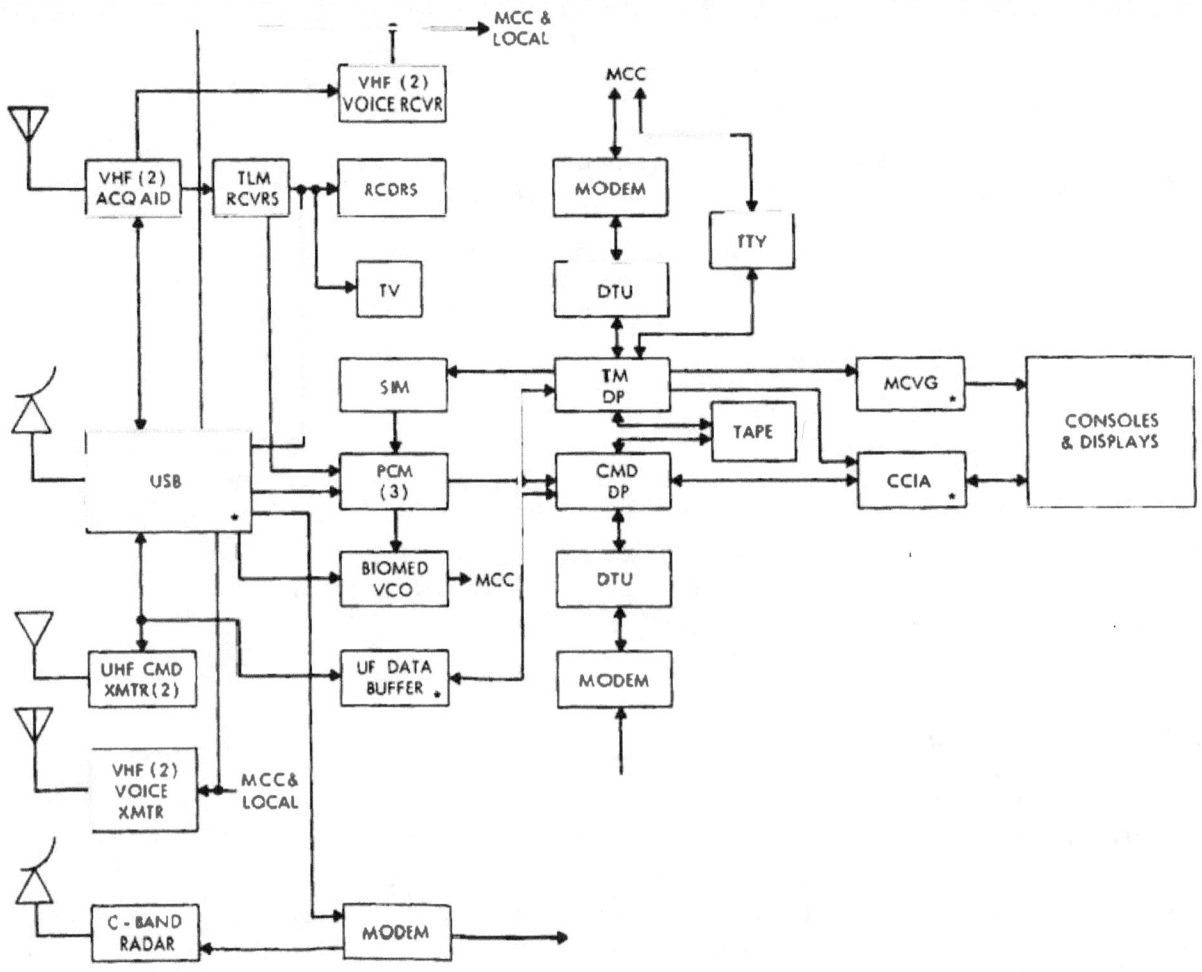

Figure 1—Typical remote site block diagram.

Telemetry

The data flow between the major subsystems at the remote sites is shown in Figure 2. Telemetry data received from the various vehicles will be demodulated by the Unified S-Band System. The data will then be decommutated and distributed by the PCM system. Selected telemetry parameters will be transferred to the telemetry (TM) data processer, and to the command data processer (CDP). Each PCM decom will contain two computer buffers to transfer TM parameters broadside to the associated data processer. The decom has the capability of transferring any selection of parameters into each of the computers.

The parameters transferred to the TM data processer are independent of those transferred to the CDP. The PCM decom also has event storage and digital/analog (D/A) converter capabilities. Data transferred from the PCM stations to the TM data processer will normally consist of parameters to be converted to engineering units, analog data to be displayed on strip/chart recorders, clock data to be displayed for time comparison, and any other parameters required for display on the consoles. Parameters required for transmission to the control center as part of the telemetry summary message will also be transferred to the TM computer.

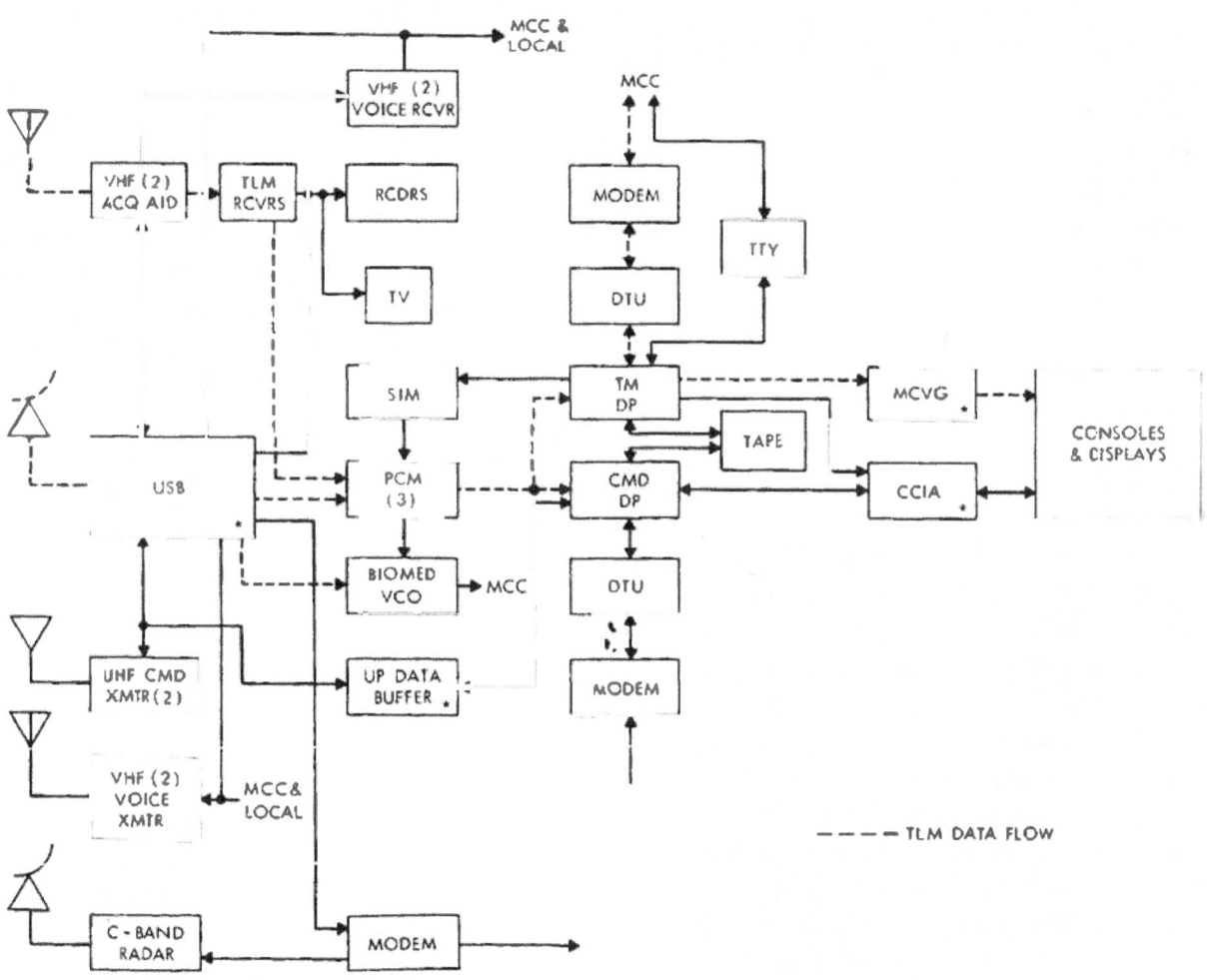

Figure 2—Typical remote site telemetry data flow block diagram.

Data transferred from the PCM stations to the CDP will include MAPS, spacecraft parameters and clock words which may be updated by command, and events for driving indications on the console command panel. Data to be displayed on the consoles will be processed and formatted by the TM data processor and transferred to the memory character/vector generator for storage. Data stored in the memory character vector generator (MCVG) will be utilized to continuously update displays on the cathode-ray tube located in the individual consoles. Data will also be processed by the TM data processor for transmission in real time to the control center over high speed lines.

In addition to the Unified S-Band (USB), the 30-foot remote sites will also be equipped with a VHF acquisition aid. During the early phases of the Apollo program, data received from the spacecraft will be VHF rather than USB. Provision has been made to provide either USB or VHF telemetry data to the PCM decoms. Biomedical parameters from the USB and PCM Systems will modulate voltage controlled oscillators (VCO's) and be multiplexed for transmission to the control center over a voice-data line.

Command

Command data will be received from the control center over high speed lines as shown in Figure 3. The data will be checked by polynomial code techniques and stored in the CDP after

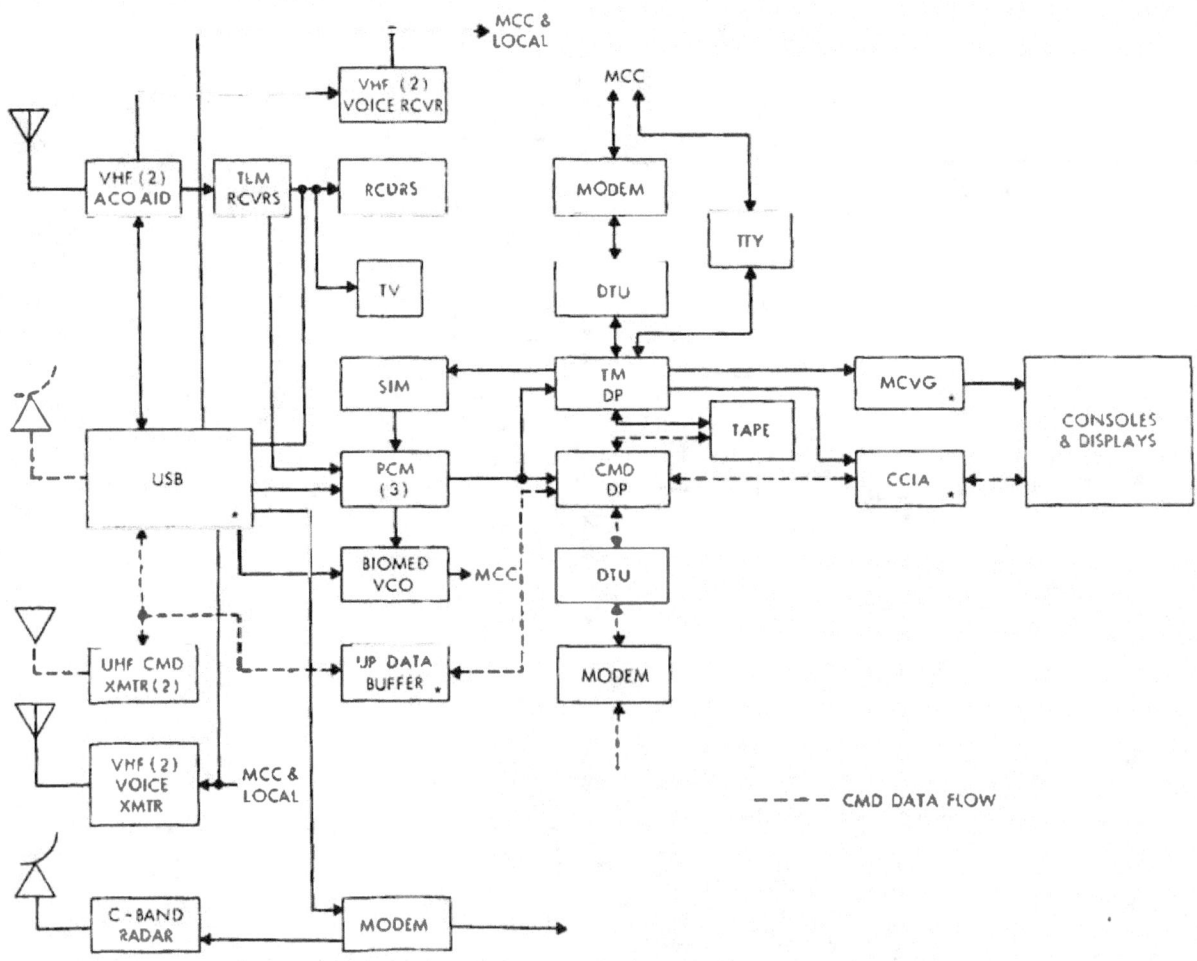

Figure 3—Typical remote site command data flow block diagram.

verification. If the data received from the control center is not valid, the CDP will generate a request for an automatic retransmission of blocks of data which were not valid. Individual commands, spacecraft clock, and up-date of command loads may be initiated from the flight control consoles or from the control center. Upon initiation of a command, data will be transferred from the command data processor to the up-data buffer. The data received by the up-data buffer will then be serialized and converted into a PSK wave form consisting of a two-kilocycle data tone combined with a one-kilocycle reference. The PSK wave form may then be utilized to modulate the UHF command system or the USB system.

Both systems are equipped with monitor receivers which will detect the data which were transmitted and convert them to parallel words for entry into the command data processer (Figure 4). These data will then be utilized in the preparation of summary messages to be transmitted to the control center over a high speed line in real time. The magnetic tape unit being provided with each data processer may be driven by either data processer. Therefore, it appears advisable to transfer all command data received from the control center to the magnetic tape unit for storage. If either computer fails, the data would then be immediately available to the remaining data processer.

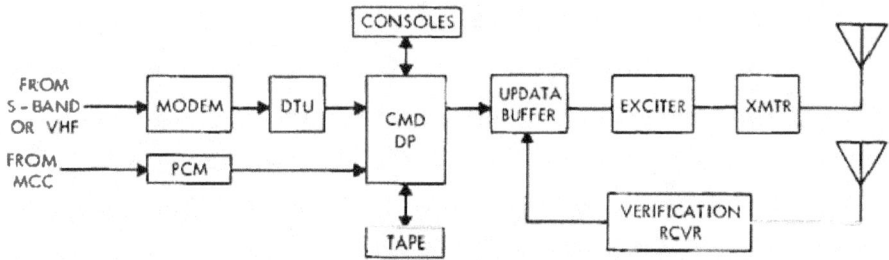

Figure 4—Uplink data block diagram.

Voice, Acquisition, and Recording (Figure 5)

Acquisition of the various space vehicles may be accomplished by one of three methods: The utilization of the VHF telemetry acquisition aid, the C-Band radar, and the USB. Air-to-ground voice capabilities will be provided on both VHF and USB. Tone remoting is being provided to permit voice modulation of the transmitters from the control center. Wide band, narrow band, voice, and chart recorders will be provided for each site to record PCM telemetry, spacecraft TV, voice, analog event, and status information. A TV monitor will be provided to display the slow scan TV from the spacecraft. Teletype input to the data processers will be provided in order that telemetry summaries from other remote sites may be stored in the telemetry computer. Summary data may then be called up for display on the cathode-ray tube by the flight controllers. PCM simulator will be provided for maintenance of the PCM system and for closed-loop tests. Data call up from the consoles and initiation of commands will be converted into computer instructions by the console computer interface adaptor (CCIA).

Sites which will not have flight controllers during missions will not have MCVG, CCIA or flight control installed.

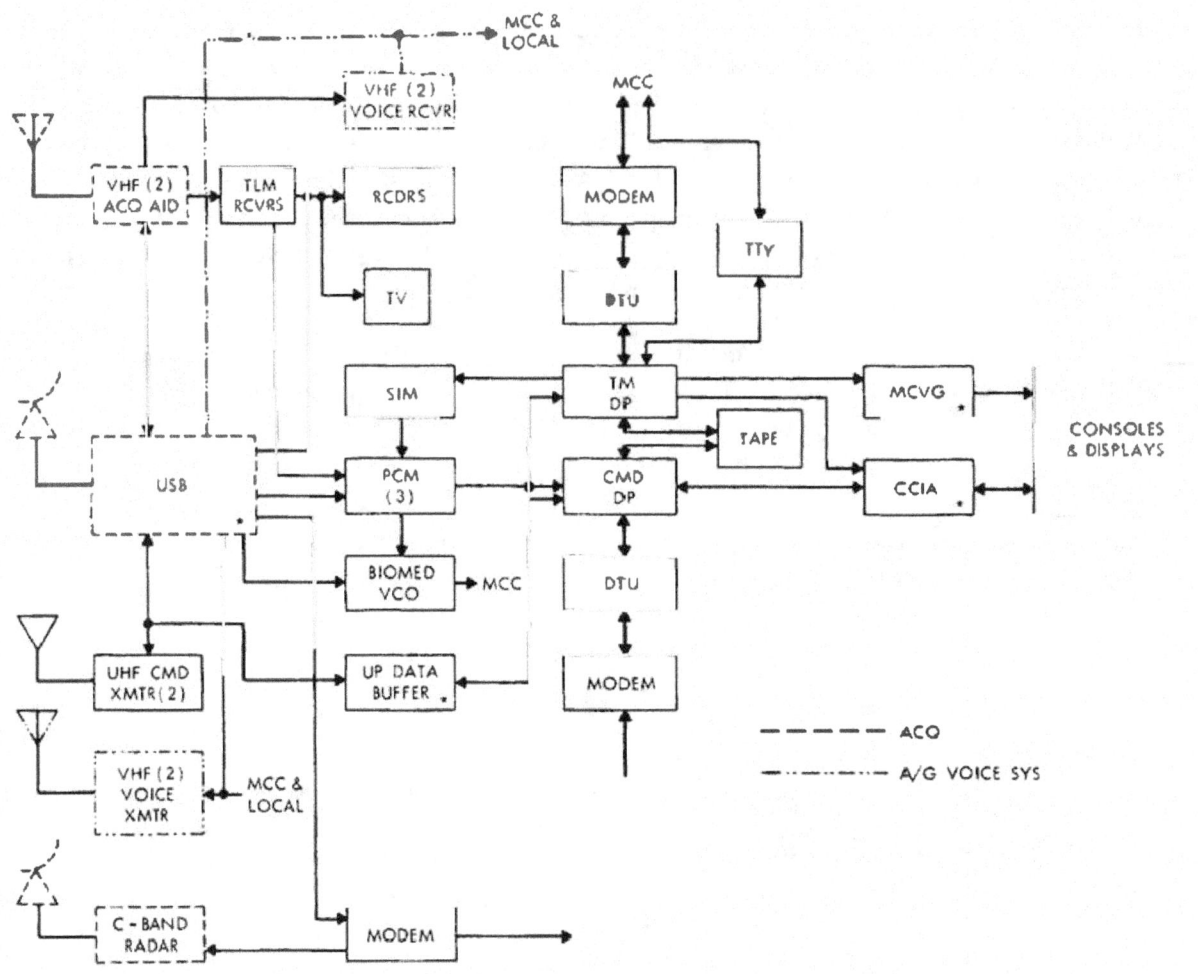

Figure 5—Typical remote site acquisition block diagram.

Redundancy has been provided throughout the major systems (Figure 6). Each USB system must be capable of handling at least two links in both receiving and transmitting modes. Dual VHF acquisition aids have been installed at most sites. Dual VHF voice receivers and transmitters are installed. All sites equipped with UHF command capabilities are dual. All primary sites will have three PCM decoms. The telemetry and command data processes are identical. In case of failure of one data processor, the remaining data processor will be utilized to process both telemetry and command information. All peripheral equipment which interfaces with only one computer will be wired through switch units to permit them to be connected to either data process.

The up-data buffer is designed to provide redundant channels. The MCVG has three identical channels. Failure of any channel will result in the loss of only four of the 12 CRT displays. The CCIA consists of two identical channels. The failure of either channel will result in the loss of control from half of the consoles.

Full utilization of digital equipment has been made throughout the USB, PCM, Data Processor, MCVG, CCIA and up-data buffer. These systems comprise the major part of the remote site instrumentation shown in Figure 7.

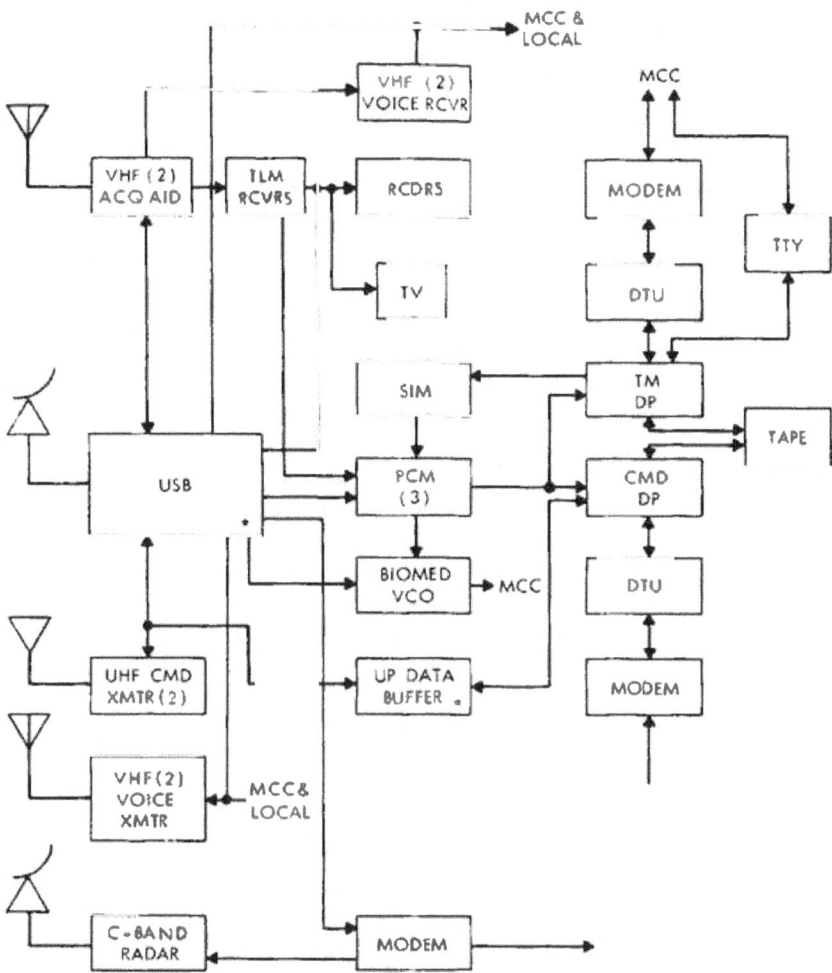

Figure 6—Typical remote site block diagram without consoles.

SYSTEM CAPABILITIES

Computer

Some of the major features of the data processor system are:

1. Two identical 642B modified computers are provided. One is for the primary purpose of telemetry data process and driving console display. Both units have an identical complement of peripheral equipment.

2. The computer memory has a two microsecond cycle time.

3. Each computer has a 32K directability, addressability memory. This memory size was selected to provide for the capability of driving CRT displays. It was determined that a 24 bit word length was desirable to permit the computer to process spacecraft computer words and ground and spacecraft time words as single units. The unit select has a 30-bit word length capability.

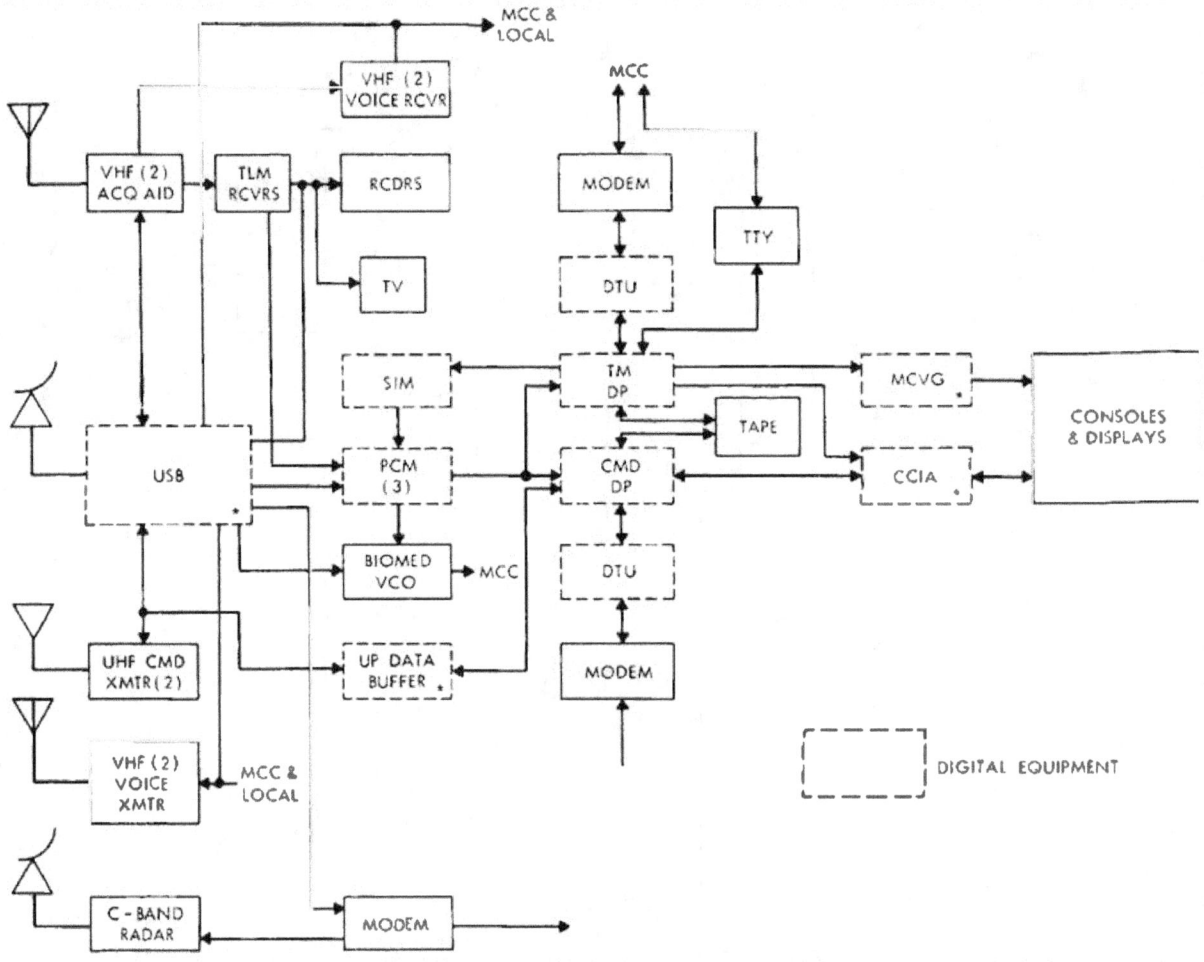

Figure 7—Typical remote site digital equipment block diagram.

4. Sixteen fully buffered input/output (I/O) channels were considered the minimum acceptable to provide for connection of all peripheral equipment without the necessity of multiplexing channels.

5. An overlap memory capability provides additional execution speed for the computer.

6. Continuous data mode capability permits the computer to transmit large blocks of data to external devices without reinitiation of the buffer.

7. As the computer forms the heart of the remote site system through which both command and display functions must pass, a militarized unit was considered necessary to provide high reliability.

8. Inter-computer transfer capability is necessary to permit both computers to be interconnected for transfer of data. This capability is also required for the Apollo ships in order that data may be transferred from the telemetry computer to the ships' central processor.

Pulse-code modulation (PCM)

The major features of the PCM system are:

1. The PCM system procured for the Apollo project utilizes program control for both acquisition and distribution formats. A 4096 by 36 memory provides the capability for routing any bit to any one of the 127 binary stores, and for the routing of any word in the format to any one of the selected 127 D/A convertors.

2. The computer load capability is provided for rapidly loading the PCM program. This feature permits the updating of the format by the computer in real time.

3. A computer output capability is provided to permit the PCM station to transfer data and status to the telemetry and command data processors.

4. The 4096 word memory of the system provides the capability of storing 10 selectable formats; six formats of the Gemini complexity can be accommodated.

5. Bit rate selection is under program control and up to 10 different bit rates may be selected. This feature permits complete control of the PCM station by the program.

6. The system has the capability of handling all existing formats with bit rates up to one megobit. Since all system parameters are under program control, the system has the flexibility to handle a wide variety of new formats.

7. Self check capabilities are provided through the use of a stored program simulator and a comparator for determining bit error.

Command

The main characteristics of the command system are:

1. Redundant channels are provided throughout.

2. Flexibility has been provided throughout the system by placing all major parameters under program control. Only software modifications are required to change word structure, word length, sub-bit encoding, transmission verification method, and input verification format. Command words transferred from the CDP to the up data buffer are in sub-bit encoded form. The only parameter of the system which may not be changed by program control is the data rate. The maximum data rate is established at one kilobit by the selection of one kilocycle as the reference for the PSK signal.

3. Continuous equipment status monitoring is provided by the utilization diagnostic routines. Equipment status may be determined even during periods when commands are not being transmitted.

4. A real time command summary capability is provided. Data transmitted from either the USB system or the UHF system will be received by a monitor receiver, demodulated, and transferred to the command processor for summary preparation.

Displays

The principal features of displays are:

1. Computer driven cathode-ray to displays were selected for the Apollo consoles to derive greater flexibility than can be obtained from analog meters.

2. A rapid reconfiguration capability was considered necessary to keep the number of consoles to a minimum. In a matter of minutes a console may be configured from S-IV-B support to Lunar Excursion Module (LEM) or Command and Service Module (CSM) support. The four systems consoles and the command communicator console are almost identical. One may be reconfigured to provide the support previously supplied by any of the others.

3. Two CRT's were provided on each system console and the command communicator console. This feature provides the capability of continuously viewing prime data while cycling through contingency data. It also provides additional reliability in that the two CRT's are connected to separate memory channels.

4. Three separate and identical channels are provided in the MCVG unit. No hardware is common to all three channels. This factor minimizes the effect of an individual failure. A failure in any of the three channels would result in the loss of four cathode-ray tubes, no two of which would be located on the same console.

5. A flexible modular design of the consoles was necessary to permit addition of displays and repositioning of existing displays in the console shell. The panels designed in standard unit sizes may be repositioned on the consoles' viewing area.

6. A maintenance monitor was provided to permit rapid maintenance of the system. The display module in the maintenance monitor is identical to the display module utilized in the consoles. If necessary it may be used as a spare unit. The maintenance monitor may be paralleled with any of the 12 CRT displays for tests of the systems performance; however, the capability of separate call-up of a new format for display on the maintenance monitor has not been provided.

7. Both character and vector display capabilities were considered necessary to provide system flexibility for displaying tabular data, analog data, trend plotting or meter formats.

8. A 3,000 word per minute printer is provided at each of the systems console locations as well as at the command communicator console. The printer is multiplexed to both the telemetry and command computers. Data from either computer may be printed by the high speed printer. It is anticipated that the printer will be utilized for read-out of command information transmitted to the remote site from the control center, read-out of the commands initiated from the consoles display, selection of tabular data and the display formats, and to display results of computer diagnostic tests.

DATA FLOW

The capability exists at the remote site for transmitting various types of data to the control center in real time (Table 1). The following data may be transmitted from the remote site in real time: air-ground voice command summary messages, telemetry summary messages, C-Band and USB tracking data, biomedical data, and administrative traffic. The remote site has the capability of accepting many types of data from the control center in real time. These include air-ground voice, commands, format select, acquisition data, remote site summaries, and administrative traffic. At present, all types of data mentioned may be processed automatically with the exception of acquisition data and administrative traffic.

Table 1
Data Flow.

From Remote Site	From Control Center
A/G voice	A/G voice
CMD sum	CMDS
TM sum	Format select
Tracking data	ACQ data
BIO-MED	Remote site sum
Admin. traffic	Admin. traffic

CLOSED LOOP TESTS

A block diagram of the remote site closed-loop test capability is shown in Figure 8. The method illustrated will permit closed-loop tests of the major portions of the system from GSFC under computer control. Test data from GSFC can be transmitted to one of the remote site processors over either high speed or teletype lines. This data can then be utilized to load the PCM simulator program. Data in the PCM bit stream may control from a remote source in real time the simulator output introduced into either the RF system or the PCM system. The output of the PCM system is then transferred to the data processer to be formatted for

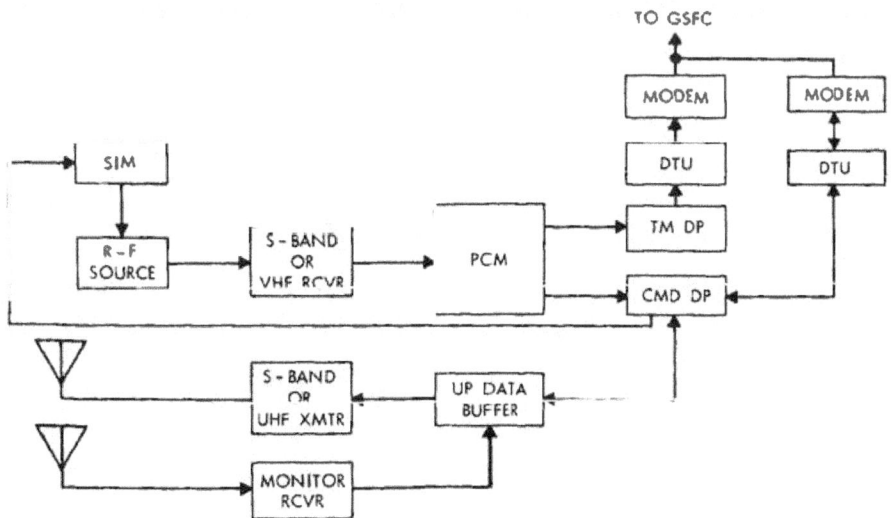

Figure 8—Typical remote site closed loop test block diagram.

transmission back to GSFC. Commands will also be initiated from GSFC through the command data processer and transmitted from the USB or UHF transmitters. The data are then sampled by the monitor receiver and transferred to the command processer through the up-data buffer. Command summaries may then be generated for transmission to GSFC to complete the closing of the command loop. Telemetry and command summary messages received at GSFC may be reduced by the computer complex to determine remote site equipment performance.

EQUIPMENT LAYOUT

The quantity of new network equipment being procured for the Apollo project (exclusive of the equipment included on the Apollo ships and aircraft contracts) is presented in Table 2. The amount of equipment provides an indication of the magnitude of the MSFN implementation program for Apollo.

Figure 9 shows how the equipment arranged in a typical remote site. The USB equipment mounted in the building has been located in one room. A large picture window has been provided to permit the antenna to be viewed from the control console. The PCM equipment, VHF receivers, acquisition aids, and recorders have been located adjacent to the USB room. The computers, peripheral and associated equipment, including the MCVG unit, have been located in one room adjacent to the operations and telemetry rooms. The communications room is located adjacent to the operations room to speed mission message handling.

Table 2

New Network Equipment for Apollo.

Site / Equip	ANA REC	W.B. REC	N.B. REC	VOICE REC	RSDP	INTER-COMM	PCM	PCM SIM	FPQ 6	CSL & DIS	T.V. MON	VHF RCVR	VHF ACAID	USB
ANT	1	2	1	1	2	1	3	1			1			S
ASC	5	2	1	1	2	1	3	1		1	1	4	1	D
BDA		2		1	2		2	1	1		1	15		S
CYI	5	2	1	1	2		1	1			1	8		S
BRA	1	2	1	1	2	1		1			1			D
BRA/JPL		2	1											D
CRO	5	1		1	2		1	1		1	1	8		D
TEX	1	1	1	1	2		1	1			1	8		S
GSFC	2	2	2	1	2	1	2	1		1	1			S
ODS	1	2	1	1	2	1	3	1			1			D
ODS/JPL		2	1											D
GBI		2		1	2	1	2	1			1			S
GUM	1	2	1	1	2	1	3	1			1	12	1	D
GYM	5	2	1	1	2	1	1	1		1	1	8		S
HAW		2	1	1	2	1	1	1			1	8		D
MAD	1	2	1	1	2	1	3	1			1			D
MAD/JPL		2	1											D
MILA	1	2	1	1	2	1	3	1			1			D
MCC-H	10				1		1	1		2	1			
Ship No. 1	5	2			2		3	1		1	1			S
Ship No. 2	5	2			2		3	1		1	1			D
Ship No. 3	5	2			2		3	1		1	1			D
Ship No. 4		2					2	1						S
Ship No. 5		2					2	1						S

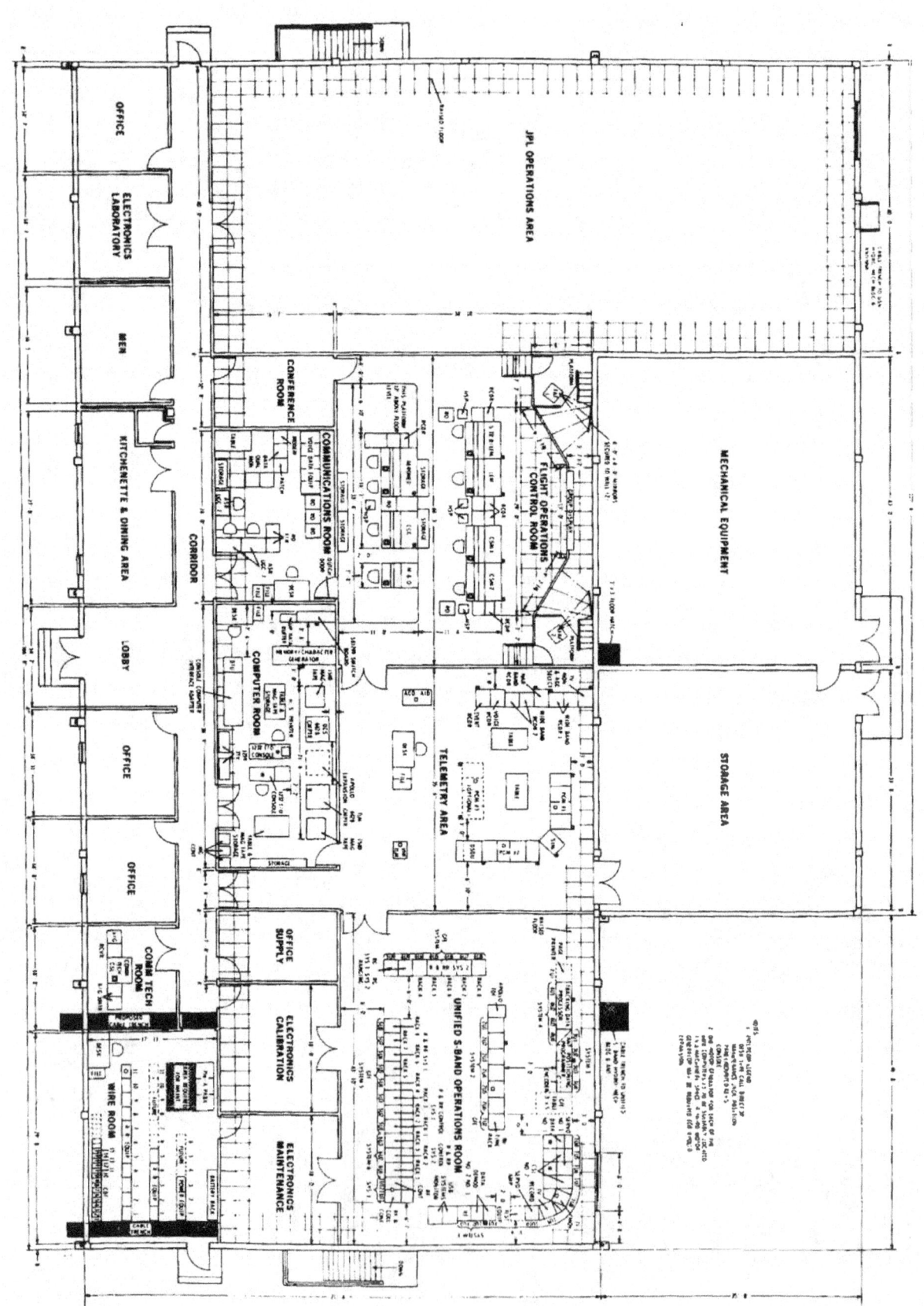

Figure 9—Typical remote site floor plan.

APOLLO NETWORK
PCM DECOMMUTATION SYSTEMS

by

W. A. Dentel

Goddard Space Flight Center

ABSTRACT

A description of the MSFTP-1 (PCM) decommutation systems currently being utilized on the Manned Space Flight Network (MSFN) is presented, including a discussion of the signal conditioner, the group synchronizer, and the telemetry output buffer and its capabilities.

A description of the new stored program PCM decommutation systems being procured for the MSFN is also presented. In addition, the added versatility of the stored program and its associated software will be discussed.

The two types of systems are discussed with respect to the interfaces with the other on-site equipment, including unified S-band (USB), VHF receivers, computers, and consoles.

INTRODUCTION

The following paragraphs describe the new pulse-code modulation (PCM) decommutation systems being procured for the Apollo missions and the differences between the new systems and the existing Gemini PCM system. The new systems will complement and increase our present capability at the Gemini sites for PCM decommutation. There will be two types of systems utilized on the Manned Space Flight Network (MSFN). The first is an existing patchboard system which was manufactured by Electrical Mechanical Research, Inc. This system is in use, and has been in use for some time on Gemini network. The second is a stored program PCM which is presently under procurement and in manufacturing at Dynatronics Inc., in Orlando, Florida. The stored program PCM will be utilized in all of the new Apollo tracking sites, as well as to supplement the Gemini site complement of PCM's for support of the Apollo program.

PCM SYSTEM INTERFACE

The block diagram (Figure 1) shows in a very simplified manner the interfaces of the PCM systems with the other network equipment. The decommutation system distribution unit (DSDU) is the major interface unit for all the PCM equipment and the other network equipment. The

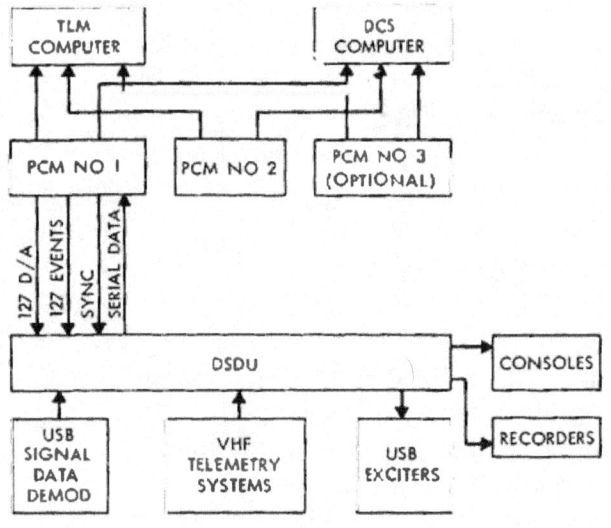

Figure 1—PCM system interface diagram.

DSDU accepts inputs from the unified S-band (USB) signal data demodulators, the VHF telemetry receivers, and the magnetic tape recorders. Serial data inputs are patched by a video patch panel into the PCM signal conditioners. These serial data inputs can be patched to PCM's 1, 2, or 3. The outputs from the PCM stations are connected to the DSDU patchboard systems. This allows the user devices at a remote site such as the consoles, the event light indicators, and the recorders to accept any event or any analog parameter from any one of the three PCM bit streams which we have the capability of receiving. There is also a direct interface at each of the PCM decommutators with each of the two computers on site. Each PCM has the capability, under program control, of selecting or stripping out any word or number of words from the format for inputting to the computer.

PCM DECOMMUTATION UNIT DESCRIPTION

The first of the two types of units being utilized on the MSFN to be covered is the new stored program decommutator. A few of the differences between the new unit and the existing patchboard type unit will be pointed out later. Figure 2 is a pictorial drawing of the new PCM decommutator which is scheduled for delivery to NASA in August. In the first rack there are two signal conditioners. Behind the door in the lower portion of the rack are several of the self-test functions which consist of a self-test panel, a paper tape reader for inputting data to memory, and other non-operational type functions. All controls for operating the station are located on the system control panel in the second rack. Below and behind the system control panel are printed circuit card gates containing the program control logic. The core memory is also located in the second rack. The third rack contains test equipment for the system. There is also an oscilloscope, electronic counter, digital voltmeter, and intercommunication panel. The fourth rack consists mostly of printed circuit gates and contains all of the output circuitry for the system. These major outputs of the system consist of 127 digital-to-analog converters, 127 on/off event outputs, and five 40-bit binary stores.

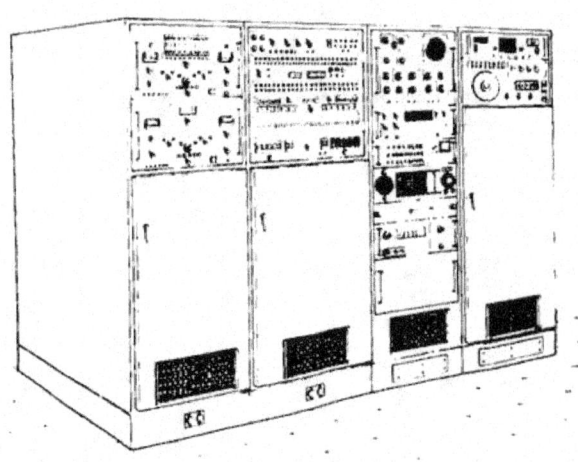

Figure 2—PCM data decommutation unit.

PCM STORED PROGRAM SIMULATOR

Figure 3 is a photograph of the PCM stored program simulator which is being procured for the Apollo. The simulator will be supplied to every site used on the MSFN whether it gets a stored program decommutator or not. The reason is with this simulator unit we can eventually perform (from GSFC) closed loop computer tests of the site utilizing the computer intput to this simulator, the output of this simulator via an RF link into a PCM decommutator, and from the PCM decommutator to the on-site computer and back to GSFC. These are at the moment in acceptance testing at the manufacturer's plant, and the first is scheduled to be delivered in July.

PCM STORED PROGRAM DECOMMUTATOR

Figure 4 is a block diagram of the new stored program PCM decommutator. The PCM system utilizes computer techniques to accomplish real time and delayed processing of serial PCM data formats under the control of an internally stored program. The system is capable of changing formats by the selection of a push button on the front panel of the system control panel, or the selection may be accomplished at a remote location. The system has the capability of storing up to 10 complete PCM formats in the decommutator.

Bit rates from 10 bits to 1 megabit per second may be accommodated in either of two bit synchronizers. Selection of either one of the two bit conditioners is provided on the system control panel. The narrow band bit synchronizer has 10 fixed bit rates associated with it which are, in turn, associated with the 10 stored programs in memory. The wide band bit synchronizer is similar, with the exception that it has a capability of manual selection of bit rate anywhere from 10 bits to 1 megabit per second.

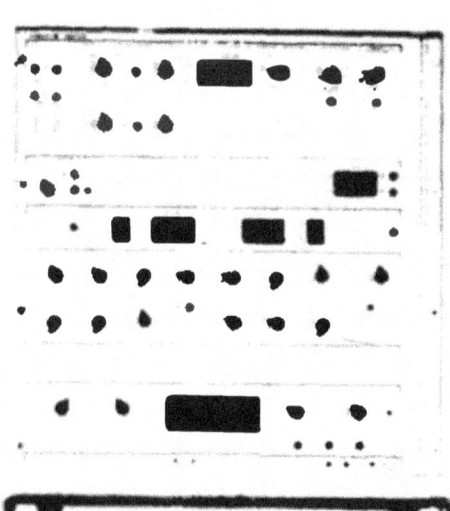
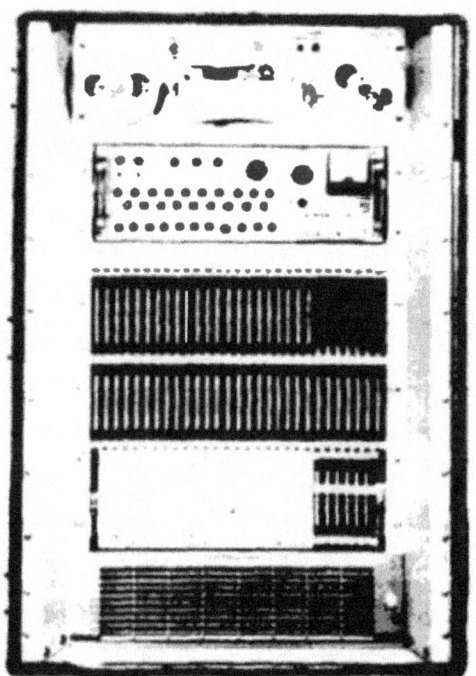

Figure 3—Stored program simulator.

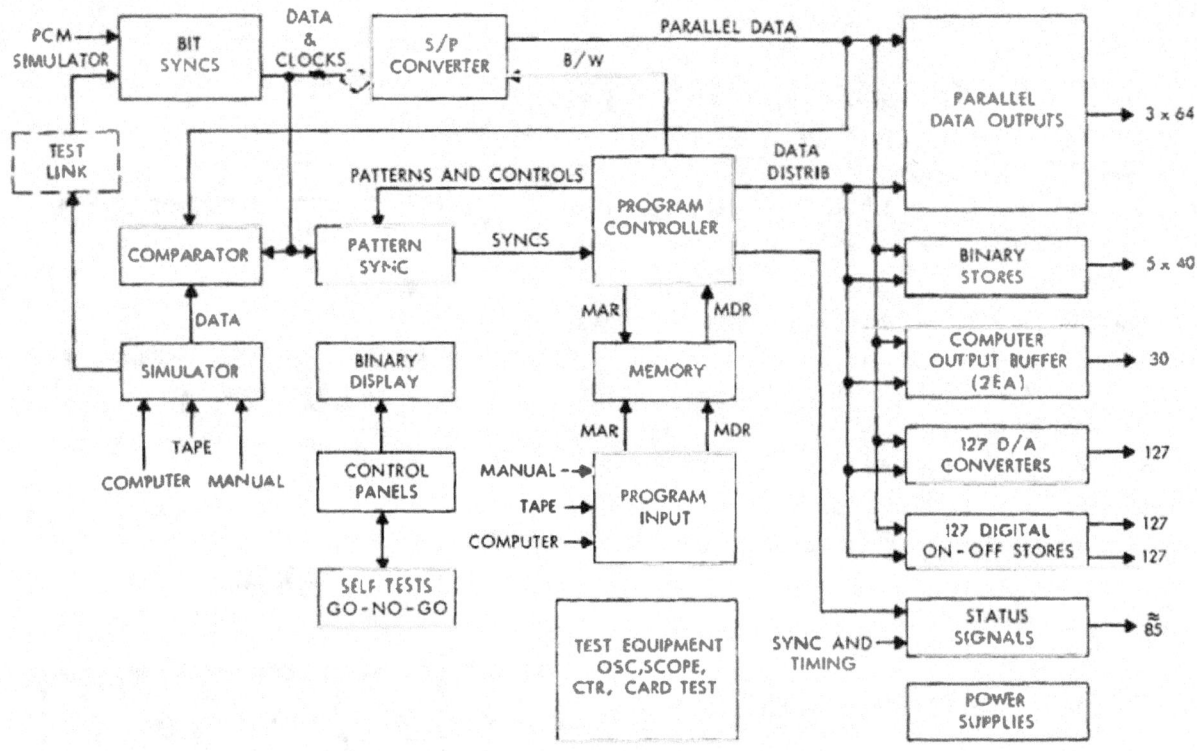

Figure 4—Block diagram of stored program PCM data decommutation system number 6000.

The parameters required for setting up the PCM system are read from memory at the initiation of a new program. At this time, memory instructions are read from the core memory to the program controller. The instructions are routed to the bit conditioner and determine what type of code this format is supposed to be receiving (NRZM, NRZS, NRZC, split-phase, or RZ), the polarity of data (negative or positive), and the bit rate (one of 10 fixed bit rates associated with this particular format).

The synchronization patterns are also read from the core memory and inputted to registers in the synchronizer for utilization in the frame synchronization, subframe synchronization, and work sync if it is utilized. At the end of the format setup procedure, the system automatically goes into a search mode and begins searching for a frame sync pattern. Once it has acquired the frame sync pattern, or word sync if it is utilized, the system goes into a check mode and then a lock mode. When in the lock mode, the decommutator is in step with the airborne commutator and the data coming from the outputs is valid.

Serial data from the bit synchronizer is coupled to the serial-to-parallel converter where it is converted to parallel form for outputting to the various output devices of the system. These consist of those shown in Figure 4. There are three 64-bit multiplex outputs and five 40-bit binary stores which have the capability of assembling, under program control, 40-bit words from any syllables within the format. The syllables need not be adjacent, and may be anywhere in length from one to ten bits.

There are two computer buffers for interfacing with the two 642B's that are being installed on the MSFN. The computer buffers have the capability, under program control, of accepting any selected number of words from the format. There are 127 digital-to-analog converters that are individually addressed by utilizing a 7 bit address in the memory instruction, and again any word in the format can be routed to any one of the 127 digital-to-analog converters. This is also true of the 127 digital on/off stores, except in this case individual bits of any word can be routed. One bit, or any number of bits from any word in the format, may be routed to any arbitrarily assigned on/off store. Status signal outputs such as synchronization status are also available.

There are three methods by which data may be inputted to the new stored program decommutator at the site: manually, from the system control panel utilizing a series of switches; by utilizing a paper tape reader which is contained in rack one; or the computer can directly interface with the PCM decommutator and in real time change a format or some instructions in an existing format. We are, of course, limited on the MSFN at the sites where we have existing patchboard systems because the computer cannot input data to the patchboard system.

Parity is used in this 4096×36 bit memory. One bit of every word is used for a parity bit. There are two modes of utilization for parity. The first is a test mode whereby the system completely runs through every memory instruction checking for parity. If a parity error is determined the system will stop and display the error (that is, the memory address and data) on the front panel displays so that it can be corrected. However, when operating on a format in the operational mode it is desired that a parity check be made without the system stopping if an error is found. In this case, parity is checked continuously as every word is brought out from memory address and data are displayed, but the program continues to operate.

NARROW BAND BIT SYNCHRONIZER

Figure 5 is a front panel view of the narrow band bit synchronizer. The controls are standard. The only things different are indicator lights on some controls. Those controls that have indicator lights are the functions which are under program control: bit rate, the type of detector being utilized, filter and sample or the integrated and dump, polarity of the incoming data, and the type of code. The wide band bit synchronizer front panel (Figure 6) is very similar, the only exception being that the bit rate itself is continuously variable, using the front panel switches, from 10 bits per second to 1 megabit.

SYSTEM CONTROL PANEL

Figure 7 is the system control panel. The top portion contains the major operating controls where format selection is controlled.

The next section is the memory control, which selects and manually inputs data to the memory or selects whether data is going to be inputted from the computer manually or from paper tape. Parity errors are displayed on this panel.

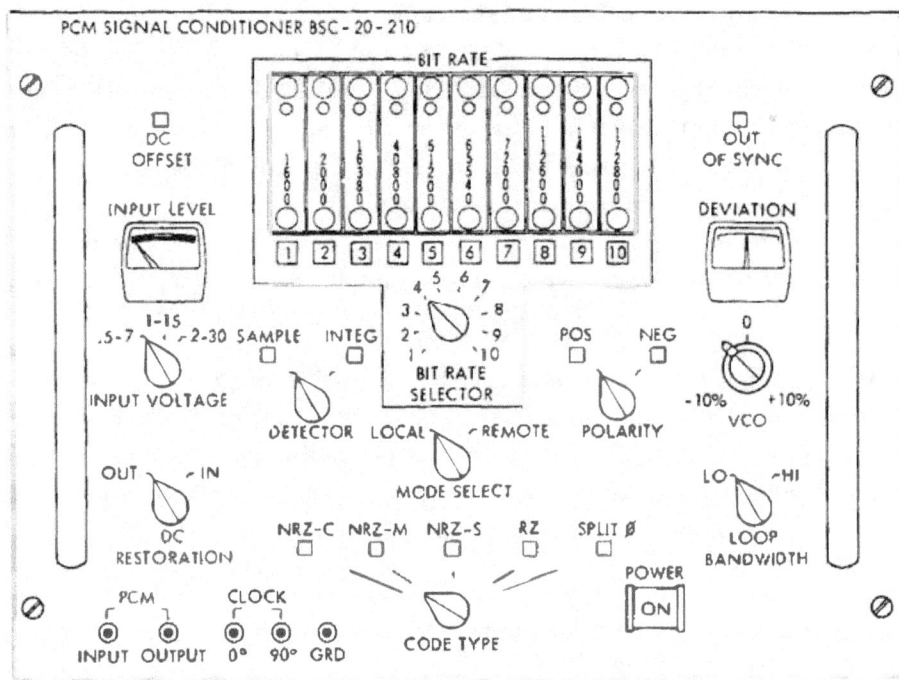

Figure 5—Narrow band bit synchronizer, front panel.

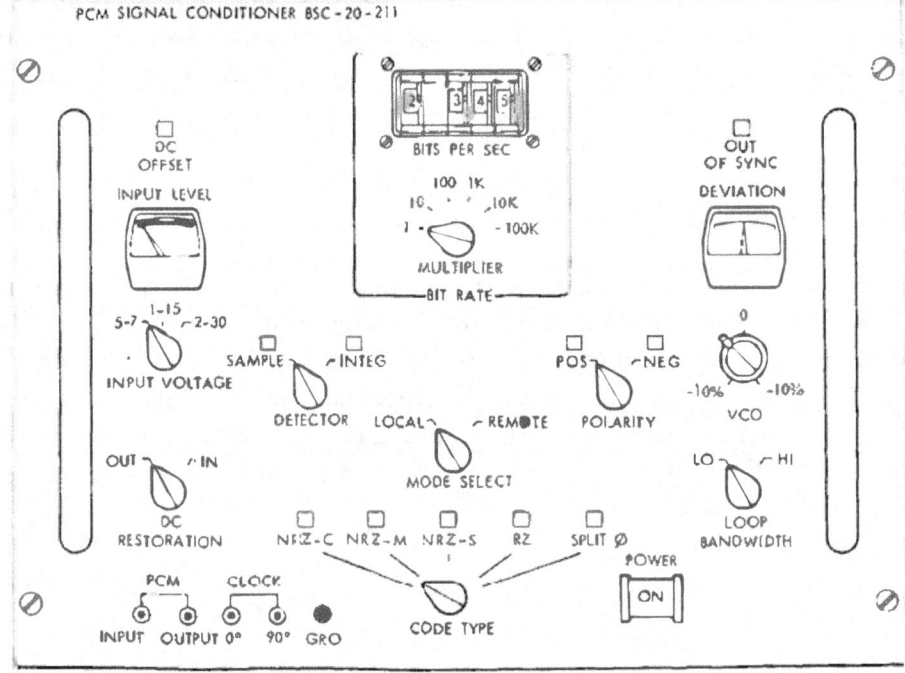

Figure 6—Wide band bit synchronizer, front panel.

The next section of panel is a monitor display. There are a 64-bit binary and two decimal displays. Any channel in any format may be selected for display on any one of the three devices located here by the channel address controls, which are associated with a unique 12-bit data tag associated with every word in the format by utilizing the memory.

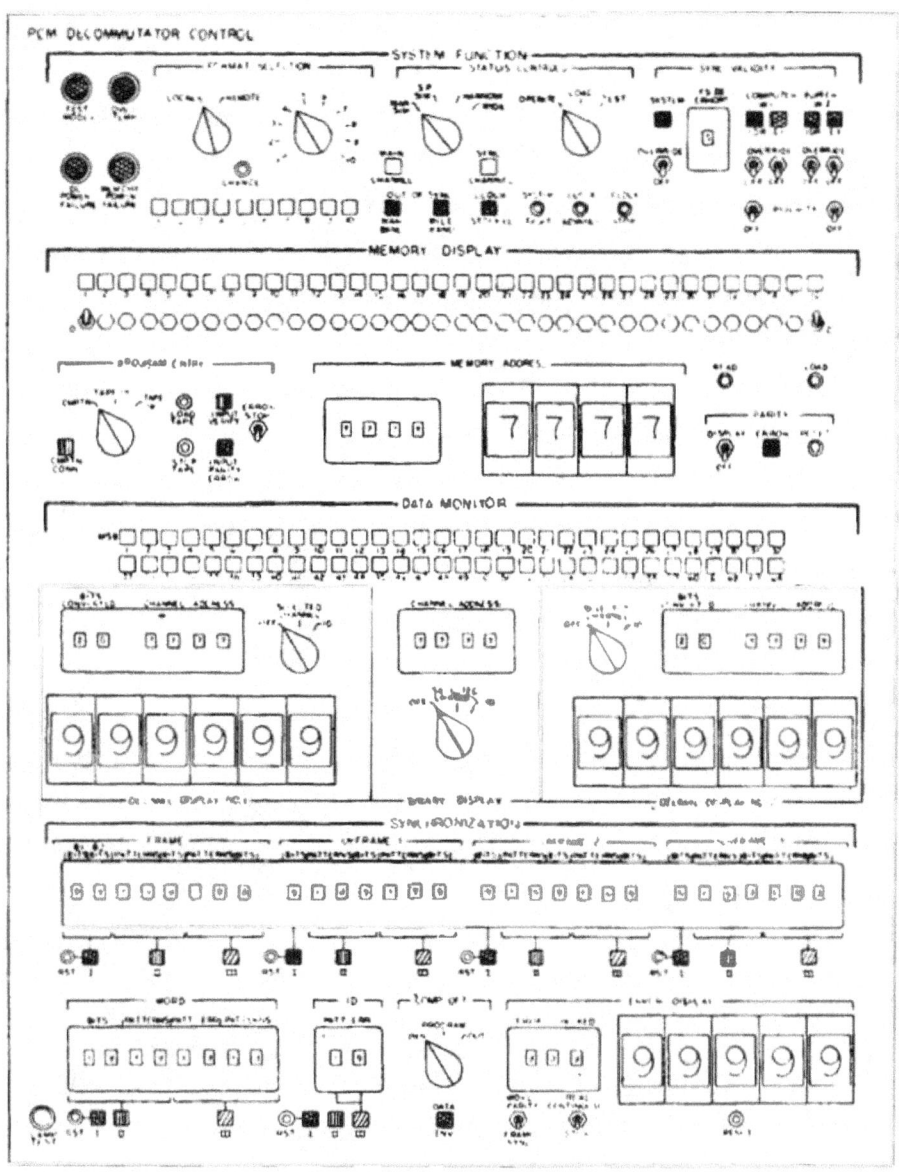

Figure 7—PCM decommutator control panel.

The lower portion is the synchronization panel, where threshold parameters for frame sync, subframe sync, and word sync are manually controllable.

D/A CONTROL PANEL

Figure 8 is a drawing of the digital/analog (D/A) control panel which is located in rack 4. It is used primarily for testing. The large switch on the left is one which will step through the 127 D/A converters. During the self-test diagnostic routine, it will check automatically for 0 and 100 percent of voltage for calibration of each D/A. Manual control of the digital voltmeter display is on this panel providing the monitoring of the voltage output of the D/A in the system.

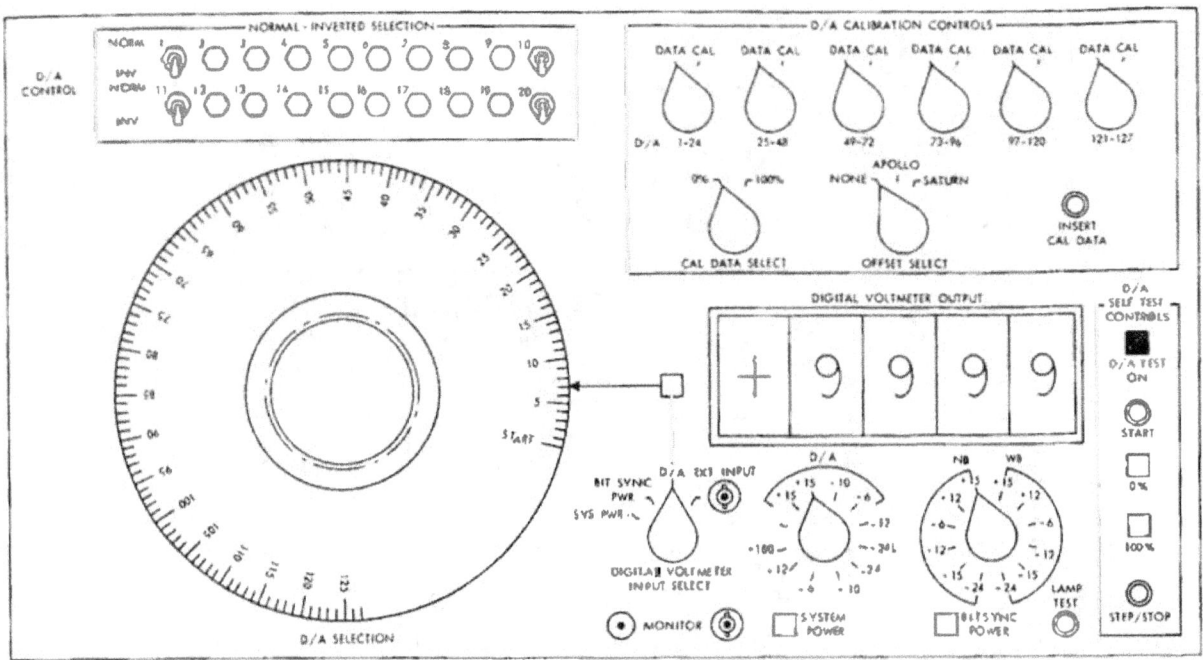

Figure 8—D/A control panel.

The capability of checking any power supply in the system or using the voltmeter for any external input is available if desired.

SELF-TEST CONTROL PANEL

Figure 9 is the self-test control panel located behind the door in rack 1. It is utilized for self-checking the system. The system can be run in several diagnostic modes, utilizing the PCM simulator in a closed loop to check the output devices, and can run bit-error-rate curves on the signal conditioners, again using the simulator in a closed loop mode. The simulator contains a bit comparator which provides a direct reading of the bit-error rate of the signal conditioner.

The core memory contains internal circuitry for self-check which will allow the input of the worst case pattern or its compliment (all 1's, all 0's), and it will automatically step through this program.

PCM STORED PROGRAM SIMULATOR

Figure 10 is a simplified block diagram of the PCM stored program simulator. Stored program is perhaps a misnomer in this case, as we are actually procuring two simulators. A manual simulator and a stored program simulator are both contained in the same rack and share the same output perturbation circuitry, where noise, jitter, or baseline variation are added to simulate some tape-recorder characteristics.

APOLLO NETWORK PCM DECOMMUTATION SYSTEMS

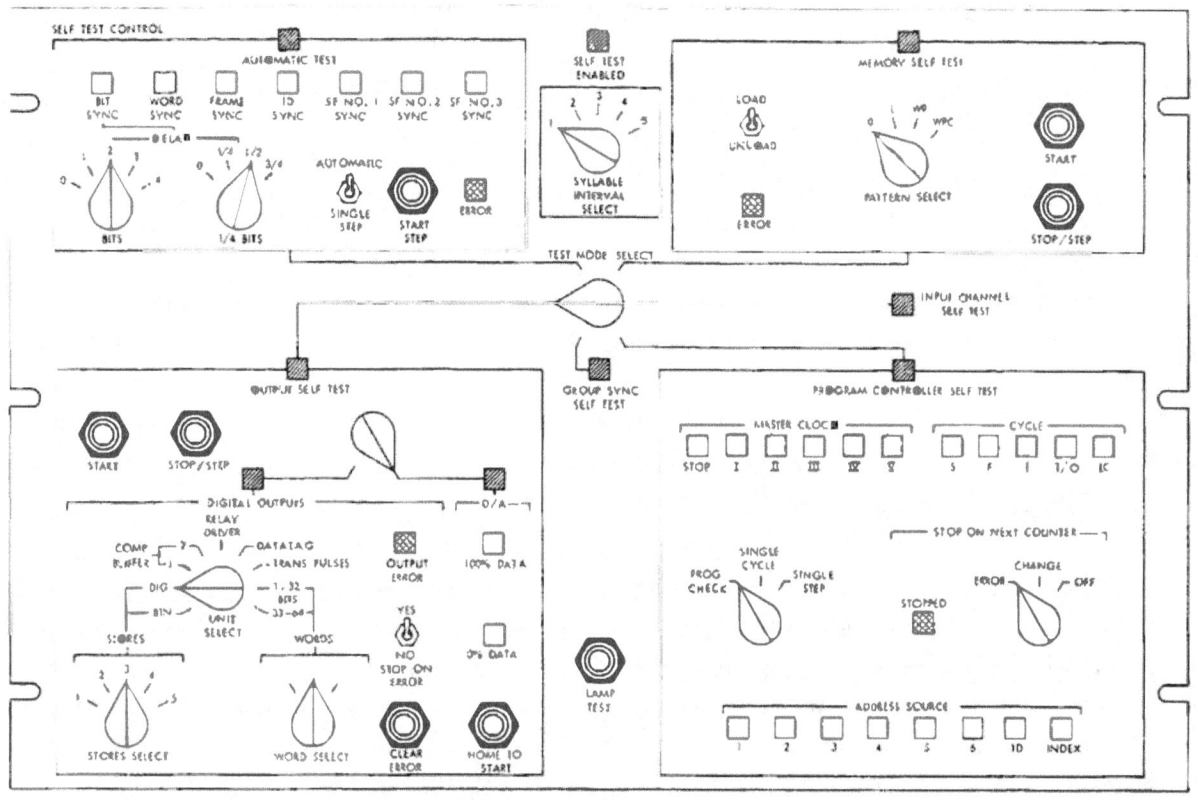

Figure 9—Self-test control panel.

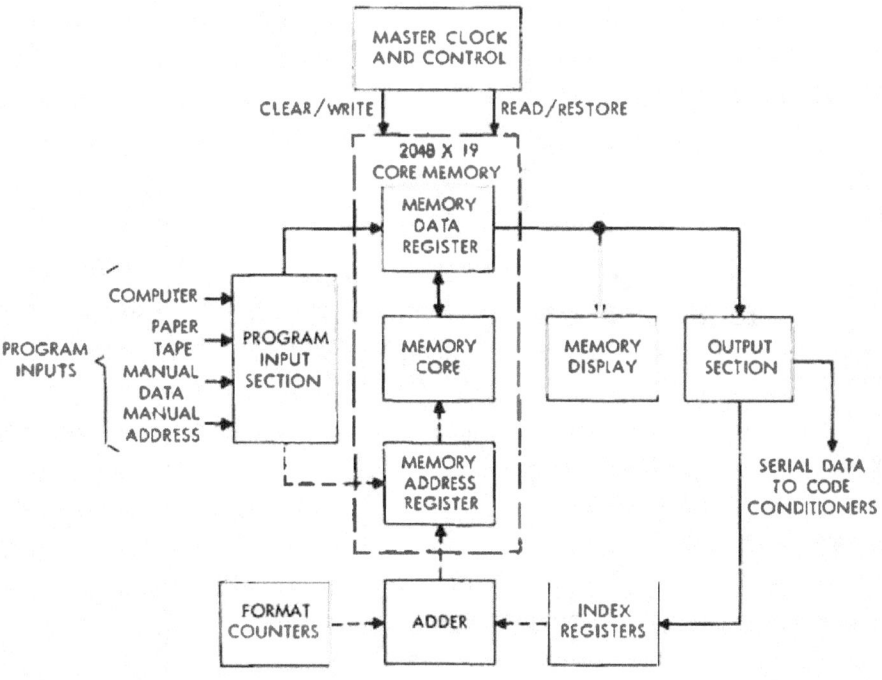

Figure 10—Block diagram of PCM stored program simulator.

The stored program simulator portion has all the characteristics of the decom with respect to core memory inputs. Data input to the memory can be manual, by computer, or by paper tape. Data are read from the memory core. Ten bits of every word in the memory contains output data. A 19 bit memory, 2,048 words, provides many unique words in the format, many more than could practically be obtained with a manual simulator. The simulator, like the decom, can contain up to ten formats. These can be controlled and selected at its front panel or remotely tied into the computer in a closed loop mode for remote checking of a site. The output portion of the simulator allows the addition of noise, calibrated with reference to a particular bandwidth to the system for running bit error rate curves. It provides for the addition of jitter, again by known amounts, and baseline variation by known amounts to simulate the various characteristics encountered when utilizing telemetry receivers and recorders in a playback mode.

SIMULATOR CONTROL PANEL

Figure 11 is the control panel of the simulator. It is quite similar to the system control panel of the decommutator. The first portion controls operational functions and data register. The next portion is utilized for blanking, which may be started anywhere in a serial bit stream and stopped at any other point, manually selectable. The remainder of the control panel is for output data control such as adding noise, jitter, or baseline variation. The last portion of the bit comparator, where bit errors are obtained when the serial bit stream (from the simulator to an RF link, into a PCM signal conditioner, and out of the signal conditioner back into the simulator) is compared with the simulator output. Any bit error or incorrect decision made by the signal conditioner is indicated on this panel, which also contains the manual simulator controls.

DECOMMUNICATION SYSTEM DISTRIBUTION

Figure 12 is a block diagram of the decommutation system distribution unit, which is the main interface box for all the PCM's and other network equipment. All the inputs from each PCM which are going to be utilized on Apollo connect into this unit. PCM outputs come into a main patch panel where any PCM may be arbitrarily selected, and in case of a failure PCM 2 can replace PCM 1 and be utilized by selecting the proper format and replacing this patch panel to output PCM 2 directly to all the outputs interfaced with PCM 1.

The primary function of this unit is the handling of the event indications from the PCM, which are utilized to switch on the event indicator lights on the consoles.

From the main patch panel the events are routed to an individual patch panel for each console and for the recorders. All 127 events from each PCM are connected to individual console patch panels in parallel. This panel provides the means for selecting any 108 of these events from each console. These 108 can come from one PCM or any mixture of 108 events from the three PCM's.

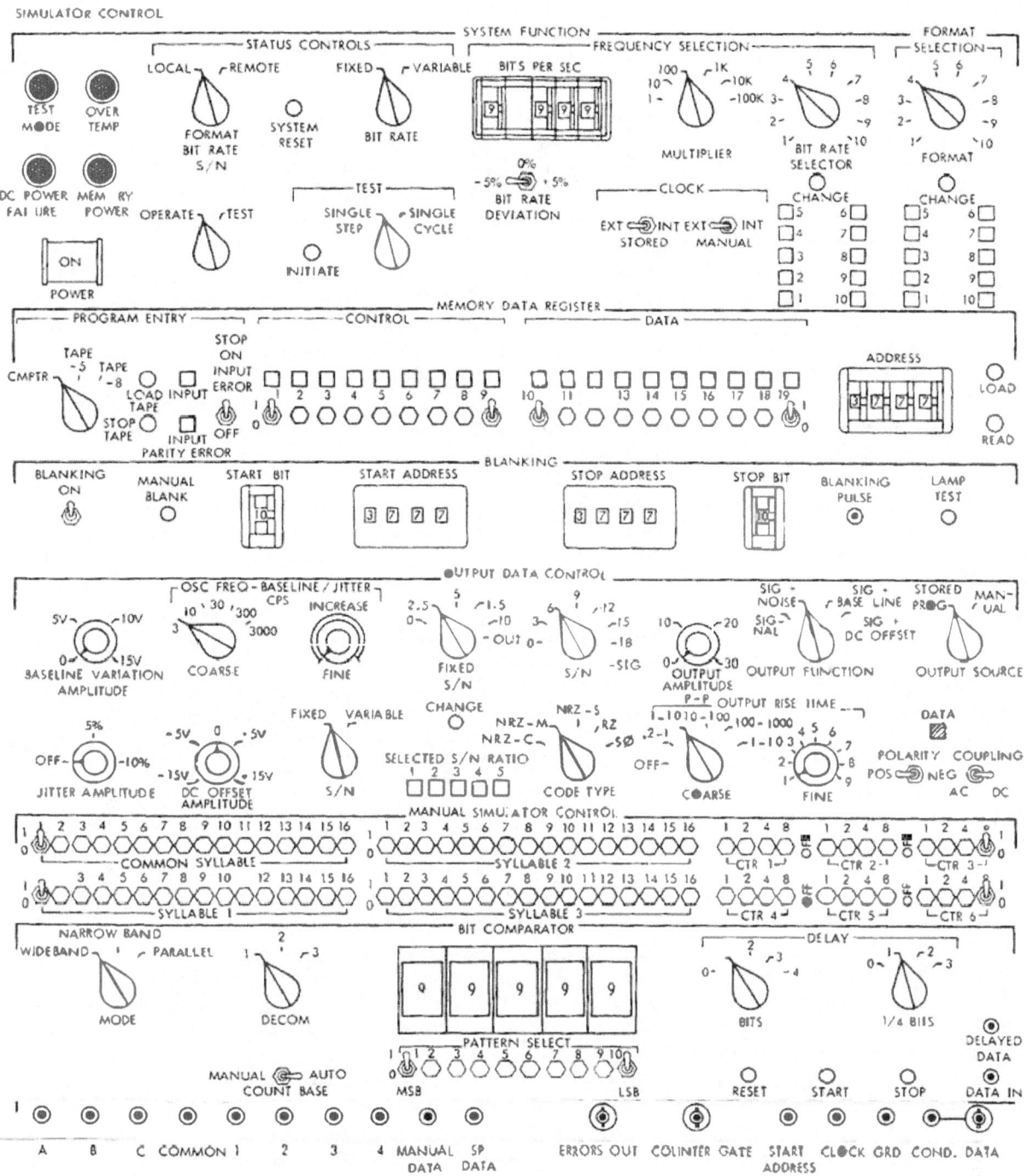

Figure 11—PCM stored program simulator control panel.

MAIN PATCH PANEL

Figure 13 is the main patch panel. On the left hand side are events coming in and going out from one, two, and three PCM's, 127 of them from each. On the right hand side are the analog functions coming in from the PCM's. These can be patched on this panel to the aeromedical

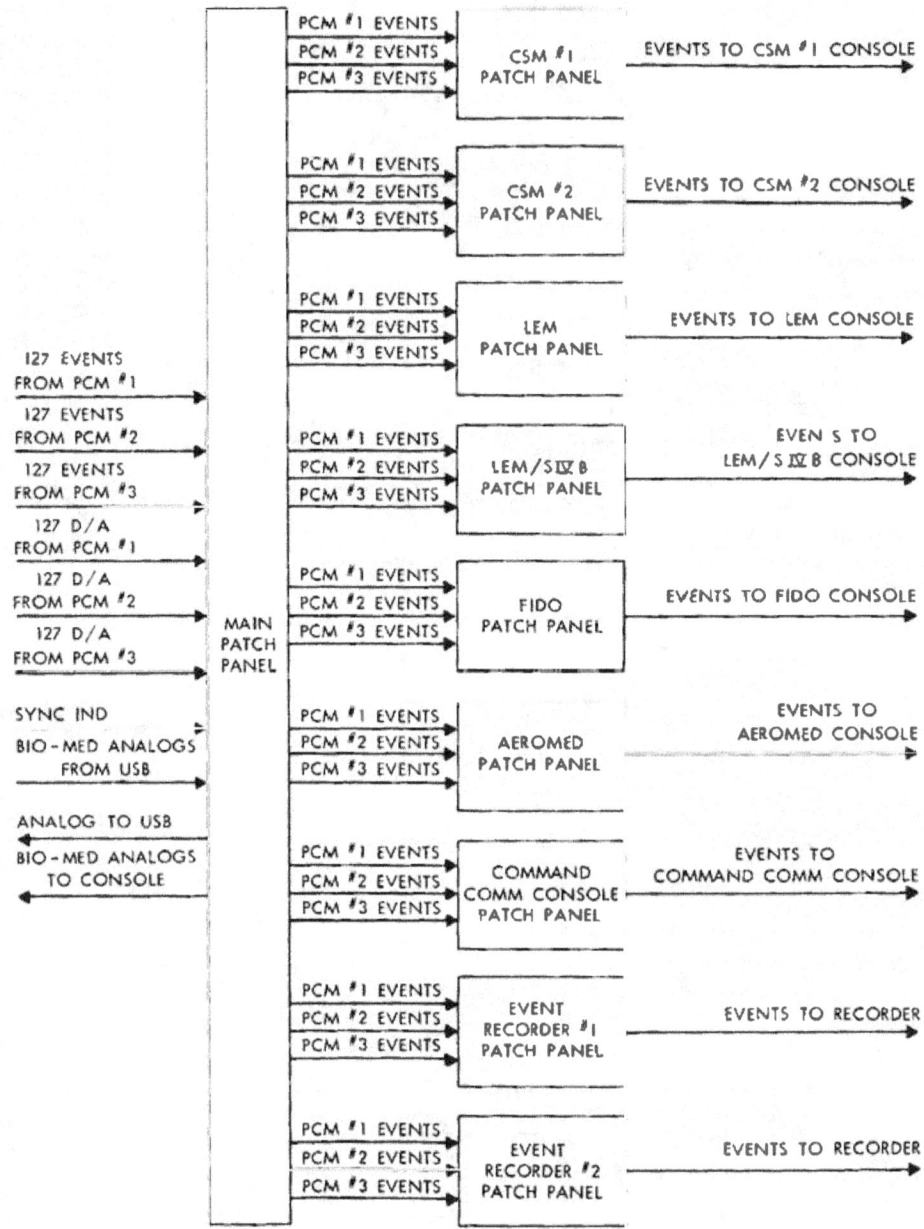

Figure 12—Block diagram of decommutation distribution system.

console for biomedical functions, and to the aeromedical recorders. Also included on this panel are a few other functions which tie the PCM's into other systems on the site. There are outputs from the PCM analogs which are required for the exciter portion of the unified S-band. These are analog spacecraft parameters indicating the static phase error and automatic gain control (AGC) range of the spacecraft receiver. These are received by the ground station, decommutated, and converted to analog form in the PCM station, then patched on this board from the PCM station back to the unified S-band exciter panel for utilization in obtaining up-link lock. This is not required, but it is utilized as an aid for obtaining up-link lock. PCM sync is also sent to the exciter with this analog to let the operator know when the data he is receiving

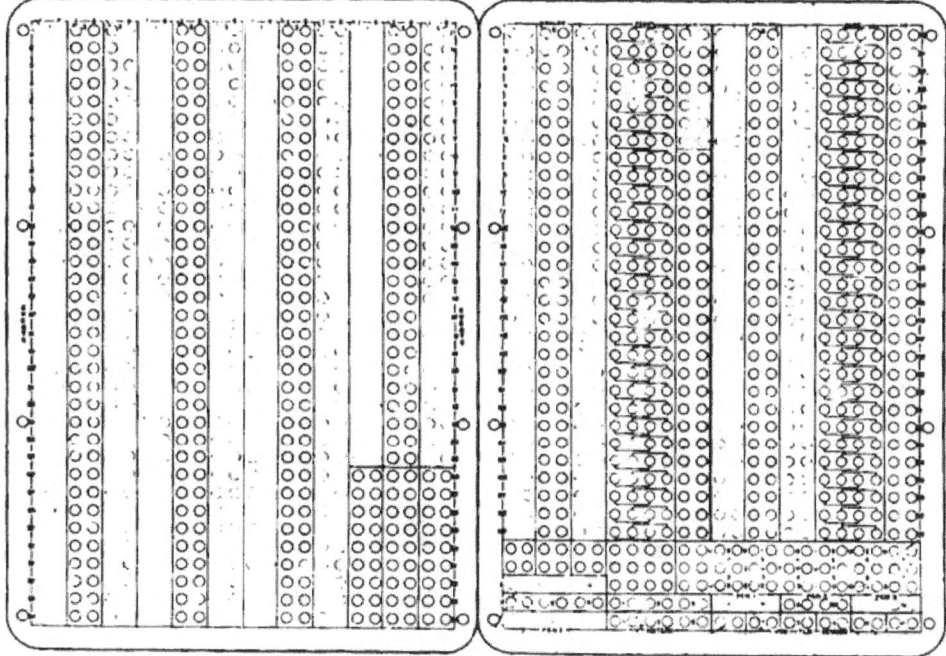

Figure 13—Main patch panel.

is valid. Also on this panel are the seven aeromedical-biomedical channels from the unified S-band. Sub-carrier discriminator outputs are brought to this panel as this unit will interface with the aeromedical consoles. There are two methods by which we can get aeromedical information. One is the unified S-band where it comes out of the sub-carrier demodulators, and the second is where it comes down on the PCM telemetry bit stream.

EVENTS PATCHBOARD

Figure 14 is the patchboard for the events. There are three sections here for inputting from the three PCM decommutators. A fourth section for the output of the events is divided into groups of 36 patches to correspond with the groups of 36 indicators on the consoles.

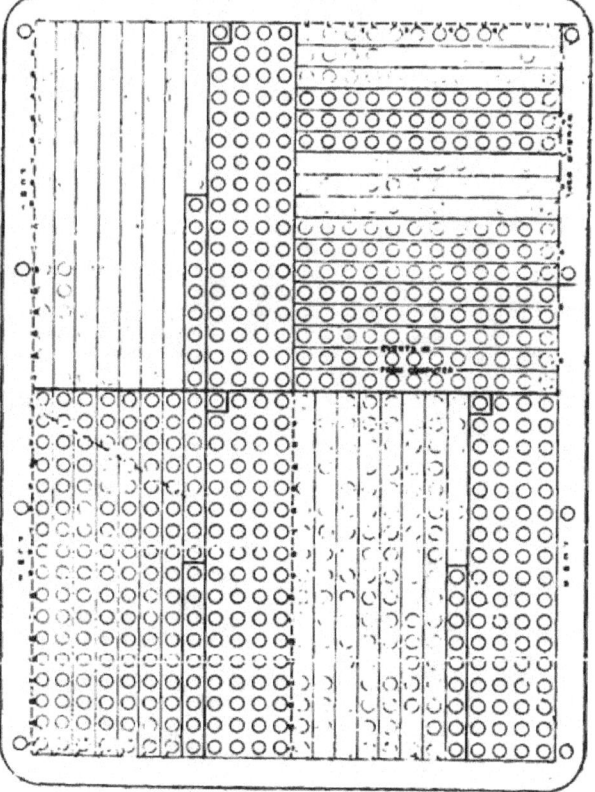

Figure 14—Events patch panel.

PATCHBOARD PCM SYSTEM

Figure 15 is the existing patchboard PCM system located on the sites. The basic difference between the present and soon-to-be-obtained systems is that the acquisition parameters of this PCM and its distribution parameters must be manually patched up (using a patch board). However, it does have the capability of containing four PCM formats which can either be selected locally or remotely. It has just about the same output capability, except that the D/A's and on/off events are limited to 100 rather than to 127. It is, of course, a larger system in physical size than what is being purchased.

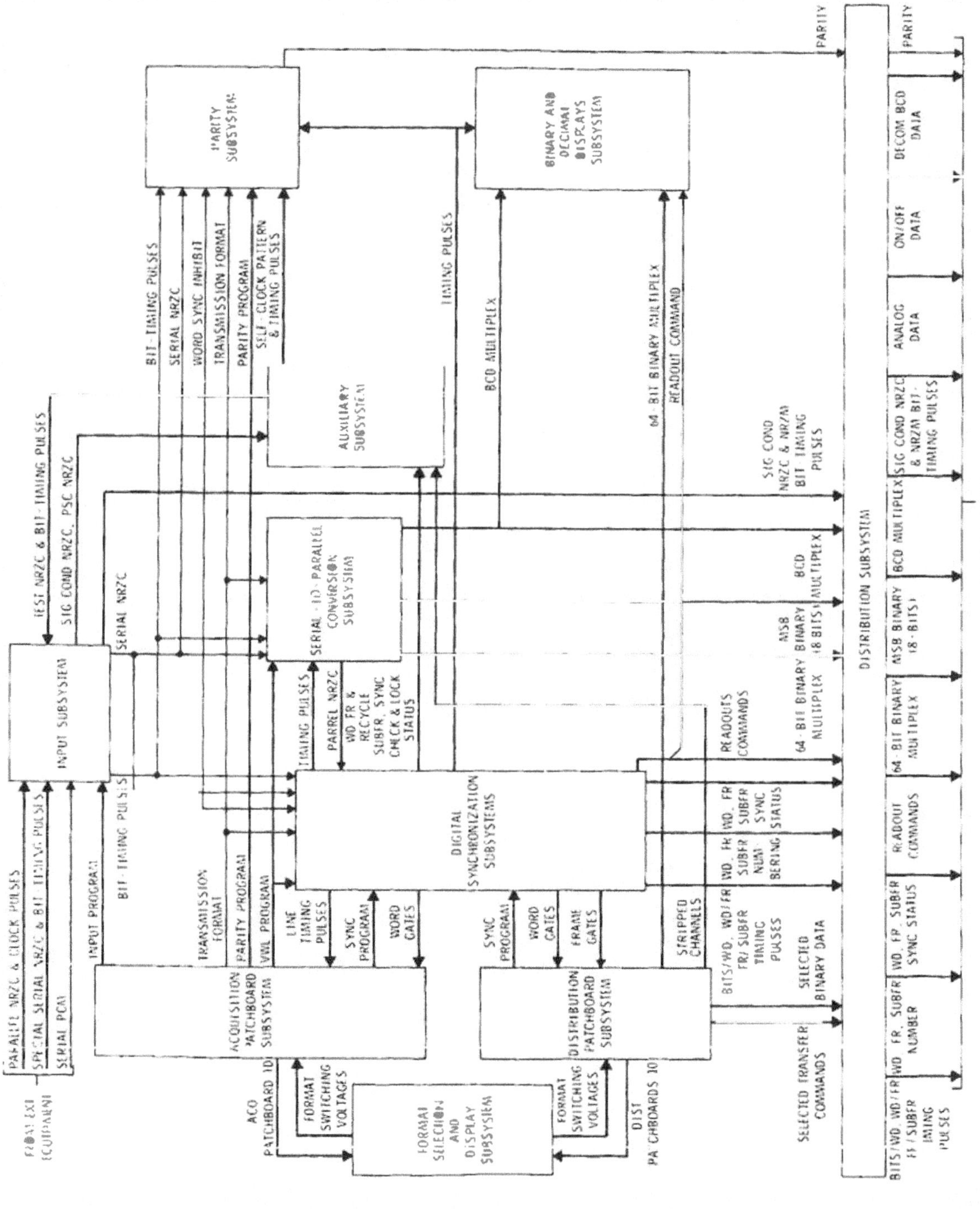

Figure 15—Patchboard PCM system.

APOLLO NETWORK
REMOTE SITE COMPUTER SYSTEMS

by
E. Willis
Goddard Space Flight Center

ABSTRACT

A brief description of the 1218 remote site data processor currently being utilized on the Manned Space Flight Network (MSFN) is presented, including brief discussion of the systems configuration and the present capabilities in Gemini missions. New computer systems which are to be installed on all of the Apollo sites are also included. This covers the:

a. UNIVAC MOD-642B computer

b. UNIVAC 1540 magnetic tape units

c. UNIVAC 1259 TTY adaptor

d. UNIVAC 2008/10 data transmission units

e. UNIVAC model 1000 CAM adapter and Greenwich mean time (GMT) buffer

f. Motorola TP-4000 3000 word per minute line printer

g. UNIVAC 1299 distribution switchload

h. Console computer interface adapter.

INTRODUCTION

The data processing system selected for Apollo remote sites is being manufactured and assembled by the UNIVAC Military Systems Division located in St. Paul, Minnesota. The computer, the UNIVAC 642B, is designed to meet military specifications. There will be two identical computing subsystems located on each of the remote sites for the Apollo network. These systems are identical in every respect with the exception of the mission requirements which will be assigned to them. One computer will process telemetry data, and the second will perform command functions.

642B MODIFIED COMPUTER

The 642B modified computer shown in Figure 1 has the following characteristics:

1. It is a general-purpose, medium-scale, solid-state, parallel, binary machine.

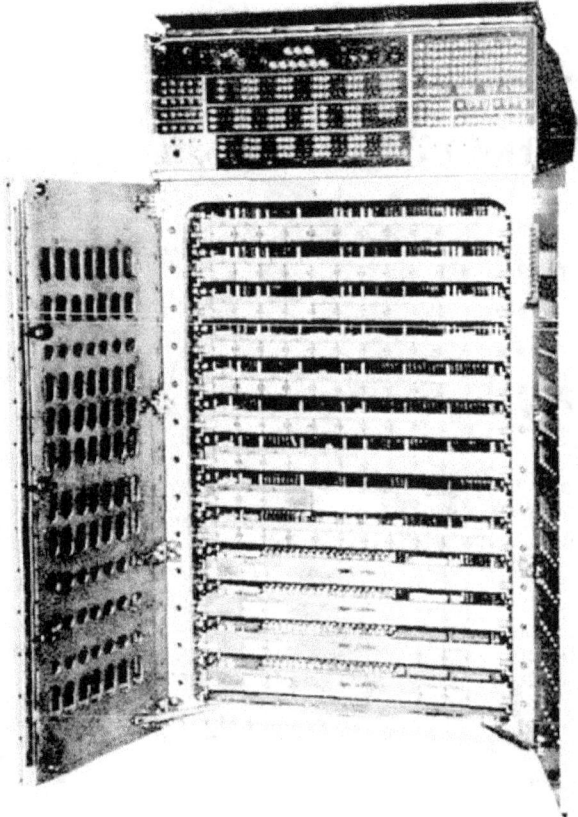

Figure 1—624B modified computer.

2. The main memory is magnetic core, capable of being randomly accessed.

3. It has overlap memory capability which can increase the execution speed of various programs.

4. It has a 2.0 microsecond read-write cycle time and a 30-bit word length.

5. There is a storage capacity of 32,768 words, directly addressable, half or full word operands which can be increased to 131,072 words, directly addressable.

6. Sixteen input and sixteen output channels are provided and all input/output transfers are under full buffer control.

7. With these sixteen I/O channels there are eighty-one unique interrupts.

Other features of the computer are:

1. Channel priority can be determined by the software with very minor adjustments made to the hardware.

2. Input/output (I/O) control, which is accomplished with ten basic instructions, provides positive control over the I/O and a high degree of sophistication in programming.

3. The continuous data mode capability will allow for automatic reinitiation of previously established buffers under program control. The termination of the buffer is also program controlled.

4. Externally specified addressing features enable a data word to be stored or read from an address directly specified by an external device.

5. The externally specified indexing capability can be used to transfer data words indirectly specified by the external device. The external device specifies the address of the buffer control word for the particular transfer.

Peripheral devices required to complete the computing system are:

1. 1540 magnetic tape unit.
2. 1232 input/output console.
3. 2010 data transmission unit.
4. Console computer interface adapter.

5. High-speed buffer, translator, and printer.

6. 1000 interface system adapter.

7. 1259 TTY adapter.

1540 MAGNETIC TAPE UNIT

The 1540 magnetic tape unit shown in Figure 2 is designed to military specifications. Significant features of the tape unit are:

1. Air-cooling requirements will operate at normal conditioned room tematures with no special ducting of the air.

2. The tape transports are shock-mounted inside the cabinet.

3. The tape transports will handle 1/2 inch mylar tape.

4. The maximum writing and reading operations are 120 inches per second. Rewind operations can be 240 inches per second.

5. The recording densities are 200, 556 or 800 frames per inch.

Figure 2—1540 magnetic tape unit.

6. The recorded tapes will be compatible from transport to transport within the system and with IBM 727, 729II, 729IV, and 729VI magnetic tape sets.

7. Each set of magnetic tape units will have duplexing capability. Either magnetic tape system can communicate with either computer during operations. The same is true for each of the computers.

UNIVAC 1232 INPUT/OUTPUT CONSOLE

The UNIVAC 1232 input/output console is shown in Figure 3. This unit consists of the tape photo electric reader, which will read 5-, 6-, 7- or 8-level oiled, or dry paper and mylar tape at 30 inches per second or 300 characters per second. The tape punch can punch 5-, 6-, 7-, or 8-level tape at the rate of 11 inches per second or 110 characters per second. The keyboard input/output capability is 10 characters per second, with 72 characters per line.

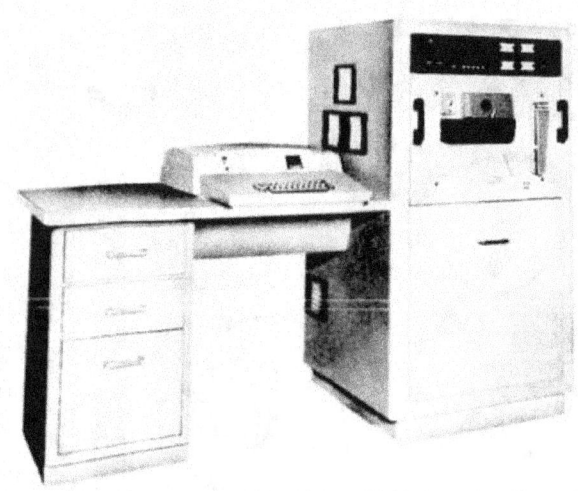

Figure 3—UNIVAC 1232 input/output console.

UNIVAC 1259 TELETYPE ADAPTER AND MODIFIED ASR-28 SET

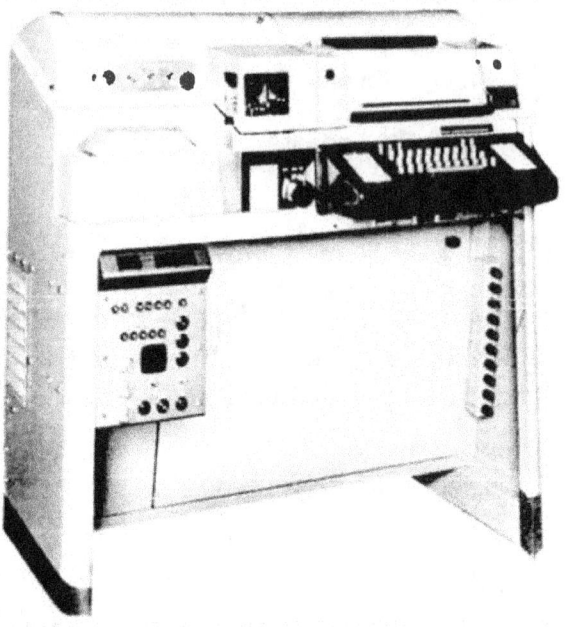

Figure 4—UNIVAC teletype adapter and modified ASR-28 set.

The UNIVAC 1259 teletype adapter and modified ASR-28 set shown in Figure 4, consists of:

Page printer which will accept TTY codes at a maximum rate of 10 characters per second.

Type reperforator (A)

Type reperforator (B)

Tape reader (transmitter/distributer)

Keyboard

Auxiliary line relay

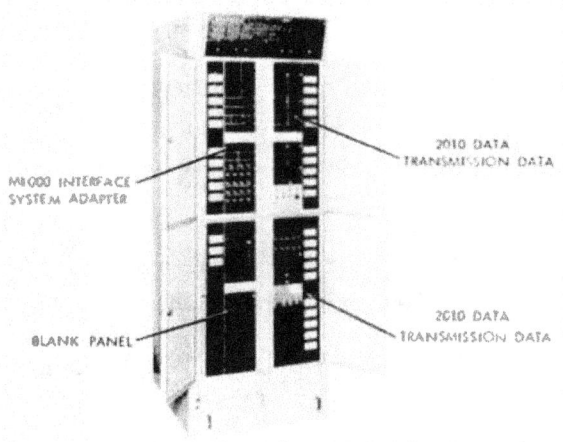

Figure 5—Peripheral communications system.

Teletypewriter adapter (which provides the necessary interface between all the 1259 units and the 642B modified computer.)

PERIPHERAL COMMUNICATIONS SYSTEM (Figure 5)

Data Transmission Units

The purpose of the data transmission unit (DTU) and the communication complex is to provide computer-controlled input and output communication between remote site communication lines and the UNIVAC 642B computer. The DTU's are each designed to permit independent operation. The DTU is also designed to provide a 8- or 10-bit parallel transfer between the computer and the DTU. The transfer rate of the data will be determined by the clock supplied from 205 modulator/demodulator. The DTU is capable of transferring serial data to the subset at speeds of 80,000 bits per second with only minor modifications consisting of adjustment of four delay cards.

Contained in the same DTU cabinet is the UNIVAC Model 1000 interface system adapter. This adapter consists of a timing buffer which will receive parallel Greenwich mean time (GMT) from the station timing system and input GMT to each of the two computers with one-second granularity.

The second purpose of this adapter is to multiplex five computer address matrix keyboards located on the flight control consoles to one of the computer channels.

Console/Computer Interface Adapter CCIA (Figure 6)

This unit provides the means of communications between seven flight control display consoles and both the telemetry and command computers. The CCIA consists of two identical sections, which are independent in all respects, including their source of power. One section will service four consoles and the second section will service three consoles. At least three consoles will be completely operational in the event of a failure in any portion of the system.

In Figure 7, it will be seen that each section of the CCIA consists of four major subsystems. First is the concentrator unit which sequentially inputs the position of all keyboard switch contacts into a 1218 computer. The 1218 computer has a four kilobit memory capacity and is programmed to detect actuation of any switch, establish communication with both the telemetry and command computers, and to format and translate data directed from these computers to the distribution and storage unit. This distribution subsystem provides the signals for the keyboard button indicators, the wall-mounted clocks, and the digital-to-analog converters. Analog voltages derived from the digital-to-analog converter unit drive eight channel pen recorders which are associated with four of the consoles.

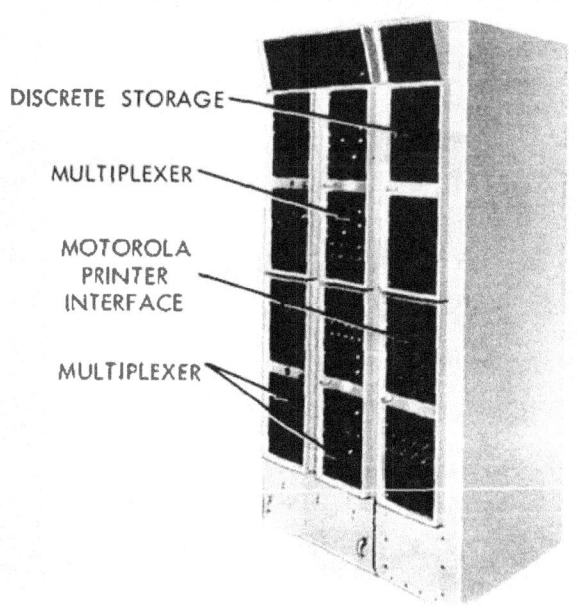

Figure 6—Console computer interface adapter.

The buffer unit and a translator unit for six Motorola TP 4000 printer units shown in Figure 8. are packaged in the CCIA unit but are not a part of it. The buffer and translator serve as the interface between the six

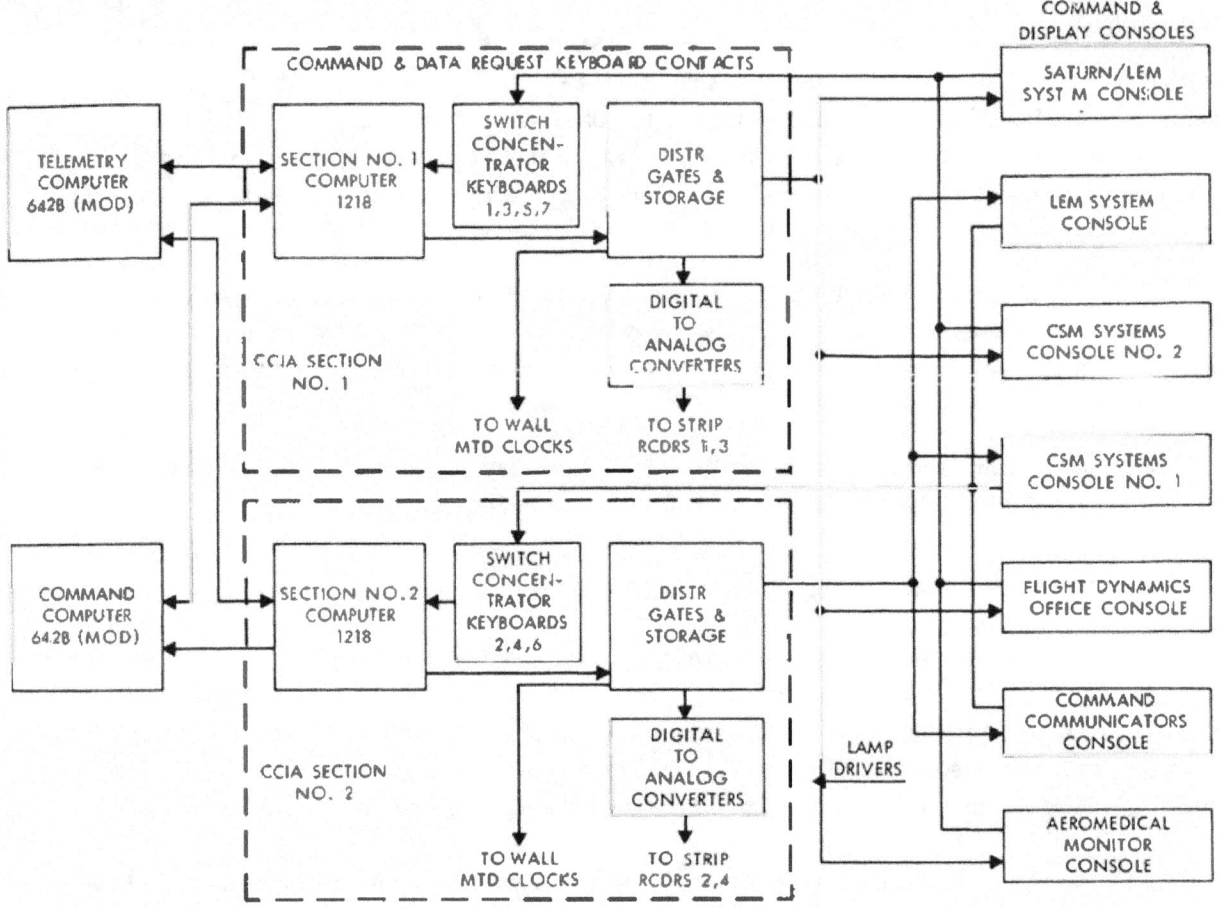

Figure 7—Block diagram - console computer interface adapter.

Figure 8—Motorolo TP400 printer set.

high-speed printers and both the telemetry and command computers. Each printer unit is small in size and is quiet, smokeless, odorless, and residue-free while operating at speeds up to 3000 words per minute. Its paper capacity is equivalent to 30,000 lines of print with each line containing 72 characters (500 feet).

The last item of equipment to be described (Figure 9) does not fall in the category of peripheral devices, but it is a vital part of the system interconnection. This is the digital signal distribution switchboard, type SB-1299. A single switchboard may contain 10 independent, manually operated, 68-pole, 3-position switches. Each site requires two of these switchboards.

APOLLO DATA PROCESSING SYSTEM OPERATION

The Apollo data processing block diagram (Figure 10) shows the interconnection between the computing system and the other remote site systems. In the normal mode of operation one 642B computer will process data for display on the flight control consoles and format telemetry data for transmission to the Mission Control Center. The second 642B computer will process, validate and store commands for transmission to the spacecraft.

If either computer fails, a third program may be loaded into the remaining computer. This program will have the capability of processing data for both display and command. At present it is anticipated that part of the display capability may not be accommodated due to the size and speed limits

Figure 9—Digital signal distribution switchboard, type SP-1299.

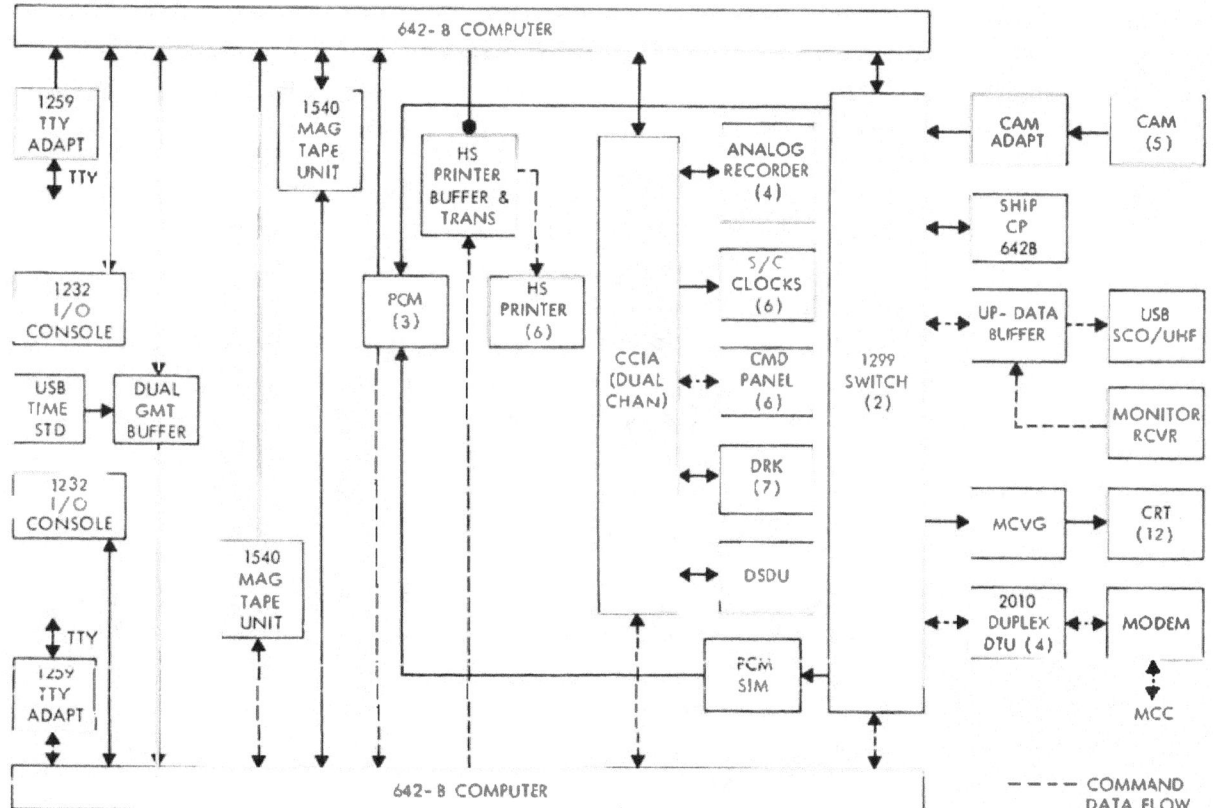

Figure 10—Apollo data processing system, block diagram.

of the computer. The limitations will be defined further after the computer program has been developed. In the failure mode, peripheral devices required for either command or display functions will be switched to the remaining computer manually at the 1299 switchboard. Units which will not require switching include the following:

a. The I/O console

b. The 1259 TTY adapter

c. The GMT buffer

d. The 1540 magnetic tape unit

Each computer has these units connected directly to it.

The I/O console will be utilized mainly for maintenance and check-out of the system. It is anticipated that the teletype input to the console will be utilized mainly as a backup to the high-speed input. The GMT buffer will input time to correlate up-link clock words and to time-tag telemetry data. The 1540 magnetic tape units will be utilized for storage of computer programs, background data, command loads, and tables for conversion of parameters to engineering units. A high-speed printer will be located in the computer room to provide a high-speed read-out to the computer technician. It is anticipated that he will be required to assist the flight control team in monitoring command loading and transmission.

Figure 11 indicates the block diagram of sites which do not have flight control consoles. These sites will not contain either a CCIA, memory character vector generator, cathode ray tube (CRT) displays or consoles with associated keyboard, analog recorders and spacecraft clocks. One high-speed printer will be provided for Management and Operations personnel.

Figure 12 indicates telemetry data flow. Data received from the pulse-code modulation (PCM) station will be processed by the telemetry computer and formatted for transfer to the MCVG. Data displayed on the CRT will be refreshed from the MCVG. Selection of data to be displayed on the CRT will be made from the data request keyboard (DRK). Parameters in the various display formats will be limit-sensed by the computer. Indications will be given on the DRK of formats which contain out-of-limit parameters. Analog functions from the telemetry data to be displayed on recorders associated with the systems consoles will be stripped from the telemetry data by the computer and routed to digital/analog (D/A) converters in the CCIA. The D/A converters will be hard-wired to the recorders. Selection of one of six groups of parameters can be made by a switch on the recorder.

The selection of a format by positioning the switch will generate instructions for the computer through the CCIA. The 642B computer will perform the selection or switching function of the analog parameters. Spacecraft clocks contained in the telemetry bit stream will be formatted by the computer and transferred to the CCIA for driving in-line digital displays. Provision has been made for transferring 16 additional off/on indications from the computer to the CCIA. These functions will be terminated on the decom system distribution unit and may be utilized similarly to off/on stores to the PCM. Data to be displayed on the high speed printer may be selected from the DRK.

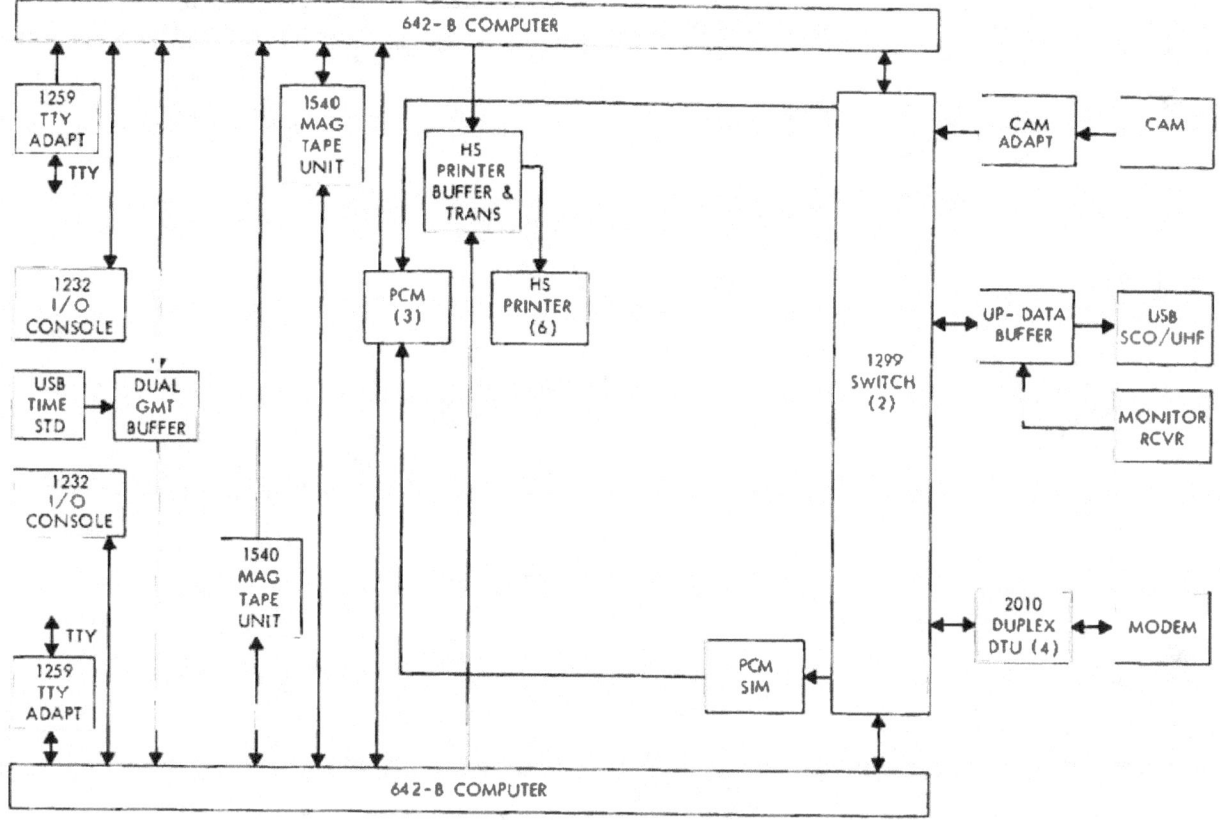

Figure 11—Block diagram of site without flight control console.

A 5X5 CAM matrix will be located on each of the systems consoles. This panel will permit the flight controllers to communicate with the computer. A three-digit number plus mode indications may be transferred to the computer from the CAM. The data format to be transferred to MSCC may be selected from the DRK. The data is then transferred to the DTU in either eight- or ten-bit bites, converted from parallel to serial format and transmitted over the high speed modulator/demodulator. Telemetry summary messages from other sites or the control center may be input to the computer for storage through the teletype adapter. This data is then available for display on the CRT at the flight controllers command.

Commands and command loads will be transmitted from mission control center (MCC) over high-speed lines to the 642B computer. The computer will perform error checking functions as the data is stored in the command computer. It will also be output to the 1540 magnetic tape unit for storage. Indication that a command load has been received will be provided to the computer technicians and to the flight controllers on the high speed printer.

Off/on stores indicating spacecraft equipment status will be stripped from the PCM format by the decom and loaded to the computer. This data will then be utilized to drive the command panel displays through the CCIA. Switch selections of the command panel will be converted to computer instruction by the CCIA. The command panel will contain modulator selection switches, load and clock selections, and real time command switches. Sixteen miscellaneous

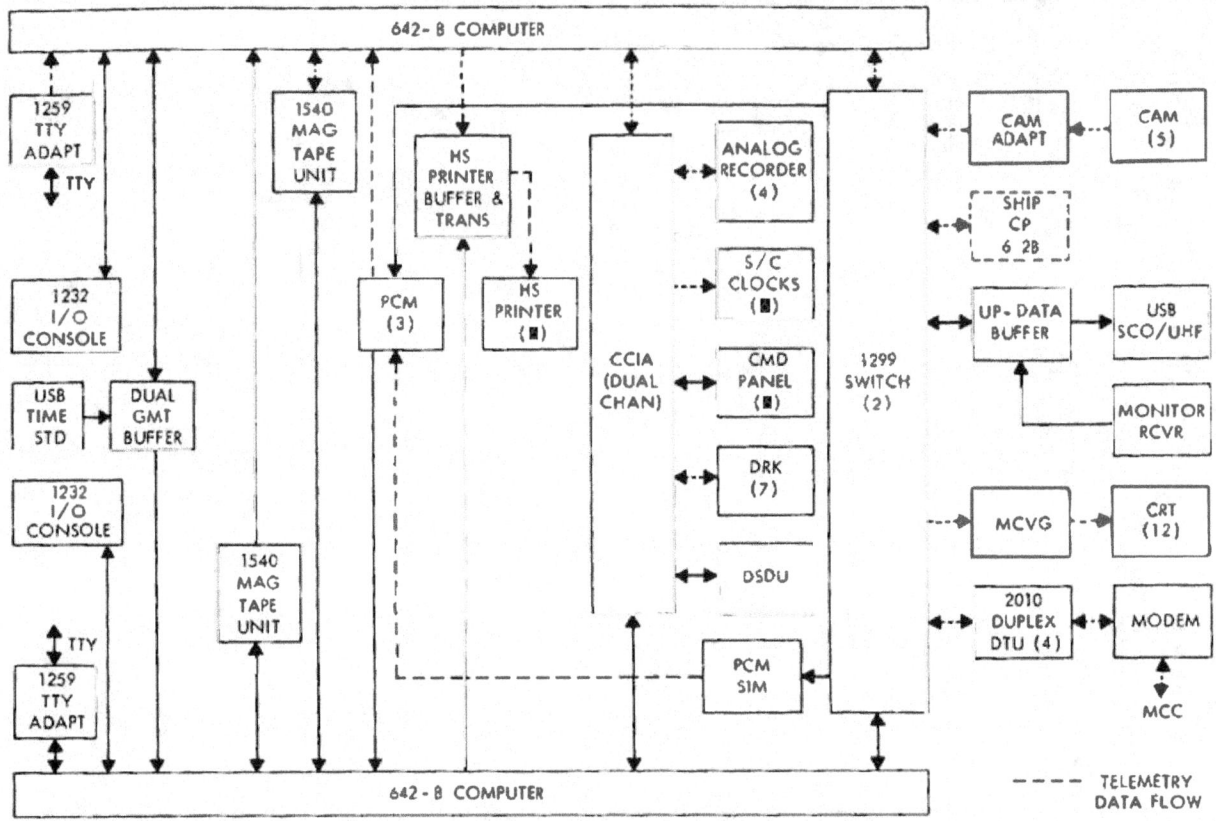

Figure 12—Telemetry data flow.

off/on functions have been provided at the CCIA. These off/on stores will be utilized to provide miscellaneous outputs from the command computer.

When the command transmission is desired by the flight controller, he will depress the assocated switch on the command panel. The selection transferred to the computer will cause the selected data to be transferred to the up-data buffer and consequently, up-link to the spacecraft via either the Unified S-Band (USB) or UHF transmitter.

Two command validation loops are available. One loop samples the transmitted command throughout the monitor receiver for comparison in the computer. The other validation loop consists of message acceptance pulse data received from the spacecraft through the PCM system. Diagnostic routine may be utilized in real time by the computer to check the status of the command-transmitting system and the ground validation loops even during times of no command transmission. This greatly increases the level of confidence in the equipment in that the malfunctions will be discovered very rapidly.

Results of the diagnostic routines may be recorded on the high-speed printer. Summary data, acknowledgment of valid data receipt, or a request for the data retransmission may be transmitted to the control center on the high-speed lines. This message could also be transmitted to the control center over the teletype adapter. This method as anticipated will be utilized strictly for backup and for simulations.

APOLLO DIGITAL COMMAND SYSTEM

by
C. B. Knox
Goddard Space Flight Center

ABSTRACT

The presentation explains the purpose of the command system, the command word structure, the on-site equipment configuration, and the operational capabilities of the command system. The purpose of the command system is explained utilizing typical commands employed during a mission. Different Apollo/Saturn command types are also itemized for each vehicle. An explanation of the command word structure demonstrates how each vehicle and vehicle system is addressed and updated. This explanation will also include the purpose and use of the sub-bit codes and error codes used in the Apollo command system. Diagrams are utilized to present the on-site equipment configurations and how they participate in the command system. The discussion will include the function, associated capability, and purpose of each unit in the system. The operational capabilities and procedures will also be presented. In conclusion, the presentation will discuss a typical data transmission from its origin at the control center to the transmission at the remote site.

INTRODUCTION

The purpose of the Apollo digital command system (DCS) is to provide a means for communicating with and controlling the spacecraft's equipment from the ground. Some commands are identified before the launch; others are developed by the computing complex at the Mission Control Center (MCC) in Houston during the mission.

The commands identified before the mission are the type that are to be executed at selected intervals during the mission. One of these commands may be sent several times from one or more ground stations. For example, a command requesting a tape playback of spacecraft recorded data may be sent by all stations once or twice during a mission.

Other commands developed during the mission by the MCC computers would ordinarily be sent only once by one station. However, if the transmission were not successfully accomplished by the selected station, it would be retransmitted by another station later during the mission. An example would be the correct time for the spacecraft computer. The time of the spacecraft computer is telemetered to the ground stations. If this parameter does not agree with the indicated ground time, a command will be sent to the spacecraft computer that will update the timing system with the correct time.

COMMAND WORD STRUCTURE

General Purpose

The command word is configured so that each command will only be accepted by the selected spacecraft if it has the correct vehicle address, systems address, bit structure, and word length. The vehicle address requirement is obvious. The system address is required to differentiate between the RTC, AGC, and CTE systems aboard the vehicle. The bit structure employs sub-bit coding for security measures.

The philosophy for secure up-data transmission is to insure that the command word will be rejected by the spacecraft if it is not the exact command word transmitted. Measures have been taken in all phases of the up-data transmission system to insure that the chances of the spacecraft accepting an invalid command is 1×10^{-9}. In the ground-to-air link each information bit is encoded into five sub-bits to insure the nonvalid command rejection ratios.

Bit Arrangement (see Figure 1)

The general bit arrangement is as follows: Bits 1-3, vehicle address identification; bits 4-6, system address identification; bits 7-remainder, system instructions. Each information bit in the transmission is encoded into five sub-bits to insure spacecraft rejection of nonvalid words. The sub-bits are transmitted at a 1 kilocycle rate. Additional precaution is taken by encoding the vehicle address bits with a different code than the remainder of the information bits. Normally, the sub-bit code for a *zero* is the complement of the sub-bit code for a *one*. Also, the CSM, LEM, and S-IV-B plan to use the same sub-bit codes. Either of these parameters, codes for sub-bits ONES/ZEROS being complementary or sub-bit codes for all space vehicles, could be different.

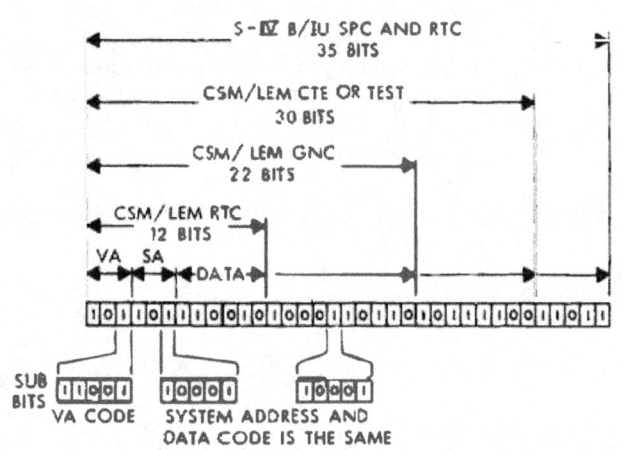

Figure 1—Apollo/Saturn command structure.

Tables 1, 2, 3, and 4 describe the bit structure of the varous commands.

In Table 1, data is MSB first - Each RTC is unique, Maximum number of RTC is $2^6 = 64$.

In Table 2, data is MSB first. The five bit data words represent the following 18 symbols: verb, noun, key release, error reset, enter, clear, plus, minus, 0, 1, 2, 3, 4, 5, 6, 7, 8, and 9. Symbols are transmitted in unique sequences to form computer up-dates.

Table 1

CSM/LEM Real Time Commands (RTC).

Bits	Identification
1-3	Vehicle Address
4-6	System Address
7-12	RTC Instruction

Every effort is made to insure that the commands generated at the Mission Control Center are transmitted safely to the remote site and from the remote site to the spacecraft. Redundant data circuits are employed between the Mission Control Center and the remote site to insure that the command reaches the site. Backup requirement is utilized at the remote site whereever possible. Also alternate data paths are used within some of the remote site equipment.

We must not only have the facilities for transmitting commands, but must also insure that only valid commands are transmitted. Various means are used throughout the system to insure that the chances of transmitting a nonvalid command are 1×10^{-9}.

Specific Types of Commands

The Command Service Module (CSM) and the Lunar Excursion Module (LEM) will use the same command format for similar commands. The S-IV-B will use similar formats but the word length is different from the CSM/LEM commands.

CSM/LEM Real Time Commands

The real time commands (RTC) are commands that are known and identified before the launch. These commands are the on/off variety that are used to control the spacecraft systems. For example, dump tape recorder playback, ON or OFF, C-band radar beacon, ON or OFF; telemetry mode select, ONE or TWO.

CSM/LEM, Apollo Guidance Navigational Computer

The Apollo guidance navigational computer (GNC) commands will provide updated information to the spacecraft computer. This will enable the spacecraft computer program to be updated or varied due to the new information developed during the mission.

CSM/LEM Central Timing Equipment

The central timing equipment (CTE) will receive correct timing data wherever there are indications that the spacecraft timing system is not accurate.

S-IV-B Real Time Commands

The S-IV-B real time commands have not been identified at this time but they will perform the same on/off functions as the CSM/LEM RTC if implemented.

S-IV-B Computer-Stored Program Commands

The S-IV-B computer-stored program commands (SPC) will incorporate timing and computer instructions. These commands will perform the same function as the CSM/LEM, GNC, and CTE commands described earlier.

Table 2

CSM/LEM Apollo Guidance Navigational Computer (GNC).

Bits	Identification
1-3	Vehicle Address
4-6	System Address
7	Parity
8-12	Symbol
13-17	Symbol Complement
18-22	Symbol Repeat

Table 3

CSM/LEM Central Timing Equipment (CTE)

Bits	Identification
1-3	Vehicle Address
4-6	System Address
7-12	Seconds
13-18	Minutes
19-24	Hours
25-30	Days

Table 4

S-IV-B Computer-Store Program Command (SPC).

Bits	Identification
1-3	Vehicle Address
4-6	System Address
7-10	Message Control Information
11	Decoder Address
12	Message Control Information
13-14	Up-data
15-18	Decoder Address
19-24	Up-data
25	Decoder Address
26-28	Up-data
29-30	Togo Bit
31-35	Decoder Address

In Table 3, data is LSB first. Bits 24 and 30 are zero fillers since five bits are sufficient to specify 24 hours or 31 days.

In Table 4, one 35 bit transmission constitutes one syllable. Each set of four syllables is followed by an execute command. The execute command is initiated following spacecraft verification of the four syllables.

Error Coding

The previous command word descriptions have pertained to the transmission of data from the remote sites to the spacecraft. The communication link between the computing complex at Houston, where the commands are generated, must also be protected to maintain the nonvalid command rejection ratio of 1×10^{-9}.

Proposed Code

The Base-Chaudhuri[*] code has been proposed as a means for obtaining the required error detection capabilities. Since many factors of the missions and the computer programs are not completely defined at this time, this proposal may vary before actual implementation. However, the final product will probably be very similar to the proposed system.

The Base-Chaudhuri code has the following characteristics:

1. Probability of an undetected error may be stated essentially independent of transmission line error statistics.

2. Detects all burst errors equal to or less than $n - k$ in length.

---

[*]Hines, S. M., Command Format for the High-Speed Data Link, Philco/MSC.

3. Probability that an error goes undetected when the burst length equals $n-k+1$ is:

$$P_1 = 2^{-(n-k+1)}$$

4. Probability that an error goes undetected when the burst length is greater than $n-k$ is:

$$P_2 = 2^{-(n-k)}$$

where n = total number of bits per word, and k = number of information bits per word.

For a Base-Chaudhuri code where $n = 57$
$k = 30$,

$P_1 = 1.49 \times 10^{-8}$

$P_2 = 7.7 \times 10^{-9}$

The parameters selected in the example are compatible with the equipment characteristics and insure the desired error detection capability.

Proposed Operation of the Command System

It has been proposed to operate the command system in three modes:

Mode 1. Used at sites with flight controllers. The command is generated at Houston, transmitted to the site, validated and stored in the command data processor. At the selected time, a flight controller at the remote site selects the command for transmission and the command is sent to the spacecraft.

Mode 2. Used at sites without flight controllers. The command is generated at Houston, transmitted to the site, validated, and stored in the command data processor. At the selected time, a flight controller at Houston sends an execute command to the remote site and the command is then sent to the spacecraft.

Mode 3. Also used at sites without flight controllers. The command is generated at Houston, transmitted to the remote site, and immediately sent to spacecraft if it is validated by the command data processor.

GENERATION AND TRANSMISSION OF A CSM/GNC COMMAND (see Figure 2)

The computing centers at Houston will be receiving telemetry and tracking data from all the remote sites. This data will be monitored to determine if any command action is necessary. If the spacecraft is to change orbits by ground command and the spacecraft computer needs new instructions, the necessary data will be generated by the mission computers. Error coding

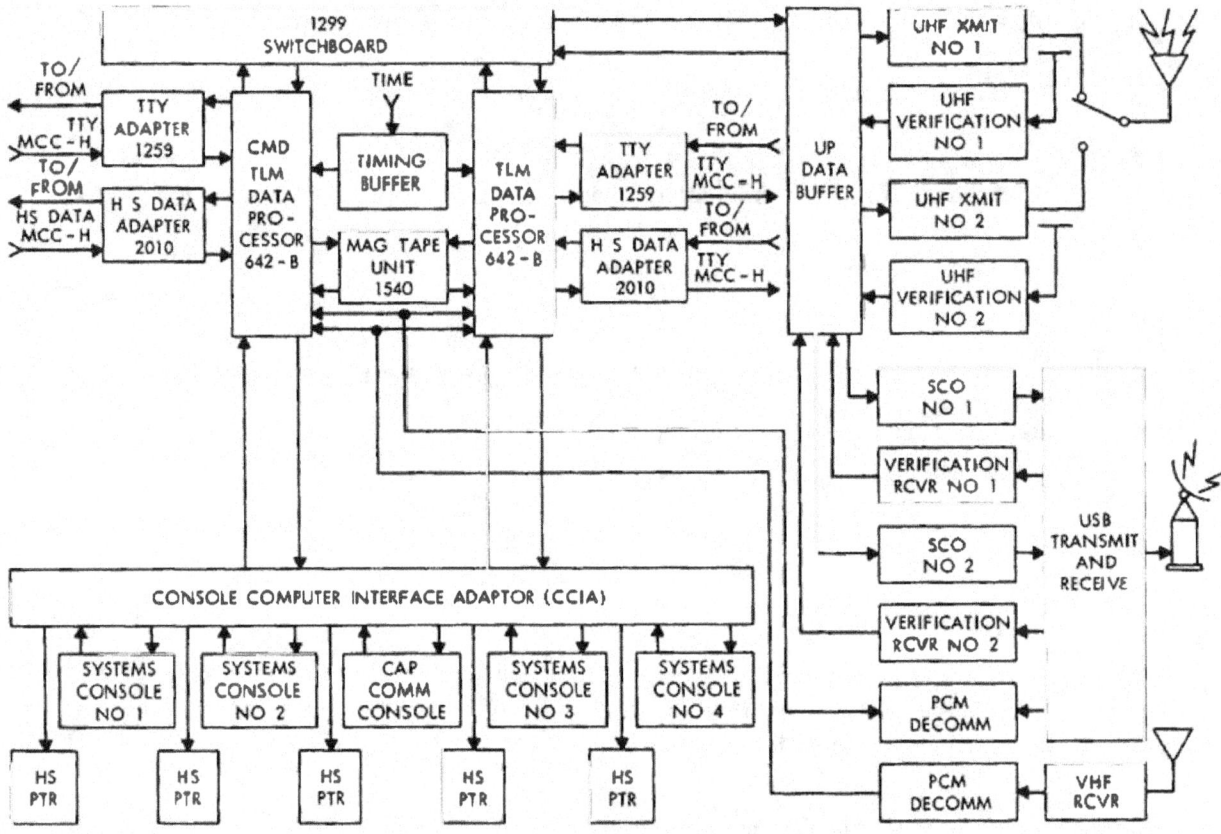

Figure 2—Apollo digital command system.

detection bits will be added to the format and the data will be transmitted to the appropriate site at 2400, 1200 or 600 bits per second. The data processor is also capable of receiving and transmitting over teletype circuits. Should the high speed circuits or equipment become unusable, the teletype circuits will be used.

Command data will be transmitted to the remote sites in data blocks. Each data block will consist of sub-blocks. The number of sub-blocks per block will vary with each transmission; however, each sub-block will contain the same number of bits, 57. (Three extra bits may be used for data synchronization which would make each sub-block 60-bits long since multiples of 30 would easily adapt to the remote site equipment).

The first sub-block of each load would be an introductory sub-block. This sub-block would contain bits as shown in Table 5.

The remaining sub-blocks of each load will vary with the type of data being transmitted; however, each sub-block will always contain 27 bits for error protection.

MCC-H Remote Site Validation

The data transmission is received at the remote site via the high speed data circuits. The command data processor accepts the data for validation if it has the correct site address. The

Table 5

Introductory Sub-Block.

Number of Bits	Identification
5	Station Address (capable of addressing 32 stations)
3	Sub-Block ID (capable of identifying 8 modes)
3	Vehicle Address (capable of identifying 8 vehicles)
3	System Address (capable of addressing 8 stations)
8	Sub-Block Count (identifies number of sub-blocks per block; capable of 256)
8	Execute ID (may be used addressing commands stored at the remote site or may contain data)
27	Error Protection

processor then performs the following checks to determine if the command is valid:

1. Checks the error coding,
2. Checks the vehicle and systems address,
3. Checks the command word structure.

If the command is validated, it is then tagged and stored in the processor's memory and printed out on the high speed printers near the consoles. If the command is not validated, a retransmission is automatically requested by the data processor. The data are then transferred to the magnetic tape unit for storage. Both the telemetry data processor and the command data processor have access to the magnetic tape unit. If the command data processor experiences a failure, the commands will be transferred from the tape unit to the telemetry data processor. The telemetry data processor will then be programmed to perform the *critical* functions of command and telemetry data processing.

Flight Control Console

The flight control consoles contain the pushbutton command matrix for initiating commands and indicators for displaying events associated with the commands. Each of the four systems consoles and the capsule communicator console has the capability of transmitting 36 real time commands, a clock command, and a computer load.

Each console can be used for initiating commands to any vehicle by interchanging the mask overlays. The mask overlay has the proper coding to identify itself as the CSM system number 1 console or any of the other systems consoles.

When the flight controllers decide to execute the command, the capsule communicator will determine that the transmit system, UHF or S-band is in the proper configuration to transmit the command. He will be able to determine the status of the system by observing the status indicators on his console, i.e., RF power, ON, S-band No. 1, and up-data mode selected. If the command is to be transmitted by the CSM systems controller, the capsule communicator will permit this console to have access to the command system. The CSM Systems controller will then press the CSM/GNC transmit control. The command request is sensed by the console computer interface adaptor, coded properly for computer use and transferred to the command data processor.

Console Computer Interface Adapter

The console computer interface adapter (CCIA) interfaces the consoles with the computers. The primary function of this equipment is to receive command data from the consoles, identify the originator, format the request in computer language and present the data to the computer; and receive status information from the computer about the various events displayed on the consoles and update the display as new data is received.

The CCIA is configured in two separate data channels. This is to prevent the command system from being incapacitated by a partial failure. If a failure occurred in part of the system, the second channel could support half the consoles. Masks would be interchanged among the consoles to allow the highest priority console to have access to the computer.

Data Processor

The command data processor calls the CSM/GNC load from memory. The processor also adds the redundant complementary bits, the symbol repeat bits, and the parity bit. The system control bits for arranging the UHF or S-band equipment are also generated by the data processor. The data bits are then sub-bit encoded in the proper sub-bit code and transferred to the up-data buffer in 30-bit parallel transfers. Twenty-five bits of each transfer are data and five are control bits.

Up-Data Buffer

The up-data buffer interfaces the command data processor with the RF transmitters. The up-data buffer receives the 30-bit parallel transfer of data from the data processor and separates the control data from the information data. The control bits are processed to control the system as instructed by the computer.

The twenty-five information bits are serialized and transmitted at a 1 kilocycle rate. Each bit controls a 2 kilocycle phase shift keyed (PSK) oscillator that is shifted 180 degrees by a change of state from a *one* or a *zero*. The output of the 2 kilocycle oscillator is added with a coherent 1-kilocycle signal that is transmitted for synchronization purposes. A *one* is defined as a composite signal where the 1 kilocycle and 2 kilocycle signals are going positive at the zero crossover point. The PSK demodulators are designed to recover the digital data from the composite PSK signal for computer verification.

The control data are transformed to relay closures to provide the data transmission path that the computer has requested. Some of the control circuits are within the buffer and some are external circuits. The buffer has a dual-data path for redundancy purposes. Some of the control bits select the correct data path when an error is detected in the ground verification loop.

There are two UHF transmitters employed in the system and one or two subcarrier oscillators of the unified S-band (USB) equipment. The buffer will have the capability for transmitting through either system sequentially. Normally, the S-IV-B will utilize the UHF transmitters and the CSM or LEM will utilize the USB equipment. Therefore, the buffer must select the UHF transmitters when processing a S-IV-B command and the S-band subcarrier oscillator subsystem (SCO) when processing a CSM or LEM command. Associated with each UHF transmitter and each SCO is a verification receiver for monitoring the effectiveness of the ground command equipment. The verification receivers are connected to PSK demodulators within the buffer where the composite 1 kilocycle and 2 kilocycle signal is converted back to a digital signal. The control and status bits necessary to coordinate the UHF operation are diagrammed in Table 6.

Table 6

Control and Status Bits for UHF Mode

Bit Number	Control From Computer to Up-Data Buffer		Status From Buffer to Computer	
	One	Zero	One	Zero
26	UHF Mode	USB Mode	UHF Mode	USB Mode
27	UHF #1 Fail	UHF #1 OK	UHF #1 On	UHF #1 Off
28	UHF #2 Fail	UHF #2 OK	UHF #2 On	UHF #2 Off
29	Demodulator Reverse	Demodulator Normal	Prime #1	Prime #2
30	Modulator Reverse	Modulator Normal	RF Power On	RF Power Off

The control bits are identified on one side of the chart and the status bits on the other. The control bits will be explained first.

Bit 26 - A *one* selects the UHF equipment for transmission to the S-IV-B, a *zero* selects the USB equipment for transmission to the CSM or LEM.

Note: Assuming that Bit 26 is *one*, the following is applicable:

Bit 27 - A *one* indicated that the number 1 UHF transmitter should be shut down (fail) because of a nonvalid transmission. A *zero* indicates the equipment is operating satisfactorily.

Bit 28 - Similar to bit 27.

Bit 29 - A *one* indicates that the PSK demodulators should be reversed because an error has been detected in the ground verification loop. A *zero* indicates that the normal (#1 demodulator to #1 verification receiver) is desired.

Bit 30 - A *one* indicates that the PSK modulators should be reversed because an error has been detected in the ground verification loop. A *zero* indicates that the normal (#1 to #1) configuration is desired.

The philosophy for switching units after an error is detected is: The PSK demodulators in the verification loop will be switched first. The PSK modulator will be switched next if an error is still present. If an error is still sensed after these two switches, the external transmitting equipment will be requested to switch. The status bits are explained as follows:

Bit 26 - A *one* indicates that the buffer is supplying data to the UHF equipment. A *zero* bit indicates that the up-data is going to the USB equipment.

Note: Assuming that bit 26 is a *one*, the following is applicable:

Bit 27 - A *one* indicates that the UHF number 1 is on and working properly. A *zero* indicates that the UHF number 1 is off.

Bit 28 - Similar to bit 27.

Bit 29 - A *one* indicates that the number 1 UHF equipment has been designed as the prime equipment. The prime equipment will provide the RF signal to the antenna while it is working properly. A *zero* indicates that the number 2 UHF equipment has been designated as the prime equipment.

Bit 30 - A *one* indicates that the RF power is on. This means that the RF power has been sampled at the input to the antenna and is at a sufficient level. A *zero* indicates that the RF power is off.

Status signals for bits 27-30 are provided to the buffer from external equipment. The buffer develops the status for bit 26 internally. The information is then transferred to the computer for evaluation of the ground transmitting equipment.

If control bit 26 was a *zero* instead of a *one*, the USB equipment would be selected as the transmitting equipment (see Table 7).

Table 7

Control and Status Bits for USB Mode.

Bit Number	Control From Computer to Up-Data Buffer		Status From Buffer to Computer	
	One	Zero	One	Zero
26	UHF Mode	USB Mode	UHF Mode	USB Mode
27	Verification Receiver #1 Select	Verification Receiver #1 Off	S-Band #1 On	S-Band #1 Off
28	Verification Receiver #2 Select	Verification Receiver #2 Off	S-Band #2 On	S-Band #2 Off
29	Demodulator Reverse	Demodulator Normal	SCO #1 Up-Data Mode	SCO #1 Off
30	Modulator Reverse	Modulator Normal	SCO #2 Up-Data Mode	SCO #2 Off

Control Bit 27 No. 1 verification receiver has been selected for verification of the data transmission. A *zero* means No. 1 verification receiver is not selected.

Control Bit 28 Similar to Bit 27

Control Bit 29 A *one* requests the PSK demodulators to reverse their normal configuration. A *zero* requests the demodulators to maintain their normal configuration.

Control Bit 30 Similar to Bit 29. Causes PSK modulators to reverse their normal configuration.

Status Bit 26 This bit has the same meaning as explained when the buffer was configured for UHF transmissions.

Status Bit 27 A *one* indicates that the number 1 SCO of the S-band equipment is on and operating properly. A *zero* indicates that the number 1 SCO is off or not operating properly.

Status Bit 28 Similar to Bit 27. Indicates status of SCO No. 2.

Status Bit 29 A *one* indicates that SCO no. 1 is capable of modulating the exciter. A *zero* indicates that the mode select switch is not in one of the up-data modes.

Status Bit 30 Similar to Bit 29. Indicates whether SCO No. 2 is selected for update transmissions.

The control bits originate in the computer and are processed by the buffer to configure the transmit loop to satisfy the up-data requirements for each up-data transmission. The information data transmitted by the RF system and monitored by the verification receivers is returned to the computer with the status bits after each up-data transmission. Therefore the computer is able to check that the data has been transmitted validly via the desired RF path.

S-Band Up-Data Transmissions

The S-band system is capable of transmitting up-data in four of its eight transmit modes, modes C, E, F, and G. The mode select switch is designed to inform the command system if it is in one of the up-data modes.

When the up-data mode is selected, the composite PSK signal frequency modulates at 70 kilocycle oscillator. The 70 kilocycle signal then phase modulates the S-band exciter which drives the S-band power amplifier. The S-band signal is sampled at the output of the power amplifier by a verification receiver. The verification receiver demodulates the S-band and 70 kilocycle signal to obtain the composite 1 kilocycle and 2 kilocycle signal. This signal is returned to the up-data buffer and to the up-data processor for ground verification.

UHF Up-Data Transmissions

The UHF transmitting system is operated for command purposes only. The system is composed of two UHF exciters, two power amplifiers, two antennas, and two verification receivers. This equipment is now being used in the Gemini Program and will be adapted to the Apollo program without any significant changes.

The UHF transmitting system employs both transmitters, power amplifiers, and verification receivers simultaneously. One transmitter-power amplifier is connected to an antenna and radiates power continuously. The second transmitter-power amplifier is also run at full power, but it is terminated in a dummy load. This arrangement provides a hot standby unit that is also monitored continuously for proper operation.

UHF Transmitters and Power Amplifiers

The UHF transmitters at the launch sites and at the Gemini sites is the AN/FRW-2 equipment. The UHF equipment aboard the new Apollo ships is not of this variety but has similar characteristics, such as:

Type of modulation:	FM
Frequency, adjustable:	406 - 549 megacycles
Power output:	500 watts

The power amplifiers are the Collins Model 240-D at the launch areas and existing Gemini sites. The new Apollo ships will have similar power amplifiers, each capable of 10-kilowatt output.

The UHF verification receivers are compatible with the characteristics of the UHF transmitters. The receiver is housed in the transmitter's cabinet and monitors the transmitter's output. The PSK is obtained from the UHF transmissions and returned to the up-data buffer for command validation.

UHF Command Antennas

There are two UHF command antennas associated with each UHF system. The antenna characteristics are:

Type:	9-Turn Quad Helix
Polarization:	Left hand
Gain:	Approximately 18db
Beam width:	Approximately 20 degrees

Remote Sites Without Flight Controllers

Many of the Apollo sites will not have flight controllers. Therefore, there will not be any systems consoles or CCIA at these sites. The command system at these sites will be remotely controlled from the MCC at Houston. The command system at flight controller sites will be operated in mode 1. The command system at nonflight controller sites will be operated in modes 2 or 3. However, once the computer receives a command execute signal, the command transmission proceeds the same at all sites.

Transmit Validation

A message acceptance pulse (MAP) is telemetered from the spacecraft to the ground station when the spacecraft receives and validates an up-data transmission. If a command word is not validated by a MAP, the word is retransmitted a preselected number of times before alarming the flight controllers that a valid transmission can not be obtained. The valid or nonvalid indication is displayed on the consoles via the CCIA.

The command history of the up-data transmissions is printed out on the high speed printers for evaluation by the flight controllers. This printout will include the up-data transmission, the time of the transmission, and an indication that the command was or was not received. The command history may also be returned to Houston to complete the command-transmit verification loop.

APOLLO REMOTE SITE DISPLAY SYSTEM

by
G. N. Georgeadis
Goddard Space Flight Center

ABSTRACT

A description of the display system which will be used to monitor and control the spacecrafts during the many phases of Project Apollo is presented. The operation and output capability of the system is discussed together with equipment located in the Apollo flight operations room such as the fast access file and display system, silenced teletypewriters, and the group display is also discussed briefly.

INTRODUCTION

Since the pulse-code modulation (PCM) systems, the data processing systems, and the digital command systems have been described in previous papers, the writer will now describe the system which provides the direct interface between these systems and the flight controllers, the Apollo display system.

The display system for the Apollo program as shown in Figure 1 will consist of one aeromedical monitor console, four spacecraft systems consoles, one command communicator console, one memory-character generator, one maintenance monitor, one maintenance and operations console and, for shipboard installations only, one flight dynamics officer's console. The memory-character generator, which is the heart of the display system, is described in detail in the following discussions. Note that the previously described systems have their inputs fed directly into the consoles.

DUTIES OF CONSOLE OPERATORS

In order to better appreciate the display system design, a definition of the duties of each console operator is in order. Their duties are:

1. *The Command Communicator* is the flight control team leader and maintains cognizance over the more general mission-oriented events and activities. He maintains control over communications and command function and has access to all spacecraft parameters to obtain an indication of overall mission and spacecraft status.

2. *The Spacecraft Systems Monitor* performs detailed spacecraft systems monitoring and analysis for the purpose of detecting malfunctions and assisting the astronauts in

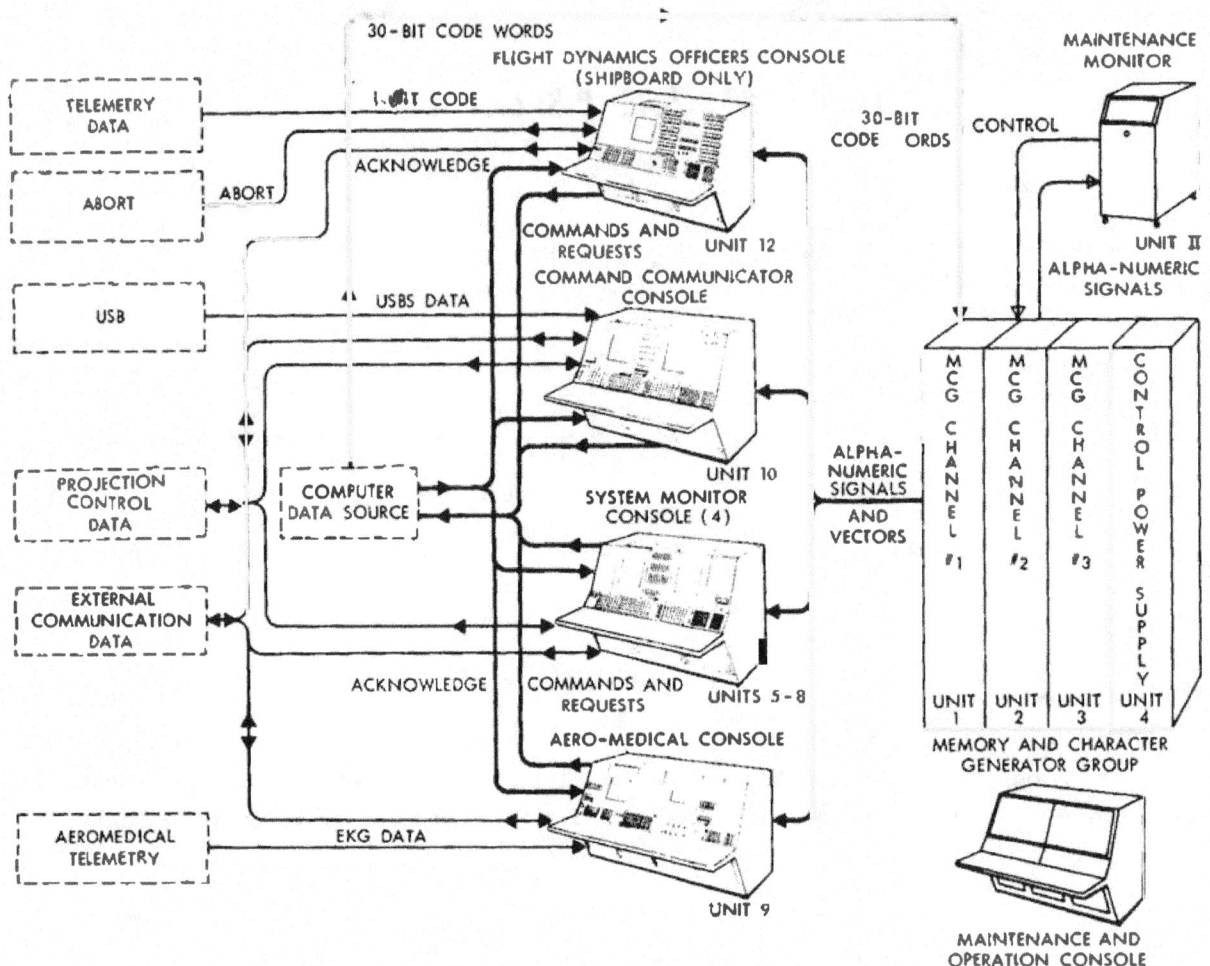

Figure 1—Display system for Apollo program.

maintaining spacecraft system integrity. Each systems monitor has access to those displays and commands contained within the jurisdiction of his designated responsibility.

3. *The aero-medical monitor console* is a two-operator position console. The aero-medical monitors perform a medical surveillance of the physiological and environmental status of the astronauts. For this purpose, they have access to special displays and a selection of telemetered parameters.

4. *The Maintenance and Operation Supervisor* exercises control over station maintenance and operations personnel during the missions. He has access to all intersite and intrasite communications circuits available during the mission.

5. *The Flight Dynamics Officer* monitors the spacecraft injection or insertion trajectory characteristics for sufficiency or abnormality and can assist in the execution of an abort maneuver if necessary.

RELATION OF DISPLAY SYSTEM TO OTHER EQUIPMENT ON SITE

Figure 2 gives an idea of where and how the display system fits in with the other equipment on site. The console/computer interface adapter provides command and display request capability to the consoles and also provides the necessary outputs to drive six high speed printers, four analog recorders, and six spacecraft clocks located on the wall-mounted group display. The group display will provide spacecraft and ground-generated times in addition to station equipment status information to all the personnel in the Apollo flight operations room. The ground-generated clocks are driven from the Apollo time distribution frame. The station equipment status inputs will come from each particular system to be monitored. The decommutation systems distribution unit (DSDU) provides inputs from the PCM stations. The two blocks labeled FAF #1 and FAF #2 are two fast access file and display system projectors containing predetermined information such as flight plans and procedures. These devices will be independently controlled slide projectors that allow random access of 500 slides each to be displayed upon request. One FAF will serve two systems console operators and the command communicator, and the other will serve the other two systems console operators and the command communicator. All consoles will have an intercom panel with local, range, and spacecraft communications.

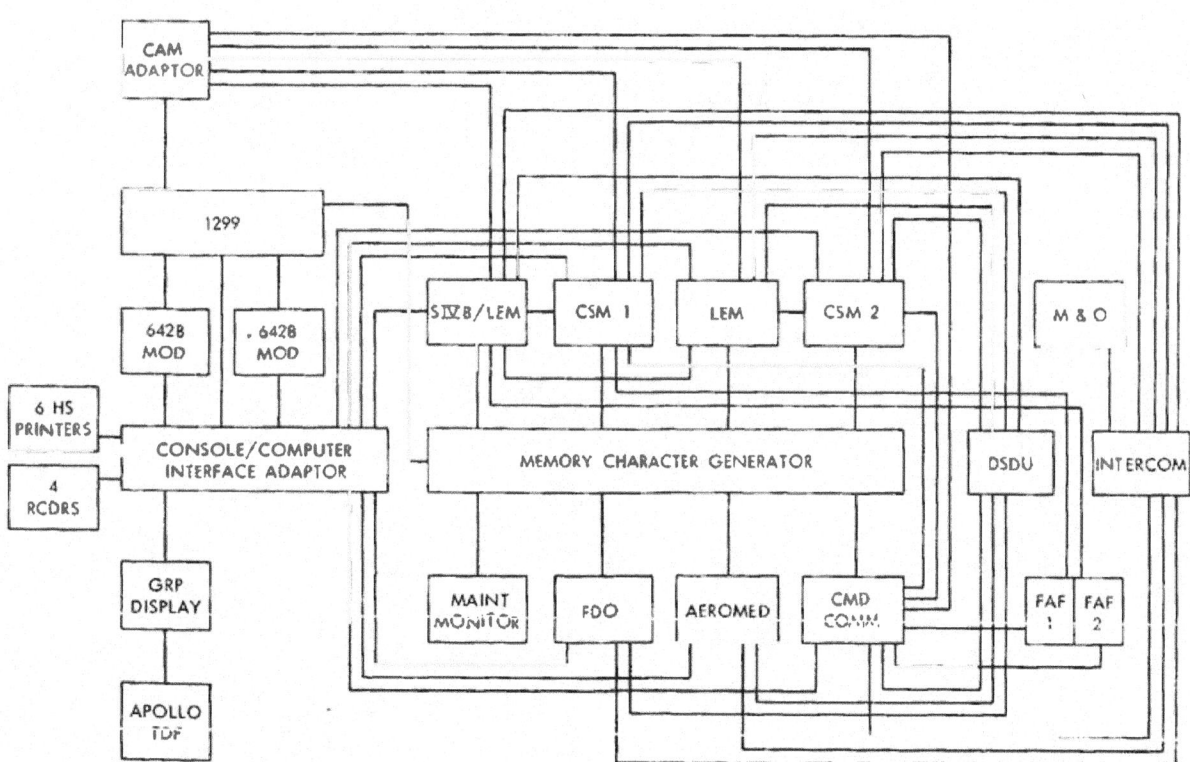

Figure 2—Apollo display and interface external signal cabling diagram.

APOLLO FLIGHT OPERATIONS ROOM LAYOUT

Figure 3 shows a typical Apollo flight operations room layout for a remote site ground installation. It should be noted that all the consoles face directly forward in order to view the group display and two rear-projection screens. The group display contains spacecraft and ground-generated clocks plus the station equipment status displays. The two rear-projection screens are part of the fast access file and display system. It should also be noted that all the systems consoles and their associated recorders are located in the front row. The high speed printers which provide the hard-copy print-outs from the computer are located in front of the systems console recorders and one between the command communicator console and the aeromedical console. The teletype RO's are located, one at each end of the front row, and one between the command communicator console and the aeromedical console. These units are Teletype Model 28 ROs which have been mounted in an enclosure which has been designed by Bendix-Pacific to silence the units and to be accessible and visible to the sealed operator.

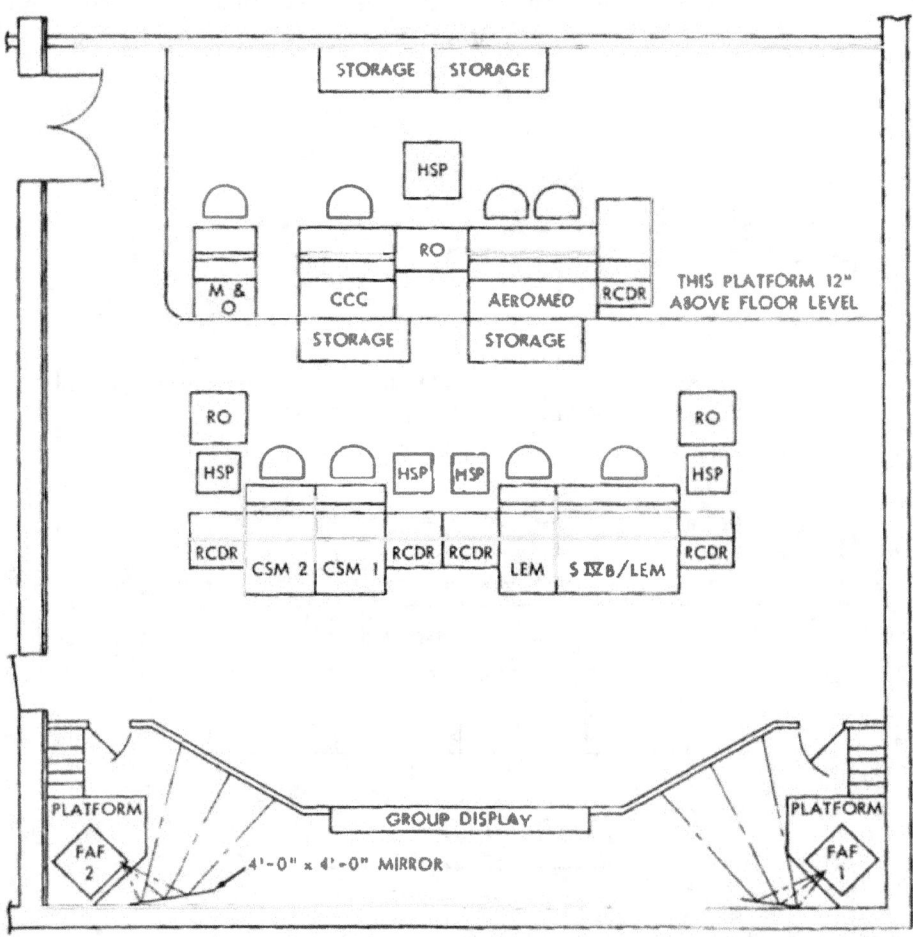

Figure 3—Apollo flight operations room layout.

COMPONENTS OF ALPHA NUMERIC DISPLAY SYSTEM

The Raytheon Company is the prime contractor for the Apollo Display System. This display system consists mainly of an alpha numeric type display system. The alpha numeric system is made up of eleven cathode ray tube display modules (or twelve in the case of shipboard installations) not including the display module in the maintenance monitor; a maintenance monitor; and a memory character generator.

CRT Display Modules

The CRT display modules are identical and interchangeable in the Apollo display system. A block diagram of this display module is shown in Figure 4. Each module contains a 17-inch rectangular CRT which uses a combination of electrostatic and electromagnetic deflection for displaying tabular data in a format of 36 lines with 72 characters in each line, in an area of at least 100 square inches. Electrostatic deflection will be used for character writing and electromagnetic deflection will be used for positioning and vector writing. In addition to the CRT, the display module includes the deflection circuits, power supplies, and controls. The two power supplies associated with each CRT display module, one low-voltage and the other high-voltage, are housed in separate containers and are removable from the rear of the console. For personnel safety, a safety glass is provided with anti-reflective coating and a tin oxide coating for RFI suppression.

Maintenance Monitor

The maintenance monitor unit will consist of a dolly, with locking wheels arranged for easy maneuverability, and a CRT display module which is electrically and mechanically identical to the other CRT display modules. In addition to the controls in the display module, the maintenance monitor is equipped with a channel selector switch which will permit the monitoring of any of the displays located in the system for maintenance purposes.

Memory Character Generator

The memory-character generator (MCG) consists of three independent logic and memory modules, two character/vector generators and independent power supplies for these. Each channel will be located in its own cabinet as shown in Figure 1 and may be independently selected by the computer for data transfers to the MCG. A single modified 642-B computer fast interface output channel connects the computer to the MCG by means of a single cable. The computer will enable the desired channel, through the use of an external function word which specifies the particular channel, to be addressed. Once a particular MCG channel is enabled, the computer will not enable another MCG channel until it has disabled the one selected. If the selected channel is in its display refresh mode, it will compute the refresh cycle. At the end of the refresh cycle, the MCG channel selected will send a data request to the computer and accept input data. The transmission of an end-of-message command word on the end of a MCG channel data block will disable that MCG channel. A new external function may then be sent.

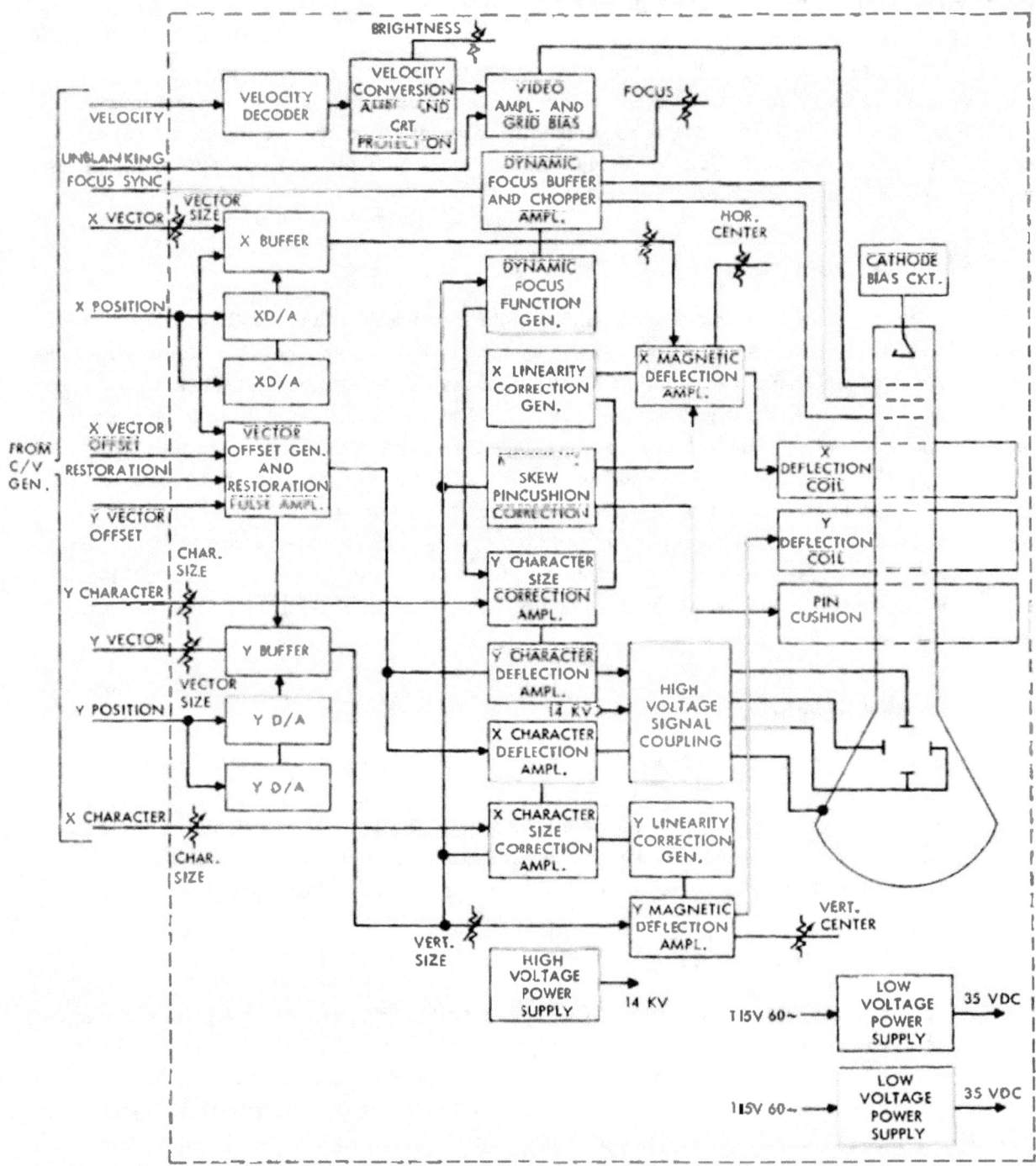

Figure 4—Digital display.

Memory Module

A memory channel block diagram is shown in Figure 5. Each MCG channel contains one 4096, 18-bit core memory with a 4-microsecond cycle time. This memory is logically divided into two memory sections of 4096 9-bit words. A memory section is further divided into two subsections under program control. Each memory subsection of an MCG channel contains the information necessary to drive a single CRT display. Therefore, the memory in each channel has the capability of storing information for four CRT display presentations since there are four subsections of memory. Since there are three MCG channels, the system is capable of operating with twelve CRT displays.

The primary function of the core memory in each MCG channel is to store the input data and refresh the four displays associated with that particular MCG channel memory at a flicker-free rate. The memory is also addressable in a random access mode through the use of certain command words. When the memory is receiving data from the computer, it is not refreshing the CRT displays. This means that the frequency of the updating of the data stored in the memory and the amount of data transferred from the computer to the MCG should be kept to a minimum since the CRT display may become dim or an objectionable flicker may occur at the face of the CRT.

Character/Vector Generator

Each memory section provides information to a character/vector generator which in turn drives two CRT displays. The character generator which is shown in Figure 6 will receive 6-bit codes and character size designation from the MCG memory and will transform these codes into character deflection and unblanking signals necessary to generate the corresponding characters. The display character codes are those shown in Figure 7. Character formation shall be based on a 32 by 32 matrix. Two character sizes are provided, 0.140 inches and 0.280 inches, with an actual character writing time of 3.16 microseconds. The characters are generated as sequential locations of memory read out. The memory refresh is read out simultaneously for two character generators. Each character generator will have the capability of displaying 4096 characters on two permanently associated CRT's, each CRT being located in a different console. Each character generator will also have the capability of driving the maintenance monitor CRT which is in parallel with either of the two permanently associated CRT's without degrading the performance of the latter. The character generators will recognize *start blink* and *stop blink* character codes. Upon interpreting a *start blink* code, that character position on the CRT will be skipped and characters between *start blink* and *stop blink* will be written or not written depending upon the phase of the blinking signal. Upon interpreting a *stop blink* code, that character on the CRT will be skipped, and subsequent characters will be written without qualification.

The vector generator shown in Figure 8 will be equipped to: accept 18-bit words; store initial position in X, Y coordinates of 9 bits each; use the final position in X, Y coordinates of 9 bits each; draw a line from the initial position to the final position in the time required to position and write two characters; and transform the old final position

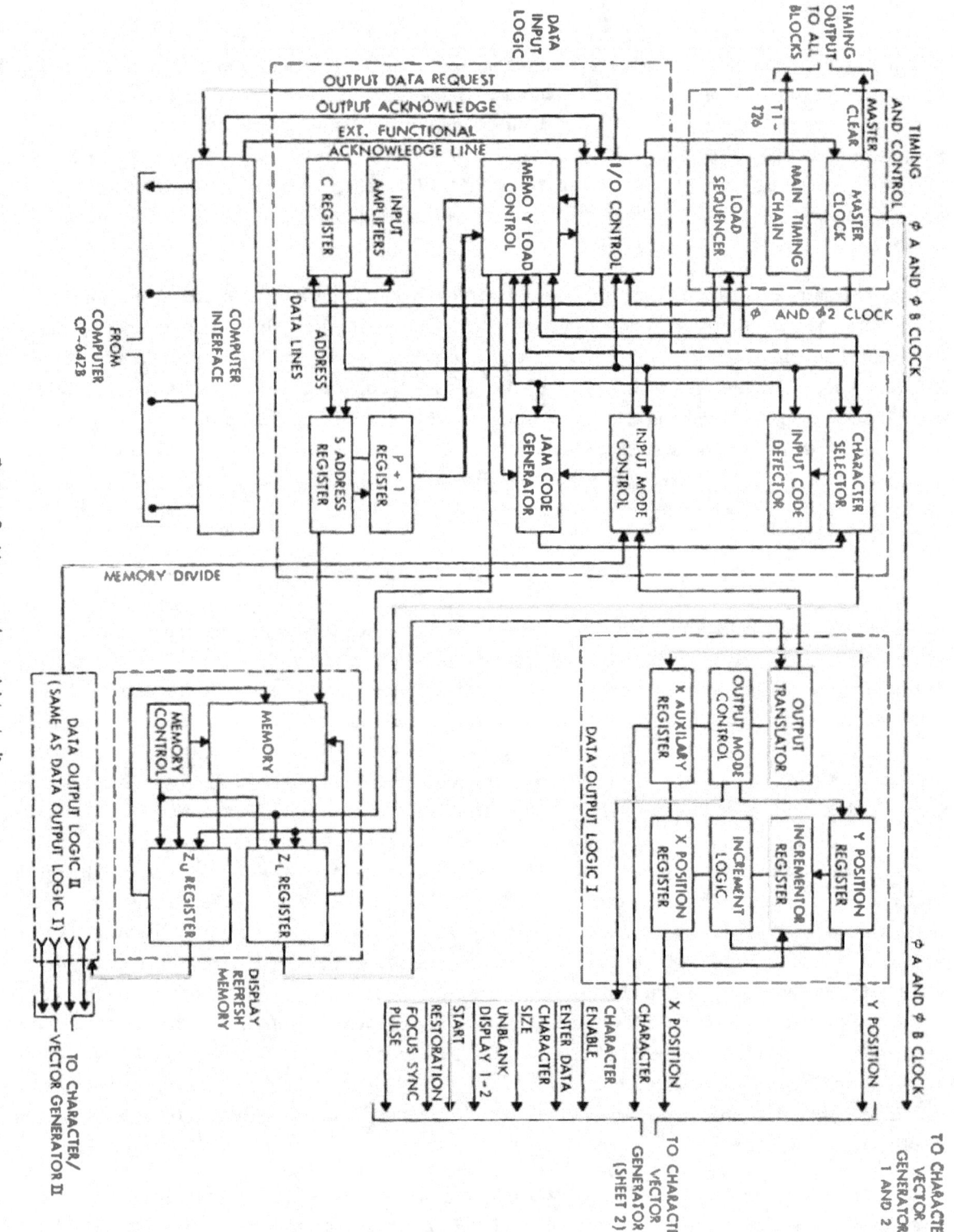

Figure 5—Memory channel block diagram.

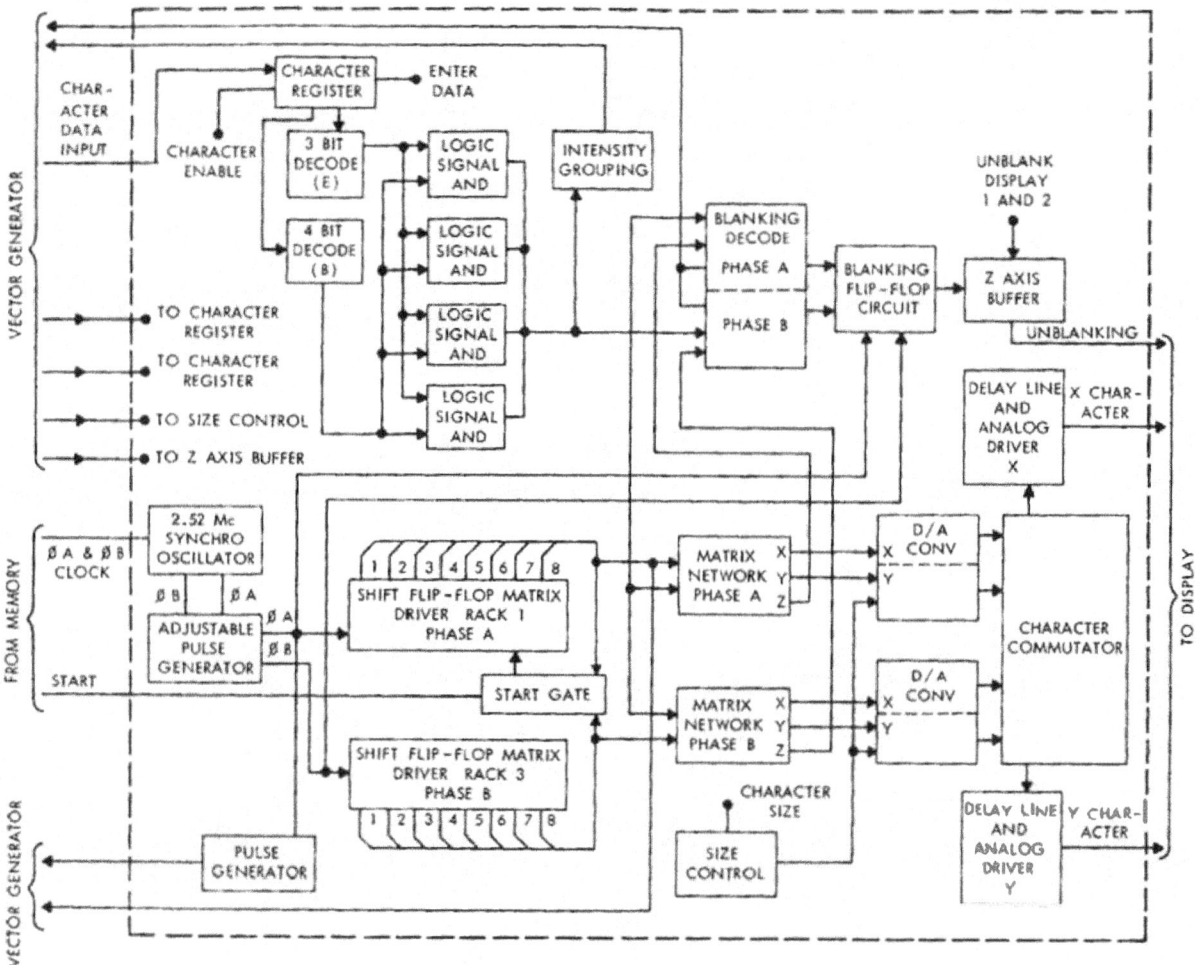

Figure 6—Character generator.

into a new initial position. By this process the vector generator will construct line segments, joined end-to-end, starting from an arbitrary position at the same refresh memory data rate used by the character generators.

When operating in the display refresh mode, the memory sections cycle all 4096 9-bit words in 21 milliseconds. The speed capability of the display system will be sufficient to continuously accept a computer word every 50 microseconds, display 22,980 characters on twelve CRT's, and refresh the data displayed on all twelve CRT's at a flicker-free rate.

The program controlled memory divider causes the information generated by the character generator to be visible or invisible on the front of its two associated CRT displays. For example, if the information for display number 1 is contained in the MCG refresh memory address ∅∅∅∅ to ∅777, this information would be visible on display number 1 while the refresh memory is cycling through these addresses and the information contained in refresh memory address 1∅∅∅ to 7777 would be displayed on display number 2 while the refresh memory is

OCTAL CODE	CHARACTER	OCTAL CODE	CHARACTER	OCTAL CODE	CHARACTER
00	SKIP*	27	X	56	Carriage Return*
01	1	30	Y	57	Skip data*
02	2	31	Z	60	+ (Plus)
03	3	32	γ (Gamma)	61	A
04	4	33	: (Colon)	62	B
05	5	34	θ (Theta, cap)	63	C
06	6	35	φ (Phi, cap)	64	D
07	7	36	Δ (Delta, cap)	65	E
10	8	37	ω (Omega, lower)	66	F
11	9	40	- (Dash)	67	G
12	0	41	J	70	H
13	* (Asterisk)	42	K	71	I
14	space	43	L	72	∠ (Angle)
15	start blink*	44	M	73	. (Period)
16	stop blink*	45	N	74	λ (lambda, lower)
17	Ω (Omega, cap)	46	O	75	Command Code*
20	No memory address increment*	47	P	76	ψ (Psi, lower)
21	/ (Slant)	50	Q	77	° (Degree)
22	S	51	R		
23	T	52	% (percent)		
24	U	53	σ (Sigma, lower)		
25	V	54	▷		
26	W	55	Increment Y, do not increment X*		

*These codes will be used as control codes

Figure 7—Character control codes.

cycling through these addresses. This is accomplished through the use of a 9-bit memory divide operation code placed in refresh memory location 1000.

The computer words entering the data input command logic of the MCG will be either command words or data words. The command words are used to signal the display system as to channel selected, whether to write small or large characters or vectors, into which portions of memory section data is to be written, which memory subsection is to be erased, where the memory sections are to be divided between two displays, what data should be modified, and when a computer transmission is ended. The command word format is shown in Figure 9. Data words are used to specify the X-Y positions of characters, the heads and tails of vectors, which characters should be written, and control data as it is being written into memory. The data word format is shown in Figure 10. Figure 11 and 12 show computer word structures and Figure 13 is a typical computer word sequence.

APOLLO REMOTE SITE DISPLAY SYSTEM

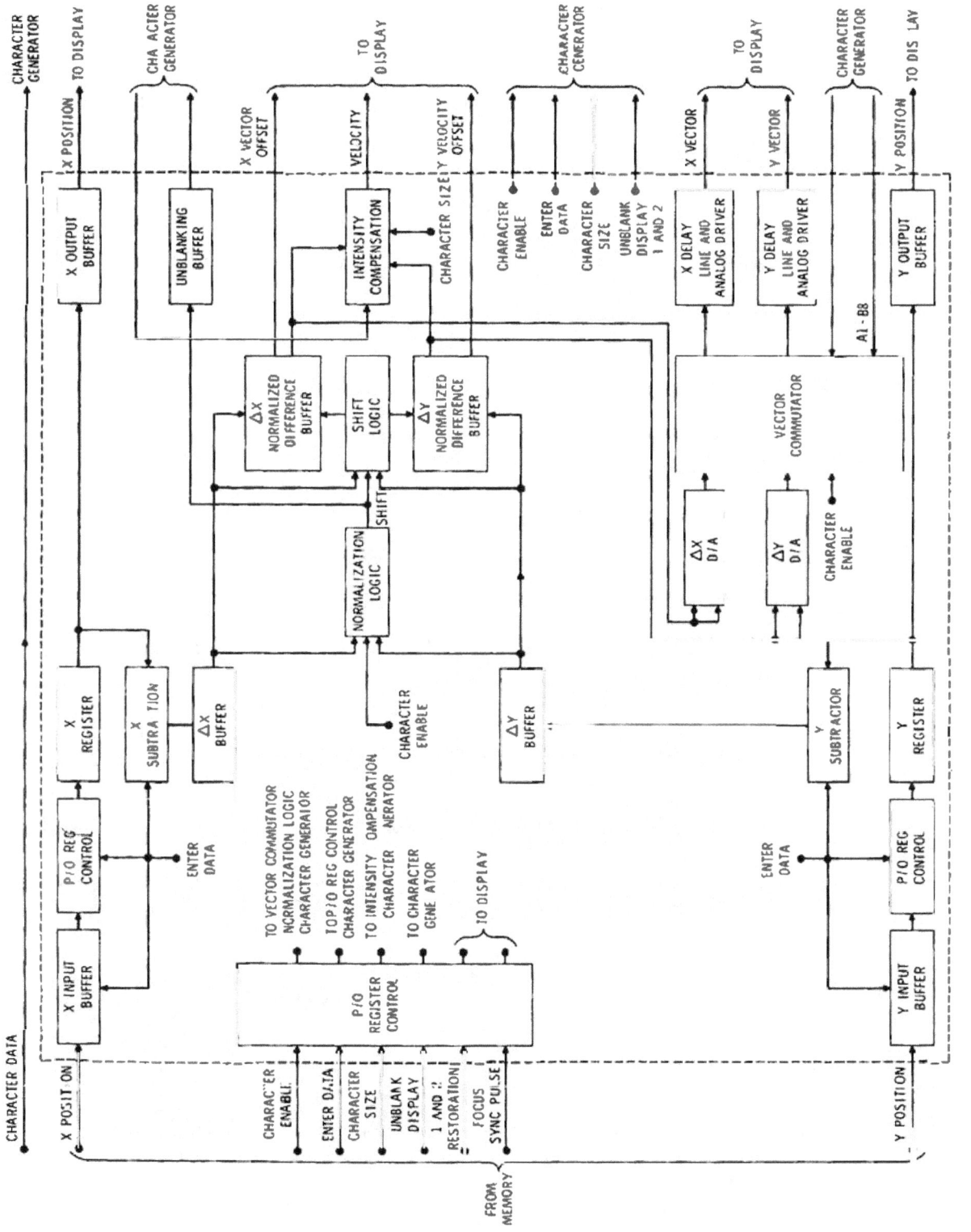

Figure 8—Vector generator.

Figure 9—Command word format.

WORD TYPE	29 28 27 26 25	24 23 22 21 20 19 18 17 16 15 14 13 12 11 10	9 8 7 6 5 4 3 2 1 0
LCCW	1 1 1 1 0 1	—— MEMORY ADDRESS ——	1 1 1 1 1 1 0 1
SCCW	1 1 1 1 0 1	—— MEMORY ADDRESS ——	1 1 1 1 1 0 0 1
VCW	1 1 1 1 0 1	—— MEMORY ADDRESS ——	1 1 1 1 1 1 0 0
EMSCW	1 1 1 1 0 1	—— MEMORY ADDRESS ——	1 1 1 1 0 1 1 1
IMDCW	1 1 1 1 0 1	—— MEMORY ADDRESS —— Not used	1 0 1 1 1 1 1 1
EOMCW	1 1 1 1 0 1	0 0 0 0 0 0 0 0 0 0 0 0 0 0 0	1 1 0 1 1 1 0 1

C5 ... C1

Figure 10—Data word format.

Character Data Word

	Character 5 (C)	Character 4 (C)	Character 3 (C)	Character 2 (C)	Character 1 (C)
bits	29 28 27 26 25 24	23 22 21 20 19 18	17 16 15 14 13 12	11 10 9 8 7 6	5 4 3 2 1 0
	Any character but 56 75	Any character but 75	Any character but 75	Any character but 75	Any character but 20 57 75

Vector Data Word

No Skip / Skip Data	Y Position	X Position	Vector Data Word Code
29 28 27 26 25 24	23 22 21 20 19 18 17 16 15	14 13 12 11 10 9 8 7 6	5 4 3 2 1 0
0 0 0 0 0 0 or 1 0 1 1 1 1	0 0 0 0 0 0 0 0 0 to 1 1 1 1 1 1 0 0 0	0 0 0 0 0 0 0 0 0 to 1 1 1 1 1 0 0 0 0	0 0 0 0 0 0

New Position Data Word

Skip / Skip Data / Start Blink / Stop Blink	Y Position	X Position	New Position Data Code
29 28 27 26 25 24	23 22 21 20 19 18 17 16 15	14 13 12 11 10 9 8 7 6	5 4 3 2 1 0
0 0 0 0 0 0	0 0 0 0 0 0 0 0 0	0 0 0 0 0 0 0 0 0	0 1 0 0 0 0
1 0 1 1 1 1	to	to	
0 0 1 1 0 1			
0 0 1 1 1 0	1 1 1 1 1 1 0 0	1 1 1 1 1 0 0 0	0 1 0 0 0 0

NOTE: C5 can contain any character but 56 or 75 for character new position word.

Figure 11—Word structures associated with characters.

	30 29 28 27 26 25	24 23 22 21 20 19 18 17 16 15 14 13 12 11 10	9 8 7 6 5 4 3 2 1 0
Command Word	1 1 1 1 0 1	0 0 0 0 0 0 0 0 0 0 0 0 0 0 0 to 1 1 1 1 1 1 1 1 1 1 1 1 1 1 1	0 0 0 0 0 0 0 0 1 0 to 0 0 0 0 0 1 0 0
	1 1 1 1 0 1		
	Command Code	Memory Address	Large character code / Small character code
New Position Data Word		Same as for Vectors	

Data Word	Character 5	Character 4	Character 3	Character 2	Character 1
	Any character but 56 75	Any character but 75	Any character but 75	Any character but 75	Any character but 20 57 75

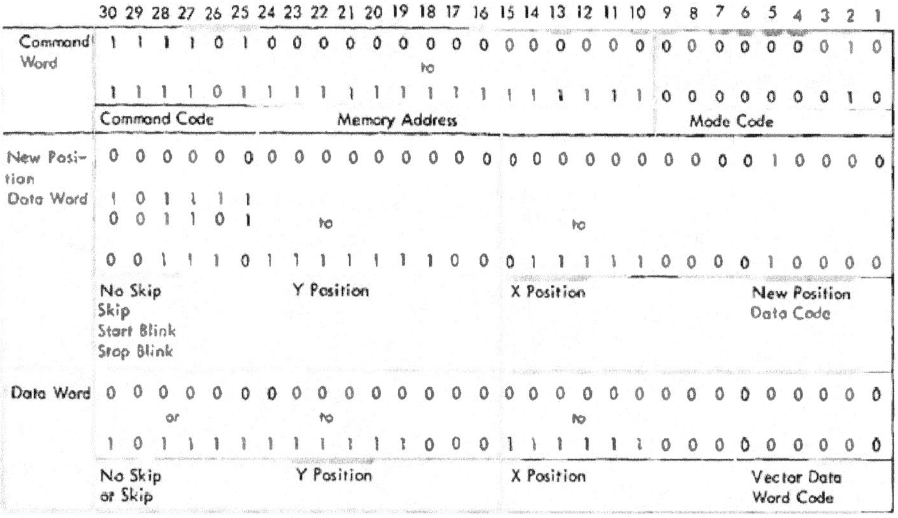

APOLLO REMOTE SITE DISPLAY SYSTEM

Figure 12—Word structures associated with vectors.

Figure 13—Typical computer word sequence.

COMMAND COMMUNICATOR CONSOLE

The command communicator console, whose display panels are shown in Figure 14, has two alpha numeric CRT display modules in it. These displays which were previously described will be used to portray spacecraft telemetered information in the form of block diagrams,

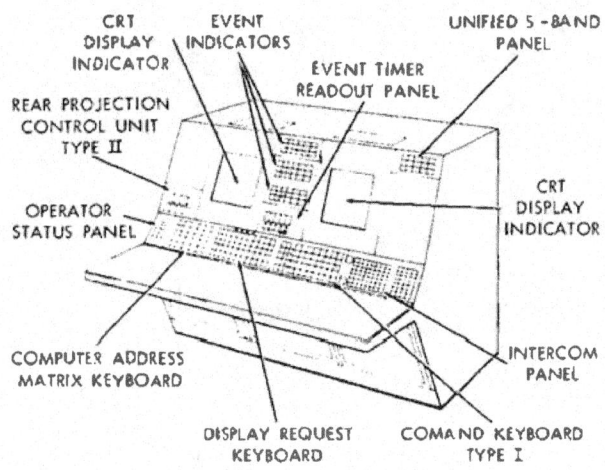

Figure 14—Command communicator console.

tabular listings, meter formats, or a combination thereof. Spacecraft telemetered time and command clock time can also be displayed on the CRT display. The information presented on an alpha numeric CRT display shall be referred to as a format. The format presented on each alpha numeric display must be capable of being changed by the console operator at any time. These changes are accomplished through the use of the display request keyboard, which is located on a sloped panel below the CRT displays. Changes will be restricted to the selection of formats previously determined and stored in the remote site data processor memory. Each display request keyboard contains fifty-five (push button indicators) PBI's, five of which are used for control functions such as designating the CRT on which information is to be presented. Each of the other fifty PBI's represent a format which may be requested by the console operator. Four coding switches are also provided on the display request keyboard which provide the capability of changing the function of the fifty format PBI's. This feature of the display request keyboard is accomplished through the use of coded plastic overlays together with the four coding switches. These coding switches allow the total capability of 15 different overlays to be used on the fifty format PBI's for a total capability of 750 formats for each display request keyboard. Each overlay contains the identifying legends for the fifty format PBI's and coding for the four coding switches. The coding switches will tell the computer which overlay is in position over the display request keyboard. Each format legend has an indicator light which will blink and illuminate red should any parameter within the format become out-of-limits.

Each display request keyboard is capable of requesting a new display presentation (format) on either CRT display located on that console. It is capable of requesting a hard copy printout of any tabular format on the keyboard whether or not the format requested is then being displayed. The high-speed printers described under the data processing system will provide the hard copy required. And last but not least, the automatic generation and transmission of selected summary messages to Mission Control Center (MCC), Houston is also initiated at the display request keyboard.

The command communicator console contains three event light panels. Each event light panel can display 36 events and also has a quick reconfiguration capability. The quick reconfiguration capability is provided by plastic overlays which are preassembled with the correct legends and desired colored lenses. No coding switches are required since this panel consists of indicators only instead of PBI's. The reconfiguration of an event light panel merely involves the replacing of the programmed patch board at the DSDU with another programmed patch board and the replacing of the overlay over the event light panel itself with another overlay.

A command panel will be provided on this console, which is divided into four sections. These sections provide the command communicator with six command enable PBI's, twelve mode select PBI's, six load command PBI's, and thirty-six real-time command (RTC) PBI's. This keyboard is also capable of quick reconfiguration due to the same type overlays and four code buttons as used on the display request keyboard. The command enable PBIs are not under the overlay since they do not change function. The matrix for RTC's will enable initiation of RTC's and will provide a means of preconditioning to aid in the rapid analysis of command status and the operation of selected commands. The load commands are required to transmit onboard computer loads such as clock times. The mode select portion is required for such commands as RCT auto/manual and setup/command.

An operator status panel is a panel on the command communicator console only, consisting of one status indicator light for each console operator position. An operator status unit is provided at each console operator position, except the command communications, which will allow each console operator to signal his state of readiness to the Command Communicator by way of the operator status panel.

An event timer is also provided in the command communicator console. This device is a five-digit, rear-projection clock which permits the console operator to program ±999 minutes and 59 seconds, with count-up, count-down, start and stop controls.

A computer address matrix is provided the Command Communicator which permits the console operator to request parameter group printout on the high-speed printer and to request automatic computer generation of summary messages.

An intercom panel is provided each operator position in the Apollo flight operations room which permits local, range and spacecraft communications. Two input jacks and a foot switch jack are provided with each intercom panel and in addition a portable speaker is provided with its output jack mounted on the upper righthand corner of the console.

The command communicator console is provided with a USB panel which indicates the mode of operation of the USB ground equipment for both Command Service Module (CSM) and Lunar Excursion Module (LEM). This is the only console in which this panel is contained.

Finally, the Command Communicator is provided with a panel to control both fast access file and display system projectors which have been previously discussed. The Command Communicator also has control of the power to each of these projectors.

It is important to note that the Apollo consoles have been designed on a modular basis for quick and easy reconfiguration of these consoles. The modules can be shuffled around or new modules added or substituted in order to support future missions, or even the present mission. This feature is better shown in Figure 15. It is also important to note that these consoles are 45-1/2 inches high and 58 inches deep with the three-bay consoles being 63 inches wide, the four-bay console (aero-medical monitor console) being 82 inches wide, and the two-bay console (maintenance and operations console) being 44 inches wide. The low-profile design permits the seated operator to maintain a full and unobstructed view of the group display (wall-mounted clocks and status lights) and the fast access file and display system wall-mounted rear-projection screens. All the controls and display devices which are mounted in or near each console must be accessible and visible to a seated operator.

SPACECRAFT SYSTEMS CONSOLE

Figure 15 shows the spacecraft systems console of which there are four. The four systems consoles are designated: S-IV-B/LEM, LEM, CSM #1, and CSM #2. The S-IV-B/LEM console is first used to monitor the S-IV-B vehicle and is then reconfigured, by use of the methods previously described, to monitor the LEM together with the other LEM console. The two CSM consoles, of course, monitor the command-service module.

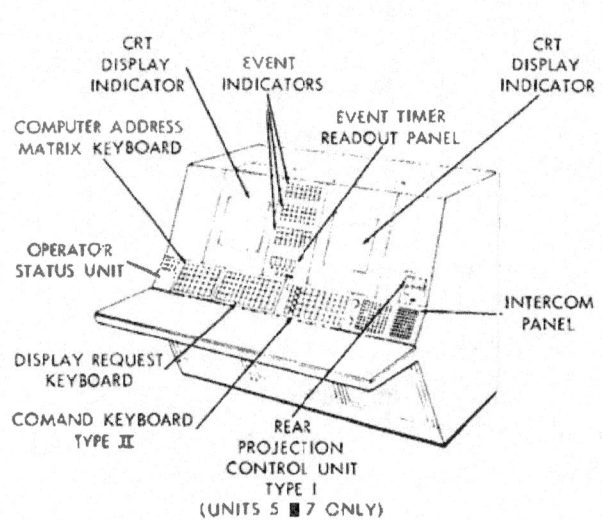

Figure 15—Spacecraft systems console.

The systems console is practically identical to the command communicator console. There are three basic differences between these two consoles. One difference is that the command keyboard on the systems console has one indicator where the command communicator command-enable PBI's are located. This indicator will tell the systems console operator whether his command keyboard is "enabled" or "disabled" by the command communicator. Another difference is the fast access file (FAF) control panel, which appears only on the S-IV-B/LEM console and the CSM #1 console. The systems console FAF control panel controls only one FAF instead of the dual control afforded the command communicator console. And finally, the systems consoles each contain an operator status unit as discussed previously.

Each systems console has a vertically mounted analog recorder associated with it which is driven from the CCIA. These recorders have eight channels, and controls are provided which permit the operator to select any one of six groups of eight parameters to record. The selection controls are on the recorder.

AERO-MEDICAL CONSOLE

Figure 16 shows the aero-medical monitor console. This console contains only one CRT display module and two event light panels. However, it also contains two intercom panels, two operator status units, two event timers, and one display request keyboard. The aero-medical console also contains a four-trace cardioscope with 10 selectable analog inputs and its associated control panels. Figure 17 shows these panels in more detail. Associated with this console is a vertical mounted analog recorder which is driven directly from the PCM stations' DSDU. The input parameters are selectable at the DSDU in the same manner the event light panel inputs were selectable. This recorder is furnished with a table extension to permit a more careful analysis of the data being recorded.

FLIGHT DYNAMICS OFFICER'S CONSOLE

Figure 18 shows the flight dynamics officer's console (FDO). This console is only

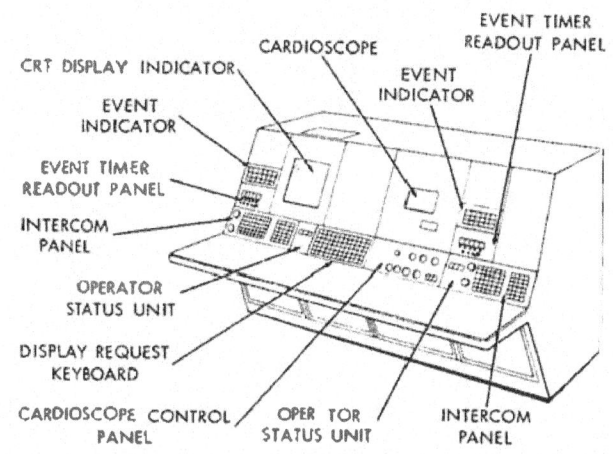

Figure 16—Aero-Medical console.

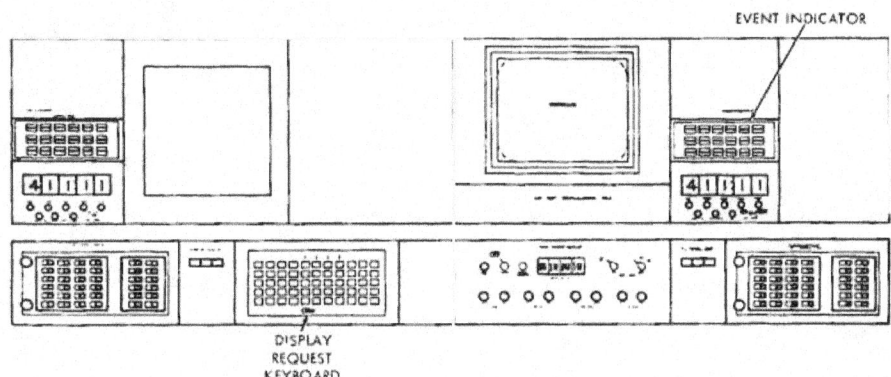

Figure 17—Panels of Aero-Medical console.

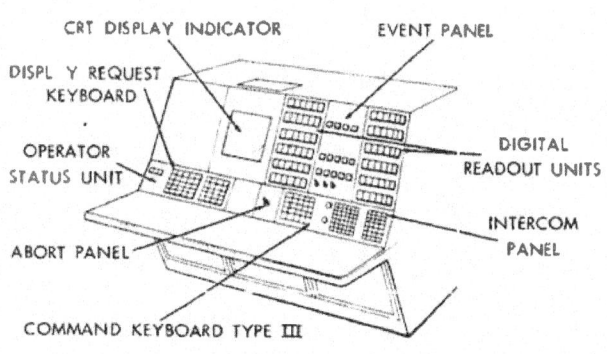

Figure 18—Flight dynamics officer's console.

present on shipboard installations. The FDO contains one CRT display module, one operator status unit, one display request keyboard, and one intercom panel. This console also contains a small command keyboard which is physically identical only to the computer address matrix (5 by 5 PBI matrix) except that the first column of this command keyboard contains 5 indicators. The other four columns contain 20 real-time commands. The FDO contains 12 digital readouts and an FDO switch panel. The digital readouts are driven from the ship's central data processor, and the switch panel provides the ship's central data processor with information on the computer mode, the programmed lift, and the abort mode. Finally, this console contains an event light/override panel which is merely a special event light panel with override capability on six of the events. Twelve event lights are provided, half to be driven from the DSDU and the other half with override capability to be driven from the ship's central data processor. Figure 19 shows the display panels of the FDO console, some of which have since been revised.

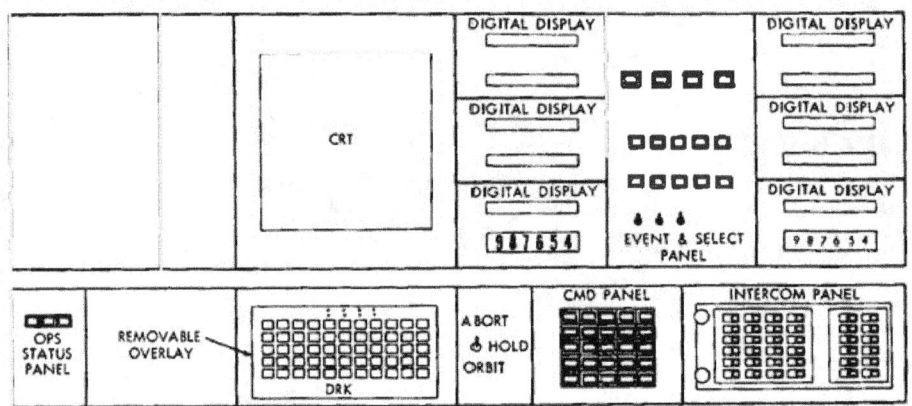

Figure 19—Display panels of FDO console.

MAINTENANCE AND OPERATIONS CONSOLE

The maintenance and operations console requirements are presently being finalized. To date it contains only an intercom panel and an operator status unit.

COMMAND AND SERVICE MODULE UNIFIED S-BAND SYSTEM

by
B. Hood
Manned Spacecraft Center

ABSTRACT

The basic communications and tracking requirements for the Command and Service Module Unified S-Band System are given along with a brief discussion of the reasoning behind these requirements. The operational modes, which include simultaneous transmission of a pseudo-random noise (PRN) range code, TV, voice, telemetry and biomedical data and system configuration that evolved from these requirements, are considered. Such things as modulation techniques, major systems, and basic data flow for these are outlined along with the modes of operation and system configuration. A more detailed description for the major systems is then presented, including block diagrams and data flow and key parameters of the premodulation processor (PMP), transponder, power amplifier and antennas.

INTRODUCTION

To introduce in this session the two "blocks" of the Apollo unified S-band (USB) will be defined. The early Saturn Apollo 200 series will carry what has been designated as a Block I system. As a result of increased operational requirements, it has become necessary to modify the functional design of the Command and Service Module (CSM) S-band system. This modified or updated version has been designated as the Block II system which will first be flown on mission SA-207 and will be the system used on the lunar mission, and is the system discussed in this paper.

BASIC REQUIREMENTS FOR CSM TRACKING AND COMMUNICATION

The basic requirements for the CSM tracking and communications are to provide tracking data, two-way voice communications, up-data from ground to spacecraft, telemetry and television from spacecraft to ground, biomedical channels, relay capabilities, and scientific data channels.

Table 1 presents the communication requirements from ground to spacecraft. There are essentially three functions to transmit: ranging, voice, and up-data. As the first requirement, the spacecraft must be capable of receiving the pseudo-random noise (PRN) range code at any

Table 1

Ground To Spacecraft Communications Requirements.

Function	Requirements
Ranging	Continuous Capability
Voice	Continuous Capability Normal - 90% Word Intelligibility Backup - 70% Word Intelligibility
Up Data	Continuous Capability Maximum of 1 Correct Message Reject per 1000 Maximum of 1 False Message Accepted in 10^9

time it is transmitted from the ground. The capability for voice reception will likewise be continuous with a requirement of 90-percent word intelligibility under normal operating conditions and 70-percent word intelligibility in the backup mode. This percentage of word intelligibility would approximate 100-percent sentence intelligibility.

The spacecraft will also be capable of continuously receiving up-data or command information. The requirements for the up-data channel are that no more than one correct message per 1000 be rejected and that no more than one false message in 10^9 messages be accepted.

The spacecraft-to-ground communications provide a more detailed list of modes, as seen in Table 2. The voice channel must have continuous capability and be transmitted with the same word intelligibilities as stated previously. The telemetry system must provide a nonreturn to zero (NRZ) PCM wave train at either a 51.2 kilobit rate or at a 1.6 kilobit rate with continuous

Table 2

Spacecraft To Ground Communications Requirements.

Function	Requirements
Voice	Continuous Capability Normal - 90% Word Intelligibility Backup - 70% Word Intelligibility
Telemetry	Continuous Capability NRZ PCM at 51.2 KBPS or 1.6 KBPS Rate Maximum of 1 Error in 10^6 Bits
Ranging	Continuous Capability Phase Coherent Turn-Around of the PRN Range Code
Television	When Convenient Near-Commercial Quality Resolution and Gray Scale
Scientific Data	3 Channels
Tape Playback	Capability to Transmit Simultaneously With Real Time Data Telemetry, Voice, and Scientific Data
Emergency Key	Continuous Capability Maximum of 25 Characters Per Minute
Relay Thru CSM	Eva Voice and Biomed to MSFN LEM Simplex Voice to MSFN

capability. The ranging channel on the transponder will be capable of continuous phase-coherent turnaround of the PRN range code and also will provide a coherent carrier on the down-link for doppler extraction.

Television is not a direct mission operational requirement; thus it will be transmitted only when convenient during the mission. When transmitted, it will provide data by which the ground personnel can monitor activity in the spacecraft or scenes from the window. The data transmission for this has been specified as near-commercial quality resolution and gray scale. There is also a capability to transmit three channels of analog data. Use of these channels will be specified by the scientific experiment office.

A tape playback mode will transmit data which has been taped on board the spacecraft during the time it is behind the lunar disc or between ground stations while in earth orbit. Stored information includes telemetry, voice and scientific data. The telemetry data will either be of low or high bit-rate command module data. There will also be a capability of receiving low bit-rate data from the Lunar Excursion Module (LEM) (over a VHF link) and recording it aboard the spacecraft for later playback to ground stations. The requirement to simultaneously transmit taped data along with real-time voice and telemetry data was one of the contributing factors in changing from Block I to Block II design.

The emergency key is a last-ditch communication mode. It will have a capability to provide a maximum of 25 characters per minute. This mode would be used in case of major failure aboard the spacecraft, such as losing both power amplifiers and/or the high-gain antenna. A relay capability through the CSM is also available. Voice and biomedical data from an extravehicular astronaut as well as simplex voice from the LEM can be transmitted to the ground stations through the CSM communication facilities.

TRANSMISSION MODES

A list of possible transmission modes or data combinations has been derived along with the appropriate modulation techniques. Table 3 shows this list for the ground-to-spacecraft transmission link. There are basically four pieces of data to be transmitted: PRN range code, up-voice, up-data, and backup voice. For simplicity, only modes 6 and 8 will be discussed. In mode 6, the PRN code is phase-modulated directly on the carrier and requires approximately 3 megacycles of bandwidth. The voice is frequency-modulated onto a 30-kilocycle subcarrier which is in turn phase-modulated onto the carrier. Similarly, the up-data is frequency-modulated onto a 70-kilocycle subcarrier and then phase-modulated onto the carrier. In mode 8, the backup voice is frequency-modulated onto the up-data subcarrier, which is in turn phase-modulated onto the carrier.

Table 4 shows a tabulation of the narrow band PM modes which will be used on the down-link channels from spacecraft to ground. The reasons for nine modes are that circuit margins may be optimized by using only those channels which are of immediate interest; and under certain contingency modes where a power amplifier or an antenna has been lost, the spacecraft must have the capability of transmitting at reduced rates.

Table 3

MSFN To CSM S-Band Transmission Combinations Summary (PM Modes).

2106.4 Mc Carrier Combinations	Information	Modulation Technique	Subcarrier Frequency
1	Carrier, PRN	PM On Carrier	-
2	Carrier, Voice	FM/PM	30 kc
3	Carrier, Up-Data	FM/PM	70 kc
4	Carrier		
	PRN	PM On Carrier	-
	Voice	FM/PM	30 kc
5	Carrier		
	PRN	PM On Carrier	-
	Up-Data	FM/PM	70 kc
6	Carrier		
	PRN	PM On Carrier	-
	Voice	FM/PM	30 kc
	Up-Data	FM/PM	70 kc
7	Carrier		
	Voice	FM/PM	30 kc
	Up-Data	FM/PM	70 kc
8	Carrier		
	Voice Backup	FM/PM	70 kc

Mode 2 is the primary high-data mode which will be used during critical phases of the mission. In this mode, the PRN code phase-modulates the carrier, whereas the voice frequency-modulates a 1.25-megacycle subcarrier which then phase-modulates the carrier. The telemetry data is a 51.2-kilobit PCM wave train which phase-modulates a 1.024-megacycle subcarrier, which in turn phase-modulates the carrier. Mode 3 has essentially the same function with only a 1.6-kilobit telemetry stream on the 1.024-megacycle subcarrier.

Mode 4 has been designed as a lunar coast mode. There is reduced activity in the spacecraft at this time and a minimum amount of data is required for monitoring. Modes 5, 6, 8, and 9 are possible transmission combinations which can be used to optimize circuit margins in contingency situations. The emergency key capability is provided by amplitude-modulating a 512-kilocycle subcarrier, which phase-modulates the carrier. Also, the backup voice capability is provided by directly phase-modulating the carrier with the backup voice signal.

The wideband FM combinations are shown in Table 5. The main reason for the wideband FM carrier is the TV data shown in mode 4. The other functions which exist on the FM mode are the tape playback channels and the real-time capability for the scientific data. The FM channel is designed to accommodate playback of various combinations of voice, CSM telemetry, LEM telemetry, and scientific data. A high-speed playback mode is used for rapid dump of low bit rate PCM at a 32 to 1 playback rate. This results in an apparent bit rate of 51.2 kilobits per second on the 1.024 megacycle subcarrier. For playback of high bit rate PCM, a 1:1

Table 4

CSM To MSFN S-Band Transmission Combination Summary (PM Mode).

2287.5 Mc Carrier Combination	Information	Modulation Technique	Subcarrier Frequency
1	Carrier		
	Voice	FM/PM	1.25 Mc
	51.2 KBPS TM	PCM/PM/PM	1.024 Mc
2	Carrier		
	PRN	PM On Carrier	
	Voice	FM/PM	1.25 Mc
	51.2 KBPS	PCM/PM/PM	1.024 Mc
3	Carrier		
	PRN	PM On Carrier	
	Voice	FM/PM	1.25 Mc
	1.6 KBPS	PCM/PM/PM	1.024 Mc
4	Carrier		
	Voice	FM/PM	1.25 Mc
	1.6 KBPS	PCM/PM/PM	1.024 Mc
5	Carrier		
	1.6 KBPS	PCM/PM/PM	1.024 Mc
6	Carrier		
	Key	AM/PM	512 kc
7	Carrier		
	PRN	PM On Carrier	
8	Carrier		
	Backup Voice	PM On Carrier	
	1.6 KBPS TM	PCM/PM/PM	1.024 Mc
9	Carrier		
	PRN	PM On Carrier	
	1.6 KBPS TM	PCM/PM/PM	1.024 Mc

dump speed is used. Playback of voice is accomplished simultaneously with telemetry or scientific data dump by modulating the FM carrier directly at baseband with either 1:1 or 32:1 voice signals. The scientific data frequency-modulates the subcarriers of 95, 125, and 165 kilocycles which then frequency modulates the carrier.

SPECTRA OF TRANSMISSION MODES

Typical spectra of the above transmission modes can be seen in Figure 1-4. Figure 1 shows the full up-link spectrum with the range code, voice, and up-data subcarriers. The range code is centered on the 2106.4-megacycle PM carrier, and the voice and up-data subcarriers are at 30 and 70 kilocycles, respectively.

A typical down-link spectrum is represented by Figure 2. It is seen that the spacecraft has the capability of simultaneously transmitting both an FM and a PM carrier. The PM carrier at

Table 5

CSM to MSFN S-Band Transmission Combination Summary (FM Modes).

2272.5 Mc Carrier Combination	Information	Modulation Technique	Subcarrier Frequency
1	Playback Voice at 1:1	FM at Baseband	
	Playback CSM 51.2 KBPS TM at 1:1	PCM/PM/FM	1024 kc
	Scientific Data Playback at 1:1	FM/FM	95 kc
		FM/FM	125 kc
		FM/FM	165 kc
2	Playback Voice at 32:1	FM at Baseband	
	Playback CSM 1.6 KBPS TM at 32:1	PCM/PM/FM	1024 kc
	Scientific Data Playback at 32:1	FM/FM	95 kc
		FM/FM	125 kc
		FM/FM	165 kc
3	Playback LEM 1.6 KBPS Split Phase TM at 32:1	FM at Baseband	
4	Television	FM at Baseband	
5	Real-Time Scientific Data	FM/FM	95 kc
		FM/FM	125 kc
		FM/FM	165 kc

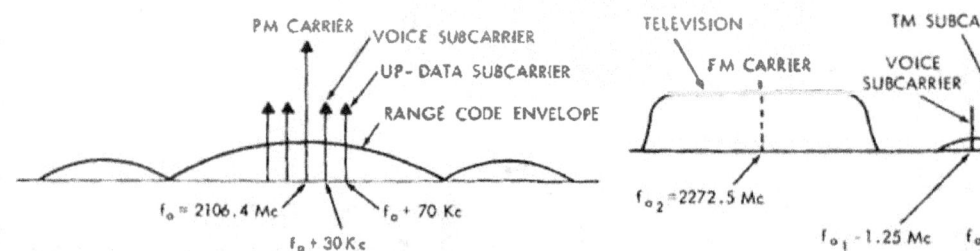

Figure 1—Full up-link spectrum with the range of code, voice, and up-data subcarriers.

Figure 2—Typical down-link spectrum.

2287.5 megacycles would be modulated with the range code, the telemetry at 1.024 megacycles and the voice at 1.25 megacycles. The FM carrier is located at 2272.5 megacycles and is modulated with the TV signal. Figure 3 indicates another typical down-link spectrum with a normal PM mode. But instead of TV, the FM channel contains the playback voice modulated directly on the carrier, the playback telemetry data modulated on its subcarrier, and the scientific data subcarriers.

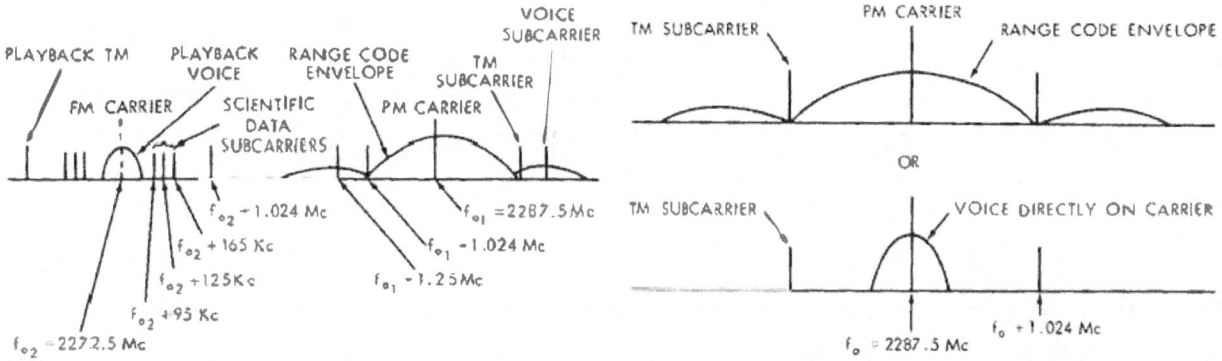

Figure 3—Typical down-link spectrum with a normal PM mode.

Figure 4—Contingency down-link spectrum.

In the event that the high-gain antenna or the power amplifier are disabled, one of the two contingency down-link spectra of Figure 4 could be employed. One spectrum shows the PM carrier modulated by the range code and a low bit-rate telemetry subcarrier (1.6 kilobits of PCM data). The other mode shows the PM carrier modulated directly by voice and a low bit-rate telemetry subcarrier. Although it is not shown, the spectrum for the emergency key would be essentially the same with a 512 kilocycle subcarrier phase-modulating the carrier.

BASIC CSM USB SYSTEM

A block diagram of the basic USB system is shown in Figure 5. It consists of four basic components or subsystems: the antenna subsystem (omnidirectional antenna and high-gain directional antenna); the power amplifier, switching, and triplexing subsystem; the transponders; and the premodulation processor. The transponder consists of a PM receiver, a PM exciter or transmitter, and an FM exciter. The premodulation processor provides the capability to demodulate the up-voice and up-data subcarrier, as well as the capability for modulating the down-link subcarriers and combining them into a composite waveform for use by the PM and FM exciters. The premodulation processor also interfaces with the VHF equipment for the communications between the LEM and command module.

The system is not fully redundant since there is only one premodulation processor; however, due to the requirements for transmitting real-time and recorded data, there are essentially two sets of modulators for each of the prime subcarriers. In the event of a failure of one of the real-time modulators, we could use a modulator which has been provided for recorded data and thereby maintain redundancy in the real-time channels.

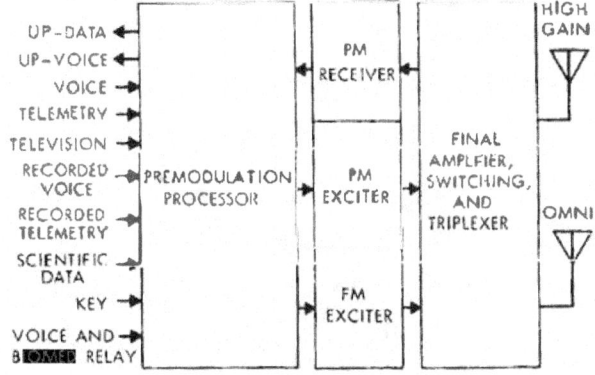

Figure 5—Diagram of basic CSM Unified S-Band System.

Figure 6 shows the basic USB receiver in the transponder. It is a dual-conversion superheterodyne PM receiver with a center frequency of 2106.4 megacycles. Following the conversion phase, at a signal level of -114dbm and a static phase error of 24 degrees, the carrier tracking loop is capable of tracking frequency ranges of ±90 kilocycles and sweep rates of 35 kilocycles per second/second. Immediately following the second mixer, the signal passes through an IF limiting amplifier into a wideband detector.

The voltage control oscillator (VCO) in the carrier-tracking loop provides a reference for the wideband detector. The output of the wideband detector consists of the range code which is coherently returned to the PM transmitter and the up-data and voice subcarriers, which are passed to the premodulation processor for demodulation and use in the spacecraft. The output of the VCO is also coupled to the PM transmitter where it provides the phase-coherent reference for this exciter. The transponder provides a frequency ratio of 240:221 for the coherent turnaround of the carrier.

The basic configuration of the transmitter exciters is shown in Figure 7. The modulators (PM and FM) receive data (telemetry and voice subcarriers) from the premodulation processor. Other inputs to the modulators include range code, a coherent reference from the carrier-tracking VCO (for the PM mode), and a reference frequency from an oscillator (for the FM mode). Once the data is modulated, it is amplified and multiplied up to the proper frequency—2287.5

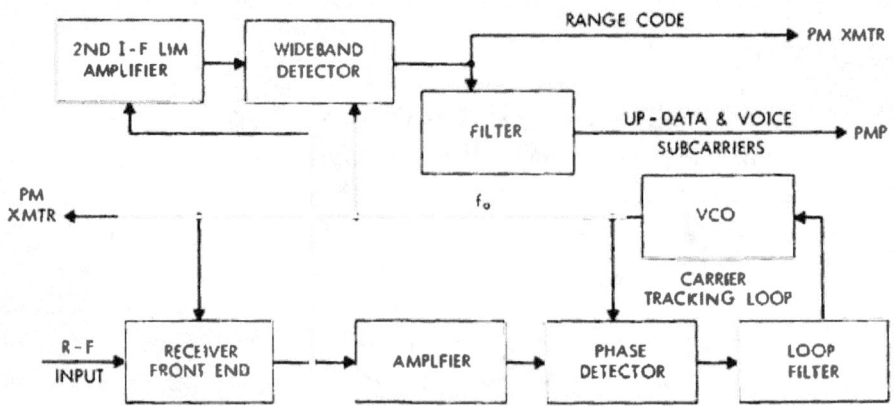

Figure 6—Diagram of CSM S-band receiver.

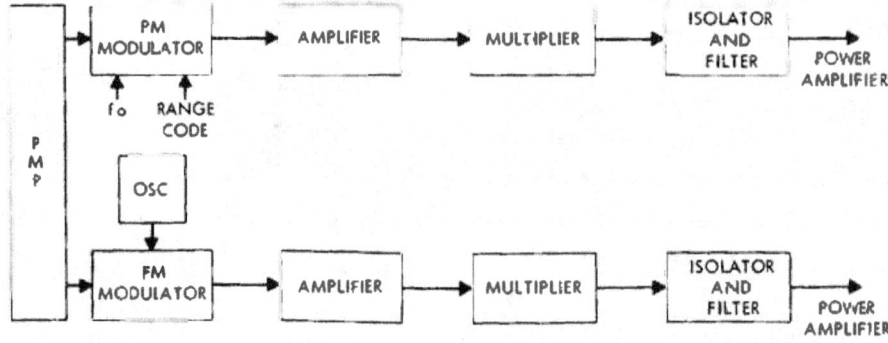

Figure 7—Basic configuration of transmitter PM and FM exciters.

megacycles for the PM mode and 2272.5 megacycles for the FM mode, and is then passed through an isolator and filter network to the power amplifier subsystem.

The power amplifier subsystem, shown in Figure 8, is composed basically of two traveling-wave tube amplifiers and the appropriate switching arrangements. The switches are designed such that either transmitter (PM or FM in Figure 7) may be connected to either power amplifier; and that the output of either amplifier may be connected to either input of the antenna subsystem.

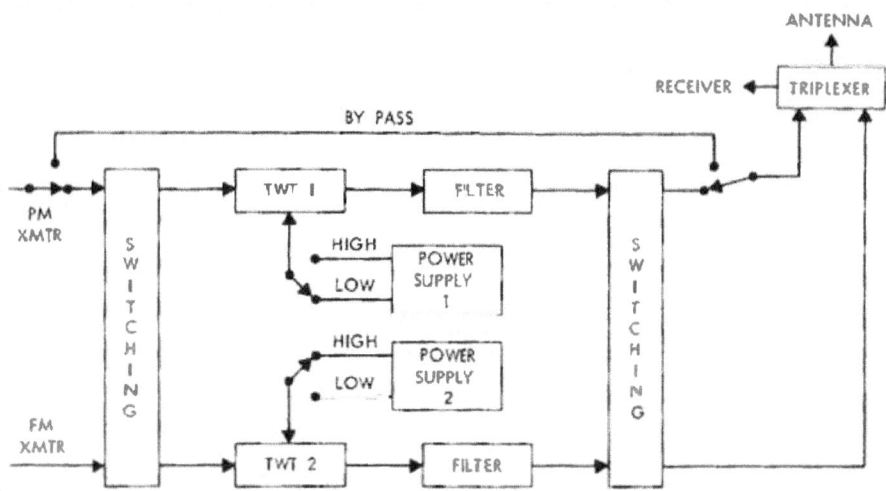

Figure 8—Diagram of CSM S-band power amplifier subsystem.

The traveling-wave tube amplifiers have two power-output levels: 5 or 20 watts. These power levels are designated as low- and high-power modes, respectively, and are controlled by the switching of the input power from the power supplies to the tubes. As can be seen, the output of the traveling wave tube is filtered and passed through switching to the triplexer arrangement and then to the antenna. The actual power delivered to the antenna is considerably less than the output power of the tube due to circuit losses between the tube and the antenna. In the event of amplifier failure, the PM exciter can bypass the power amplifier and still provide a capability of 1/4 watt of RF power.

The antenna subsystem (Figure 9), consists of a high-gain antenna and a set of omnidirectional antennas. As can be seen, PM and/or FM carriers are fed into the triplexer, which in turn supplies the power to the selected antenna. The high-gain antenna is mounted on the service module and has a selectable set of gains and beamwidths. The omnidirectional antennas (low-gain) are

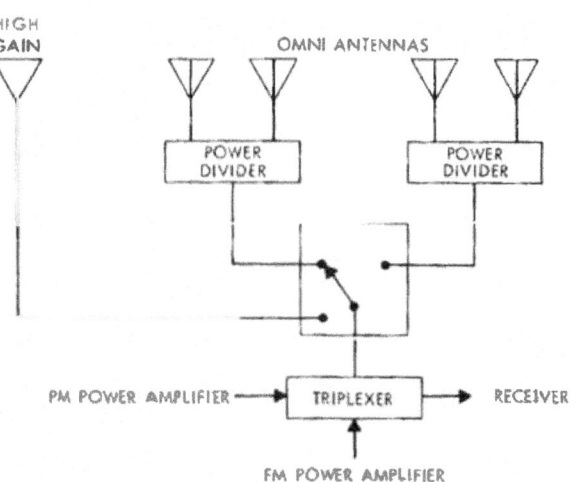

Figure 9—CSM S-band antenna system.

located around the base of the command module and are spaced at 90 degree intervals.

There are plans to use a number of combinations of antennas and power amplifiers in order to optimize the circuit margins and conserve power during different mission phases. The omnidirectional antennas will be used in earth orbit and in the first stages of translunar injection up until the high-gain antenna can be deployed. The omnidirectional antennas will also provide limited capability at lunar distances when communicating with the high-gain antennas on the ground.

The basic characteristics of the antenna system are tabulated in Table 6. As stated above, the high-gain antenna has a selectable set of gains and beamwidth. It has a capability of 28.4db at a beamwidth of 4.6 degrees, 22.9db at a beamwidth of 11.1 degrees, or 7.4db at a beamwidth of 68 degrees. In order to cover the earth's surface during early phases of injection, the beamwidth must be quite large. As the spacecraft nears the moon, beamwidth is reduced. The high-gain antenna is a parabolic radiating system and radiates a right-circularly polarized wave. It is pointed at the earth by an IR sensor device. The omnidirectional antennas are flux-mounted, right-circularly polarized, and provide an overall gain of approximately -3db over 80 percent of the sphere.

Table 6

CSM Antenna Characteristics.

Function	Omni	High Gain	
Type	Flush Mounted	Parabolic	
Polarization	RCP	RCP	
On-Axis Gain and Corresponding 3 DB Beamwidth	-3 DB Over 80% Sphere	28.4 DB 22.9 DB 7.4 DB	4.6° 11.1° 68°
Pointing		IR Sensor	

LUNAR EXCURSION MODULE UNIFIED S-BAND SYSTEM

by
W. Kuykendall
Manned Spacecraft Center

ABSTRACT

A description of the basic S-band communication and tracking requirements for the in-flight and lunar phases of the LEM operation is given. Modulation techniques and associated RF spectra for various communication modes involving voice, telemetry, TV, and ranging are presented. RF systems, including the transponder, power amplifier, diplexer and antennas (omni, steerable, and erectable) are discussed. Current estimates of system gain and loss parameters are presented along with predictions of system performance at lunar distance.

INTRODUCTION

In discussing the Lunar Excursion Module (LEM) Unified S-Band System, four essential areas are covered: mission requirements, spacecraft system configuration to meet these requirements, spacecraft signal design for the up-link and down-link, and expected system performance margins.

LEM COMMUNICATIONS AND TRACKING REQUIREMENTS

Table 1 gives the communication and tracking requirements for the LEM. These requirements can be divided essentially into two areas: the inflight area and the lunar stay. Table 1 shows that the up-link requirements for the inflight phase include for carrier, range code and voice. Note that there is no up-data link to the LEM. The down-link inflight requirements consist of carrier, voice, biomedical data, PCM telemetry, and range code. During lunar stay, the up-link requirement is for voice only; the down-link

Table 1
LEM S-Band Communication and Tracking Requirements.

Mission Phase	Requirements	
	MSFN→LEM (Up-Link)	LEM→MSFN (Down-Link)
In-Flight	Carrier Voice Range Code	Carrier Voice Biomedical Data PCM Telemetry Range Code
Lunar Stay	Voice	Voice Biomedical Data PCM Telemetry Television

requirement, for voice, biomedical data, PCM telemetry, and television. We might also note that there is no requirement for a tape-playback capability from the LEM. There is, however, an indirect tape-playback capability present since the LEM is capable of transmitting low-rate telemetry to the Command and Service Module (CSM) where it can be recorded and subsequently played back to the Manned Space Flight Network (MSFN). Another Unified S-Band (USB) requirement that will not be treated in this discussion, is the lunar surface experiment package, S-band link for communication with the MSFN.

The modulation schemes used for the up- and down-link transmissions are shown in Table 2. The up-link is always PM as is the down-link during inflight phases. During the lunar-stay period, when the LEM is not transmitting TV, the PM mode will normally be used. When transmitting TV during the lunar stay, the FM mode will be used.

Table 2
Carrier Modulation Techniques.

Link	Modulation
Up-Link	PM
Down-Link	
In-Flight	PM
Lunar Stay, Without TV	PM
Lunar Stay, With TV	FM

LEM S-BAND SYSTEM CONFIGURATION

Figure 1 shows the system block diagram for the onboard S-band system in the LEM, illustrating the several major elements in the system. At the left of the figure is the premodulation processor (PMP) which functions much like the PMP for the CSM. To the right is the S-band transceiver, or transponder. Between the transceiver and antenna system are the two power amplifiers; the antennas and associated switching are depicted in the upper right-hand part of the figure.

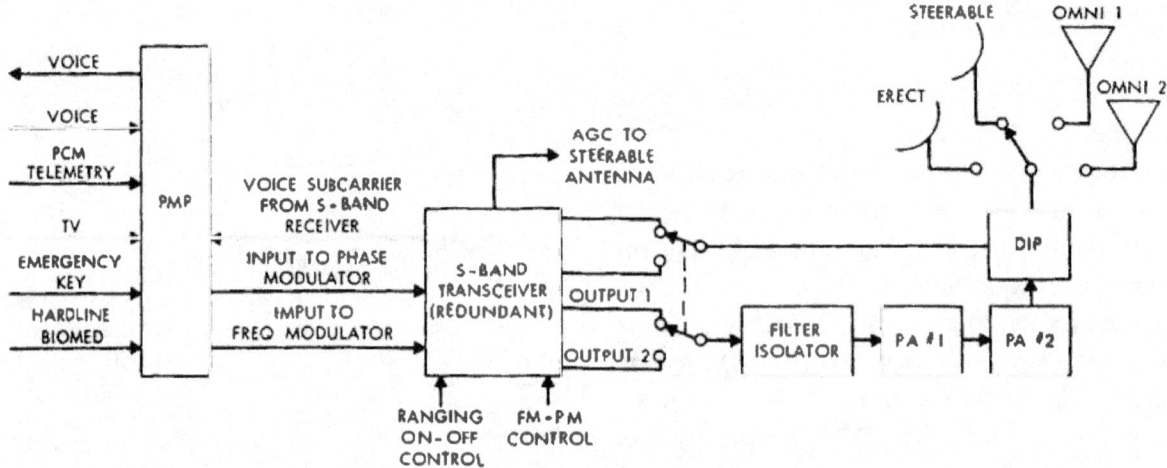

Figure 1—Block diagram of LEM S-band system.

MODULATION TECHNIQUES AND ASSOCIATED RF SPECTRA

Figure 2 shows the full up-link PM spectrum. Notice the absence of the 70-kilocycle (up-data) subcarrier. During normal flight operation the range code will be used intermittently; therefore, the spectrum without range code represents the normal inflight up-link mode. The center frequency is 2101.8 megacycles (plus or minus doppler and transmitter frequency offset).

The full down-link spectrum for normal PM modulation is shown in Figure 3. Again the range code will be used intermittently; therefore, the normal down-link mode will include the carrier and the two subcarriers.

The two contingency modes are illustrated in Figure 4. Backup voice is modulated directly at base band with low-rate telemetry transmitted on the 1.024 megacycle subcarrier. In the worst case, multiple failures such as antenna, power amplifier, and possibly PMP failure could lead to the use of the emergency-key mode which utilizes a 512-kilocycle subcarrier keyed on and off by the astronaut.

Figure 5 shows the FM down-link spectrum, which will be used during the lunar-stay mode if there is television transmission. Also shown in the figure is the spectrum containing TV and the telemetry and voice/biomedical subcarriers.

Table 3 shows the three possible combinations of up-link information: carrier and pseudo-random noise (PRN), carrier and voice, and PRN and voice. Note that the *combinations* is synonymous with *modes*. Combination 2 will be the primary up-link mode for the LEM.

The various combinations of down-link transmission for PM and FM modes of operation are shown in Table 4. For down-link transmissions there are a few keyed modes which will probably be used during most of

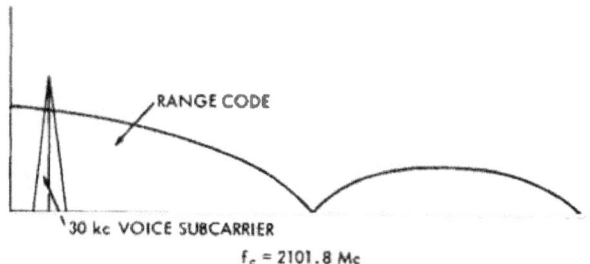

Figure 2—Full up-link PM spectrum.

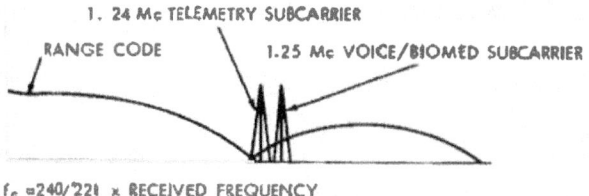

Figure 3—Full down-link PM spectrum.

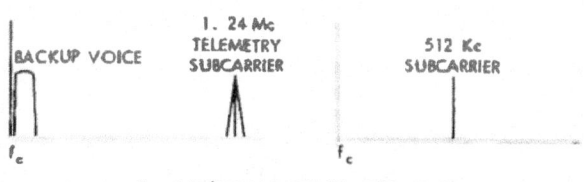

Figure 4—Down-link RF spectra for contingency modes (PM).

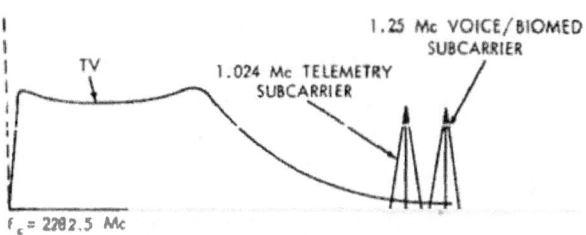

Figure 5—FM down-link spectrum.

Table 3

LEM Up-Link S-Band Transmission Combinations Summary.

Combination	Information	Modulation Technique	Subcarrier Frequency	Carrier Phase Deviation
1	Carrier, PRN	PM on Carrier	-	0.37 Radians
2	Carrier, Voice	FM/PM	30 kc	1.4 Radians
3	Carrier			
	PRN	PM on Carrier	-	0.37 Radians
	Voice	FM/PM	30 kc	1.4 Radians

the operation and additional modes to be used in the event of various failures. The primary modes are combinations 1, 7, and 10.

Combination 1 is the down-link carrier with voice and 51.2 kilobit telemetry on their respective subcarriers. Combination 7 is commonly referred to as the lunar-stay mode. Voice and biomedical information are transmitted on the 1.25 megacycle subcarrier while low-rate telemetry is on the 1.024 megacycle subcarrier. In order to optimize the circuit margins, the phase deviations for the two subcarriers modulating the PM carrier are reversed for combinations 1 and 7. In combination 1 we are transmitting a wide-band high-rate telemetry signal. In combination 7, it is not necessary during lunar stay to transmit the same quantity of information required during inflight and we are able to reduce that bit rate to 1.6 kilobits per second. The power in the two subcarriers may then be balanced, enabling combination 7 to perform well with low-power transmission and the erectable antenna during lunar stay, contrasted to another alternative of operating with full power (20 watts). This is the basic reason for a change in the modulation indices in the right-hand column.

The advantage of the reversal of the modulation indices in combination 7 is evident if we consider operation with low power during lunar stay while keeping the same phase deviations as used in combination 2. The result is a -3.9db margin for the voice and biomedical channel and about a +12.9db margin for the low-rate telemetry. However, if by reversing modulation indices, results yield positive margins for both channels operating in the low-power mode. These margins are over 3db for the voice and biomedical channel and 6db for the telemetry channel. The last primary mode here of course is number 10 which is the FM transmission of TV with the voice, biomedical and extra-vehicular mobility unit (EMU) signals. The EMU is the device carried by the astronaut on the lunar surface. The biomedical subcarriers are transmitted from the EMU to the LEM by the VHF link and are subsequently placed on the S-band for transmission to the earth.

Table 4

LEM Down-Link MSFN S-Band Transmission Combination Summary.

2282.5 Mc Carrier Combination	Information	Modulation Techniques	Subcarrier Frequency	Carrier Phase Deviation
1	Carrier Voice 51.2 kbps TM	FM/PM PCM/PM/PM	1.25 Mc 1.024 Mc	0.7 Radians 1.3 Radians
2	Carrier PRN Voice 51.2 kbps TM	PM on Carrier FM/PM PCM/PM/PM	1.25 Mc 1.024 Mc	0.2 Radians* 0.7 Radians 1.3 Radians
3	Carrier 1.6 kbps TM	PCM/PM/PM	1.024 Mc	1.3 Radians
4	Carrier BU Voice 1.6 kbps TM	PM on Carrier PCM/PM/PM	1.024 Mc	0.8 Radians 1.3 Radians
5	Carrier Backup Voice	PM (24 db clipping)		0.8 Radians
6	Carrier Key	AM/PM	512 kc	1.4 Radians
7 (Lunar Stay Mode)	Carrier Voice/Biomed 1.6 kbps TM	FM/PM PCM/PM/PM	1.25 Mc 1.024 Mc	1.3 Radians .7 Radians
8	Carrier Voice/EMU/ Biomed 51.2 kbps TM	PM on Carrier (no clipping) PCM/PM/PM		TBD

				Carrier Deviation Ratio
9	Voice/EMU/ Biomed TM	FM/FM PCM/PM/FM	1.25 Mc 1.024 Mc	0.17 0.37
10	TV Voice/EMU/ Biomed 1.6 or 51.2	FM at Baseband FM/FM/FM PCM/PM/FM	1.25 Mc 1.024 Mc	2.0 0.17 0.37

*Down PRN ranging phase deviation is to be set with up voice and up ranging modulation (Table 1 Comb 3) and with a high signal to-noise ratio in the turn around channel.

EQUIPMENT CHARACTERISTICS

Returning to Figure 1 the PMP performs the demodulation of the up-voice subcarrier and contains the voltage controlled oscillator (VCO) for the down-voice channel and the bi-phase modulator for the PCM telemetry. It interfaces with the TV equipment, emergency key, and the hardline biomedical channel. The hardline biomedical channel is a 14.5 kilocycle subcarrier on which 0 to 30 cycles per second biomedical data is modulated when LEM is transmitting low bit-rate telemetry. Due to the lower capacity in the low-rate telemetry format, certain necessary biomedical data cannot be handled by PCM; therefore, a separate subcarrier was provided.

The other equipments with which the PMP interfaces on the spacecraft include the audio center, the PCM telemetry equipment, central timing equipment, and TV camera. The S-band transceiver (transponder) block as shown here includes two fully redundant transceivers as well as an FM modulator. The two RF outputs are selectable, as are the two RF inputs to the receivers. Differing slightly from the CSM, the LEM utilizes amplitron power amplifiers operated in cascade. Normally, power amplifier, PA #2 would be used when operating in the 20-watt mode because there is additional insertion loss in the circuit with power amplifier PA #1 in operation. The three types of antennas, shown in Figure 1, are the erectable antenna, which is used during lunar stay operations, the steerable antenna, which is the primary inflight antenna, and two omnidirectional antennas, which may be used as backups. These antennas are switchable by the astronaut.

With reference to the control functions which the crew exercises over this system, the astronaut has control over the telemetry bit rate through selection of either 51.2 or 1.6 kilobits. He also may turn the ranging channel on or off by a manual control on the panel which allows the ranging channel to be turned off when PRN ranging is not being used, thus eliminating the turnaround of the up-link subcarriers and noise which is detrimental to the down-link performance. Another control selects PM or FM such that the LEM can transmit either FM or PM, but not both. The crew can also select transceiver one or two and the power amplifier (PA #2) or backup power amplifier (PA #1). In addition, the crew has manual control for pointing the steerable antenna for initial acquisition.

Table 5 lists the characteristics of the three types of antennas used. The steerable antenna, a 2-foot parabola which is the primary inflight antenna, angle-tracks the MSFN station automatically. The erectable antenna, a 10-foot parabola, is used to provide the additional gain needed for the TV transmission from the lunar surface. It is stowed in flight and is erected by the astronaut on the lunar surface. The low-gain omnidirectional antenna serve as a backup to the steerable antenna. In lunar orbit the astronaut can switch between either of two omnidirectional antennas located on opposite sides of the spacecraft for optimum communications.

As shown in Table 6, the steerable antenna is a 2-foot parabola with on-axis transmitting gain of 20.3db and receiving gain of 16.5db. The noticeable difference in the transmit and receive gains is due largely to the RF tracking technique which degrades the receive performance. The RF angle tracking involves a feed system for sampling four quadrants and deriving the error signals. The 3db beamwidth is between 12 and 14 degrees and polarization is right-circular with an elipticity of less than 1db within plus or minus 4 degrees of the boresight axis.

Table 5

Characteristics of LEM S-Band Antennas.

Name	Type	Use
Steerable	2' Parabola Auto-Tracks MSFN Station	Primary In-Flight Antenna
Erectable	10' Parabola Stowed In-Flight Erected on Lunar Surface	Primary Lunar-Stay Antenna
Omni (Backup)	Low-Gain Omnidirectional (Can Manually Switch Between 2 Antennas)	Backup to Steerable

Table 6

Characteristics of LEM Steerable Antenna (S-Band).

Type	2' Parabola
Gain (On-Axis)	Transmit: 20.3 db Receive: 16.5 db
Beamwidth (-3 DB)	12° - 14°
Polarization	RCP
Ellipticity	<1 db Within ±4° of Boresight Axis
Pointing	RF Tracker Lobe on Receive Only Utilizes S-band transceiver AGC to derive error signals

The pointing of the antenna is accomplished by RF angle tracking using a lobe-on-receiver-only technique. Original LEM plans called for an IR tracker much like that in the CSM tracker; however, it was found that during the LEM descent phase, there are certain maneuvers which could place portions of the LEM structure within the line-of-sight between LEM and earth. The infrared system could then lock on the LEM itself requiring the flight crew to manually re-acquire earth. Since this was an unacceptable crew task, several alternatives were considered such as extending the boom, providing auxiliary antennas, and using an RF tracker. The latter was the method chosen.

Table 7 gives the characteristics of the LEM erectable antenna. This is a 10-foot parabola with an on-axis transmitting gain of 34.0, receiving gain of 32.5db and a beamwidth of 3 degrees at the 3db points. Polarization is right-circular with elipticity less than 1db within 1.3 degrees of the boresight axis. This antenna is erected on the lunar surface by the astronaut

Table 7

Characteristics of LEM Erectable Antenna (S-Band).

Type	10' Parabola
Gain (On-Axis)	Transmit: 34.0 db Receive: 32.5 db
Beamwidth (-3 db)	3°
Polarization	RCP
Ellipticity	<1 db Within ±1.3° of Boresight Axis
Pointing	Manual by Astronaut Aided by Sighting Device
Dimensions When Stowed	3' Long x 10'' Diameter

Table 8

Characteristics of LEM Omniantennas (S-Band).

Type	2 Conical Spirals
Gain	Not Less Than -3db Over 85% of Spherical Area (assumes switching)
Polarization	RCP

Table 9

LEM S-Band Equipment Weight and Power Summary.

Item	Weight (lbs)*	DC Power (Watts)*
S-Band Transceiver	20.2	29
Power Amplifier and Diplexer	13.0	65.2
PMP	9	17.0
Steerable Antenna (Including electronics)	20.0	20.0
Erectable Antenna	10.0	
Backup Antennas	2.3	

*Approximate values.

and pointed by a visual alignment with the center of the earth. The astronaut will be able to align the antenna accurately enough to achieve good communications. The size of this 10-foot antenna while stowed in the LEM is to be less than three feet in length and 10 inches in diameter.

The LEM omnidirectional antennas, shown in Table 8, consist of two conical spirals with the specified gain not less than -3db over 85 percent of the spherical area. The spirals are switchable for optimum coverage. The polarization for the omniantenna is also right-circular.

Table 9 gives the present power and weight estimates of the LEM S-band equipments.

Table 10 shows the LEM S-band circuit losses. Circuit losses vary with transmission power, transmit or receiver channel and the antenna selection. The difference between the high-power and the low-power loss values is due to the insertion loss of the power amplifier system. The non-operating power amplifier system is in the transmit leg even when transmitting in the low-power mode. Thus, an additional insertion loss appears in the low-power mode which does not appear in the high-power mode.

In the case of the erectable antenna, high losses are basically due to long cable runs. A 65-foot cable run is necessary to allow the astronaut ample room to select a suitable place for erecting the antenna. Therefore, in the low-power mode with the erectable antenna there are some very high circuit losses.

Table 11 gives the expected circuit margins for normal down-link communications. No margins are given for the up-link, since the 10 kilowatt transmitter power on the ground minimizes the problem of up-link communication with the LEM at lunar distance. The first mode uses the steerable antenna at 20 watts while transmitting the carrier, range code, voice, biomedical data and telemetry. The resulting circuit margins are based on the circuit loss values given in Table 10. These margins are calculated assuming an 85-foot antenna, a system temperature of 326 degrees kelvin and operation at lunar distance. The system temperature is quite high and a 2 or 3db improvement in these margins can be expected by using the cooled parametric amplifiers now being implemented for the ground stations. In the first mode, the voice/biomedical channel has about a 6.6db margin, and high bit rate telemetry has a 5.2db margin.

The erectable antenna normally uses a transmitting power of 0.75 watts for lunar stay mode when not transmitting TV. The carrier margin is 24.6db, the voice/biomedical margin, 3.2db and the low bit-rate telemetry margin, 5.8db. If the modulation indices had not been

Table 10
LEM S-Band Circuit Losses.

Antenna	Losses (db)		
	Transmitting High Power	Transmitting Low Power	Receiving
Steerable	5.5	7.9	6.3
Erectable	9.1	11.5	10.0
Omni	5.3	7.6	6.1

Table 11
Expected Circuit Margins for Normal LEM Down-Link Combinations.

System Configuration	Channels	Margin (db)
Steerable Antenna 20 Watts	Carrier	32.2
	PRN Ranging	20.3
	Voice/Biomed	6.6
	Telemetry (51.2 KBPS)	5.2
Erectable Antenna 0.75 Watts	Carrier	24.6
	Voice/Biomed	3.2
	Telemetry (1.6 KBPS)	5.8
Erectable Antenna 20 Watts	TV, Voice/Biomed, TM on FM Channel	2.3

*Assuming 85' ground station, T_{system} = 329° K, lunar distance.

reversed, the voice/biomedical margin, which is 3.2db in this case, would be minus 3.9db and telemetry would be plus 12.9db.

The margin for the FM mode is low, but with a low system temperature, the margin should increase enough to assure good quality TV from the moon.

The above margins have assumed that the antenna is pointed directly at the center of the moon; however, during the inflight modes, the antenna could be boresighted at a spacecraft near the edge of the lunar disc. Thus another slight improvement in signal performance may be expected.

UNIFIED S-BAND RF SYSTEM COMPATIBILITY TEST PROGRAM

by

A. Travis

Manned Spacecraft Center

ABSTRACT

The background, history, and present status of the Unified S-Band RF System compatibility program are presented. A diagram of the test configuration is furnished and the equipment utilized in testing is outlined. The tests which have been accomplished are enumerated and the general results are given.

The responsibility of verifying spacecraft-ground-systems compatibility through systems analysis and tests has been assigned to the Information Systems Division of the Manned Spacecraft Center. To fulfill a portion of this responsibility, the Unified S-Band RF System Compatibility Test Program is being conducted, and the background, history, and present status of this program will be presented. Approximately one year ago it was decided that the ground and spacecraft system development status was such that it was possible and desirable to make an intensive effort to plan and conduct the S-Band Electronic Systems Test Program (ESTP). The initial objectives were to:

1. Insure basic signal compatibility between the Block I USB spacecraft system and the S-band ground equipment.

The tests which have been accomplished include: Gross compatibility tests as a first overall look at the quality of the operational modes for Block I, data channel performance tests which consisted of a detailed examination of each communication channel and signal combination, and detailed investigation of the problem areas noted during the other tests.

General results of the tests show that Block I systems are compatible with the mission requirements under some constraints, which have been defined; and several problem areas have been delineated.

Some of the problems represent system constraints, others have affected the design of spacecraft and ground equipment, and the remainder require further investigation. Tests which are intended to investigate the latter problems are presently being planned and will be conducted in the near future. In addition to the written reports, periodic briefings and reviews are held for interested parties.

Test results are documented in monthly activity reports, test-review reports, and a final report. The final report on the Block I tests will be published in the near future.

The new receiver/exciter subsystem, which was furnished by GSFC, and a Block I transponder are presently being modified to Manned Space Flight Network (MSFN) and Block II specifications.

2. Establish performance limitations and communications circuit quality for the early Apollo flights by experimental means and evaluate the results by comparison with theoretical analysis.

3. Provide a high degree of the assurance that the first Apollo USB flight tests would be successful.

4. Provide data as soon as possible on design improvements required for future development of spacecraft and ground systems to meet mission requirements.

The test planning has been a joint effort of the Manned Spacecraft Center (MSC), NASA Headquarters, Goddard Space Flight Center (GSFC), Jet Propulsion Laboratory (JPL), Marshall Space Flight Center (MSFC), and the spacecraft contractors. MSC awarded Motorola Communications and Electronics, Inc., the contract to work with MSC and the other participants in conducting the system tests at MSC.

GSFC, JPL, North American Aviation, Inc., and Collins Radio Company have provided on-site assistance by operating and maintaining equipment, as well as by handling information distribution to and from their home office.

Between July and December 1964, the facility at MSC was prepared and furnished with a modified deep space network receiver/exciter subsystem, ranging subsystem, engineering models of the data demodulators and subcarrier oscillators, and engineering models of the Block I Command and Service Module spacecraft equipment. These, along with special supporting equipment, were assembled, carefully checked, and calibrated so that controlled tests could be conducted.

A block diagram of the test configuration is given in Figure 1, and the operational equipment utilized in performing these tests includes: Receiver/exciter subsystem, ranging subsystem, signal data demodulator, subcarrier oscillators, and spacecraft equipment. Equipment shown at the right of the chart is for the spacecraft systems, while the equipment at the left is for the ground systems. The spacecraft equipment is separated from the ground equipment by an RF-shilded enclosure, to insure that both the spacecraft and the ground equipment are sufficiently isolated an that the received signal strength can be accurately controlled by the RF path. Tests utilizing the modified transponder, Command and Service Module Block I D-Models, Lunar Excursion Module production equipment, Command and Service Module Block II equipment, S-IV-B production equipment, and the up-dated ground station will be initiated in the near future.

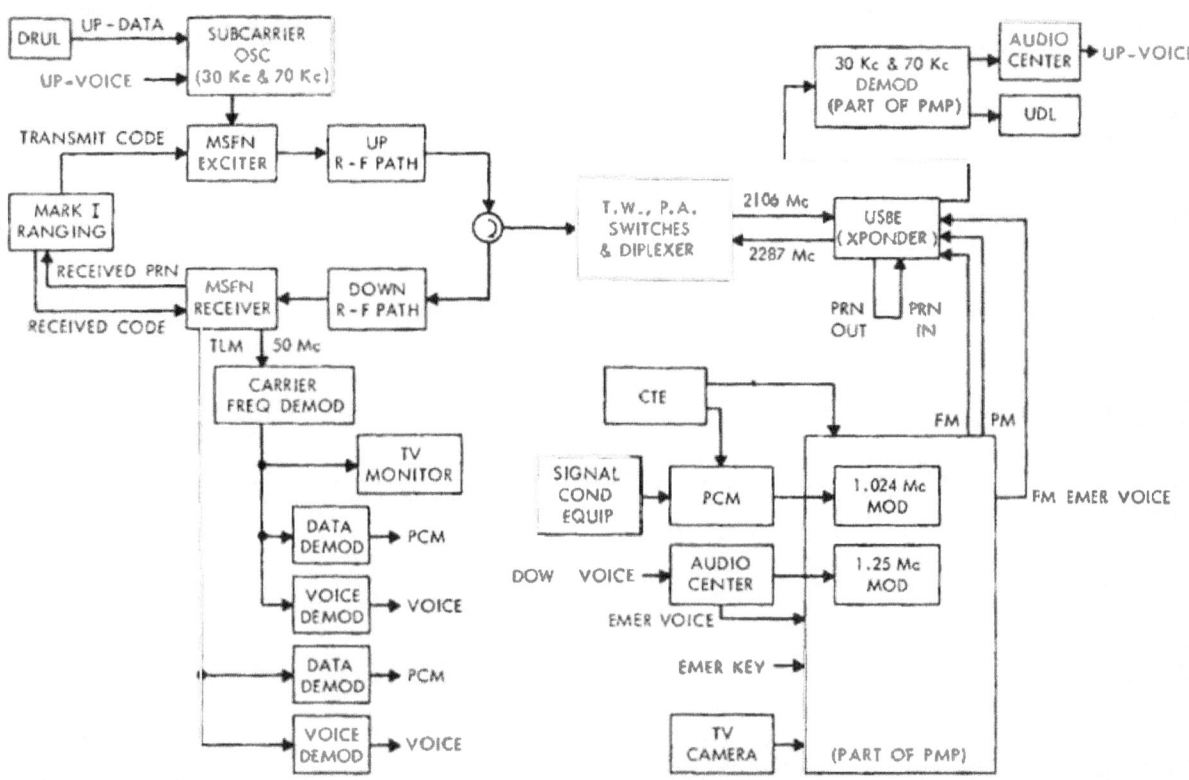

Figure 1—Block 1 system test configuration.

COMMAND AND COMMUNICATION SYSTEM

by
B. Reed
Marshall Space Flight Center

ABSTRACT

The Command and Communication System (CCS) consists of a transponder, power amplifier and antenna system. The CCS is located in the S-IVB instrument unit and is capable of up-data command, down-link telemetry and turn-around ranging. A brief description of each stage of the Saturn V vehicle is presented. The S-IVB is the third powered stage of the Saturn V vehicle and separates from the Apollo spacecraft at a distance of approximately 20,000 nautical miles from the earth. The operational requirements of the CCS during the period up to and including separation are described. This consists of the launch, parking orbit insertion, and injection phases of the Apollo mission. Electrical and environmental specifications of the CCS are presented. The errors in the transponder due to doppler shift and doppler rate are tabulated versus input signal level. Loop bandwidth and suppression factor versus signal level are also included. The CCS omni- and directional antenna system and their switching and coupling networks are described. A system analysis is presented and gain margins are tabulated for the up-data link, down telemetry link and the RF carrier.

INTRODUCTION

The S-IVB command and communication system (CCS) is a phase-coherent receiver-transmitter capable of establishing a communication link between the unified S-band (USB) ground stations and the instrument Unit (IU) of the Saturn V launch vehicle. Specifically, the CCS will: receive and demodulate command up-data for the guidance computers in the IU; transmit pulse code-modulated (PCM) mission control measurements originating in the S-IVB and the IU to the USB ground stations for processing; and coherently retransmit the pseudorandom noise (PRN) range code that is received from the USB ground stations. The CCS physically consists of a transponder, power amplifier, and antenna system. Each of these will be described later.

SATURN V VEHICLE

We will now examine the Saturn V vehicle to establish its relation to the CCS. The Saturn V consists of three powered stages and is shown in Figure 1. The first stage is designated the S-IC and is the stage normally called the booster. The S-IC is built by The Boeing Company. It has a cluster of five F-I engines which burn liquid oxygen and RP-I kerosene. Each F-I

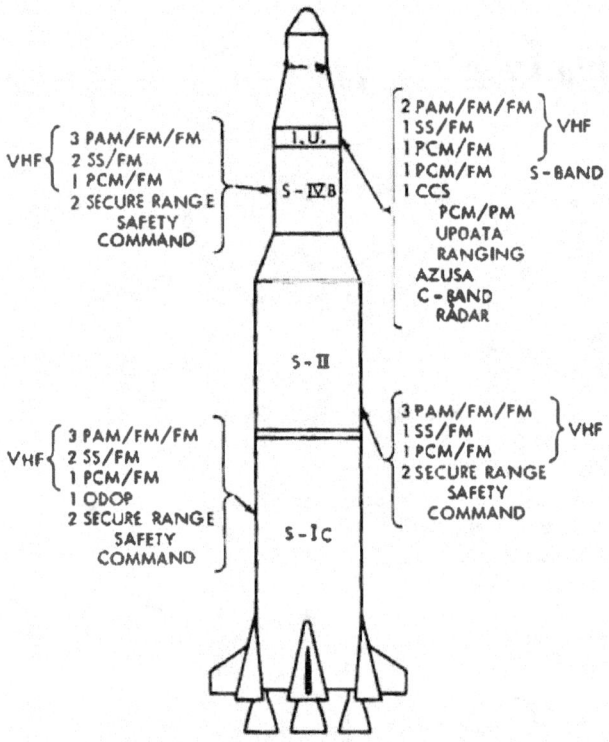

Figure 1—Saturn V instrumentation systems.

engine produces a thrust of one and one-half million pounds for a total S-IC stage thrust of seven and one-half million pounds. The F-I engine was developed by the Rocketdyne Division of North American Aviation.

The second powered stage, designated the S-II, has a cluster of five J-I engines which burn liquid oxygen and liquid hydrogen. Each J-I engine produces a thrust of 200,000 pounds for a stage thrust of one million pounds. The S-II is built by North American Aviation and the J-I engine, by Rocketdyne.

The third powered stage, designated the S-IVB and built by Douglas Aircraft Company, has one J-I engine for a stage thrust of 200,000 pounds. In addition, the S-IVB J-I engine has a restart capability which will be explained later. The CCS is contained in the IU, which is rigidly attached to the S-IVB. The Apollo spacecraft is then attached to the IU.

The total height of the Saturn V vehicle including spacecraft is 110 meters. The S-IC is 42 meters high, the S-II, 25 meters, and the S-IVB, 19 meters. The Saturn V as described has a capability to put a 200,000-pound payload into earth orbit or to thrust a 90,000-pound payload to escape velocity.

A nominal moon mission will consist of a 72 degree launch azimuth. The S-IC will burn first. After the S-IC fuel is expended, the S-IC and S-II stages separate. Separation is achieved by retro-rockets on the S-IC thrusting in the flight direction to slow the S-IC down and ullage rockets on the S-II thrusting opposite to the flight direction to speed the S-II up. After sufficient separation time, the S-II engines are ignited. After depletion of the S-II fuel, the S-II and S-IVB separate in a manner similar to the separation of the S-IC and S-II. Upon separation, the S-IVB engines will ignite and thrust the S-IVB and the spacecraft into a 185-kilometer (100 nautical mile) earth parking orbit.

Once the orbit is established, the S-IVB engines are shut down until an injection opportunity is presented. This may take from one to four orbits. At the time for injection, the S-IVB engines restart and the S-IVB and the spacecraft are injected into a lunar trajectory. The S-IVB provides attitude stabilization for approximately two hours after its second burn and separates from the spacecraft at a distance of approximately 20,000 kilometers from the earth.

CCS OPERATIONAL REQUIREMENTS

The operational requirements of the CCS include command up-data, down-link telemetry in which ranging is desirable but not mandatory. Command up-data is not required on a continuous basis throughout the entire moon mission. The requirement for command varies according to the particular phase of the flight profile. During the launch phase, there is no requirement for up-data, the logic being that the launch is of such short duration that there will not be sufficient imte to analyze and make an action decision on unpredictable situations. The guidance computer is triple-redundant and is programmed for all predictable situations.

After insertion into an earth parking orbit, command up-data will be required on a non-continuous basis for crew and equipment checkout. If all systems are functioning properly, the computers will be corrected if necessary and updated with the data required to accomplish injection.

During the injection burn (second S-IVB burn), command up-data will be required on a continuous basis to take advantage of any injection opportunity. During this period, the ground stations will monitor the vehicle and determine the first injection opportunity. At a distance of approximately 20,000 kilometers, the S-IVB/IU separates from the spacecraft and is no longer needed for mission success. The CCS requirement is then complete.

TELEMETRY

The telemetry (TM) systems on the Saturn V are shown in Figure 1. Each stage has its own self-sufficient telemetry link. During the launch and earth orbit phases, VHF transmitters

Table 1

Down-Link Telemetry VHF-UHF.

Parameters	VHF	UHF
Transmitted Power (20 w)	+43 dbm	+43 dbm
Modulation, Polarization & Cable Gain	-5.1db	-8.1db
Transmitter Antenna Gain	-3db	+12db
Space Loss (260 mc - 20,000 km)	-167db (2.2 gc 20,000 km)	-185.5db
Receiver Antenna Gain	±18db	±44.0db
Receiver Gain	-0.44db	-0.2db
Received Input Power	-114.1 dbm	-94.8db
Noise Density (KTD)	-173.4 dbm/cycle	-173.4 dbm/cycle
Noise Bandwidth (150 kc)	+51.7 db	+51.7db
Receiver Injected Noise (NF)	+4.0db	+2.5db
Noise Power	-117.3 dbm	-119.2 dbm
Actual S/N	+3.2db	+24.4db
Required S/N	+13.0db	+13db
Circuit Margin	-9.3db	+11.4db

are the prime TM source making CCS TM unnecessary at this time. At distances of approximately 10,000 kilometers and greater, VHF transmission is inadequate, primarily due to spacecraft and ground antenna gains. Table 1 is a comparison between the VHF and UHF transmission links. After injection, the CCS transponder and a UHF transmitter become the prime TM links.

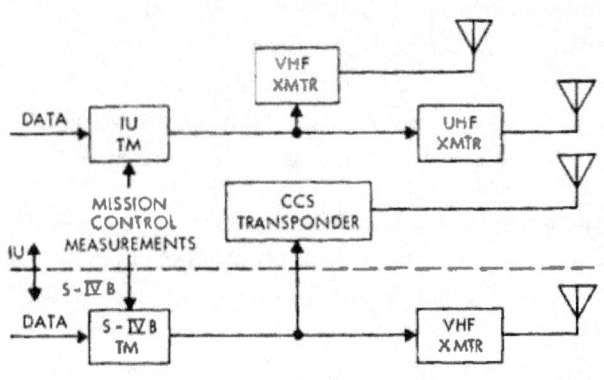

Figure 2—Telemetry systems.

The mission control data from the S-IVB stage and the IU are interconnected such that both sources of data are available at either source. This makes the mission control data double-redundant during the launch and earth orbit phase via a UHF transmitter and the CCS in the IU (Figure 2).

The TM data from the S-IVB and IU have different PCM formats. However, the mission control data occupies the same time slots in both formats making the recovery of the mission control data simple regardless of which receiver is demodulating the PCM stream.

TRACKING

Under normal circumstances, there is no mandatory tracking requirement of the CCS. During launch, tracking is provided by C-Band beacons and surface radar with the Command and Service Module (CSM) transponder providing tracking during other phases. The CCS transponder will have PRN turnaround capability and may be used as a backup to the CSM transponder in case of failure or desire for a cross-check.

CCS TRANSPONDER OPERATION

The CCS transponder consists of a double-conversion phase-coherent receiver and a continuous wave (CW) transmitter in which the received and transmitted signal are integrally related in both frequency and phase. Figure 3 illustrates the operation of the transponder.

The received frequency is designated 221f and is nominally 2101.8 megacycles where f is nominally 9.51 megacycles. The first mixer converts the 221f input to the first intermediate frequency (IF) of 5f using a local oscillator of 216f. The first IF amplifier has a bnadwidth of 4 megacycles and a gain of approximately 100db with an automatic gain control (AGC) range of 120db. After amplification, the 5f signal is converted to f in the second mixer using 6f as a local oscillator. The second IF amplifier is at a frequency of f and has a fixed gain of +54db and a bandwidth of 10 kilocycles, which is determined by a crystal bandpass filter. The second IF output is then limited to a constant signal plus noise (S+N) level of +15dbm before being presented as one input to the loop phase detector.

The other input to the loop phase detector is from a crystal voltage-controlled oscillator (VCO) which has been divided by two. It will be noticed that both local oscillators (216f and 6f) are the 2f VCO multiplied by 108 and 3. This makes the 216f and 6f local oscillators frequency and phase-coherent with the 2f VCO. Since the loop phase detector is phase-comparing the converted input and the divided VCO, any phase difference between the two will cause an output from the phase detector related to the phase difference.

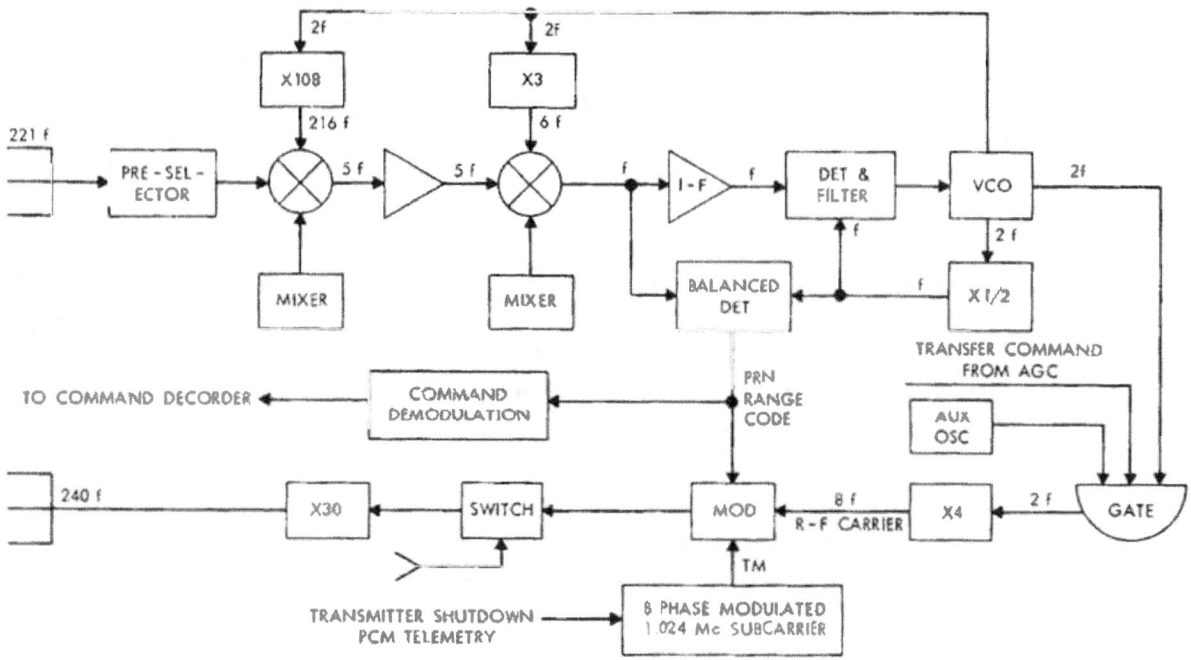

Figure 3—Transponder block diagram (AGC loop not shown).

The output of the phase detector is then processed by the loop filter and applied as a control voltage to the VCO. This control voltage is such that the phase difference between the converted signal and the divided VCO will be reduced. The amount and rate of reduction are dependent upon the loop filter, the limiter, phase detector, and the VCO. This will be covered in the analysis section later. The receiver is now considered in a "locked" condition and the VCO will follow the input signal in phase and frequency and will thus contain the up-link doppler shift.

The command up-data is frequency-modulated (FM) on a 70 kilocycles sub-carrier and 70 kilocycle sub-carrier is then phase-modulated onto the RF carrier. The PRN range code is directly phase-modulated onto the RF carrier. The second mixer has a second output of f, or approximately 9.51 megacycles. The f output is then amplified and limited before wide-band demodulation using the divided VCO. The output of the demodulator will be the 70 kilocycles sub-carrier and the PRN range code at baseband. The demodulated signal is then passed through a 70 kilocycles bandpass filter with a bandwidth of 20 kilocycles to remove the command sub-carrier. The sub-carrier is limited and FM-detected using a pulse-averaging detector. The FM-detected output is video-amplified and presented as an output to be further processed by the command decoder. The baseband PRN range code is one input of the transmitter modulator to be described later.

The transmitter section of the transponder receives 2f from the VCO and multiplies it by 4 to 8f. The 8f is then phase-modulated and multiplied by 30 to 240f. It is thus seen that the received and transmitted frequencies are integrally related by a ratio of 240 to 221. The phase modulator of the transmitter has two modulation inputs, the PRN code and the down-link telemetry sub-carrier. Either modulation input may be separately set to modulate the down-link R-F carrier to 2 radians for a peak modulation capability of 4 radians.

The down-link telemetry sub-carrier consists of a 1.024 megacycle crystal oscillator that is bi-phase modulated by 72 kilocycles nonreturn to zero PCM data which contains mission control measurements. This sub-carrier is compatible with the USB ground stations although its bit rate is different from the Lunar Excursion Module (LEM) and Command and Service Module (CSM) PCM bit rates.

An auxiliary oscillator is included so that when the receiver is unlocked, the auxiliary oscillator will provide a noise-free carrier for the transmitter. Upon lock, the auxiliary oscillator is gated out by the receiver AGC.

AGC is provided by a second phase detector which phase-detects the unlimited second IF and the divided VCO phase shifted 90 degrees. This detector will therefore have an output directly proportional to the amplitude of the input signal. The AGC system is designed to maintain the signal input into the limiter a constant power within 3db over the operating signal dynamics.

Figure 4 is a gain, bandwidth power distribution of the transponder. It relates signal and power levels and signal-to-noise ratios at various points in the transponder under locked and unlocked conditions. In the unlocked condition, all amplifiers are at maximum gain since there is no AGC voltage developed.

Since the CCS and LEM transponders have identical frequencies, there exists a possibility of interference if the two are operated simultaneously at close range. The CCS transponder has the capacity to shut down the transmitter upon external command. This shut down command will be initiated before the LEM transponder is used. The CSM receiver will remain alive so that commands may be received at all times. It is desirable but not essential that the CCS transponder transmitter be turned back on after sufficient separation from the LEM transponder. This will allow turnaround ranging to more completely define the trajectory of the S-IVB after separation from the spacecraft.

Tables 2 and 3 show the main electrical and environmental specifications of the CCS transponder

At the data of publication, power amplifier proposals were under evaluation so the power amplifier will not be described. Its main characteristic is to raise the CCS 500 milliwatts to 20 watts.

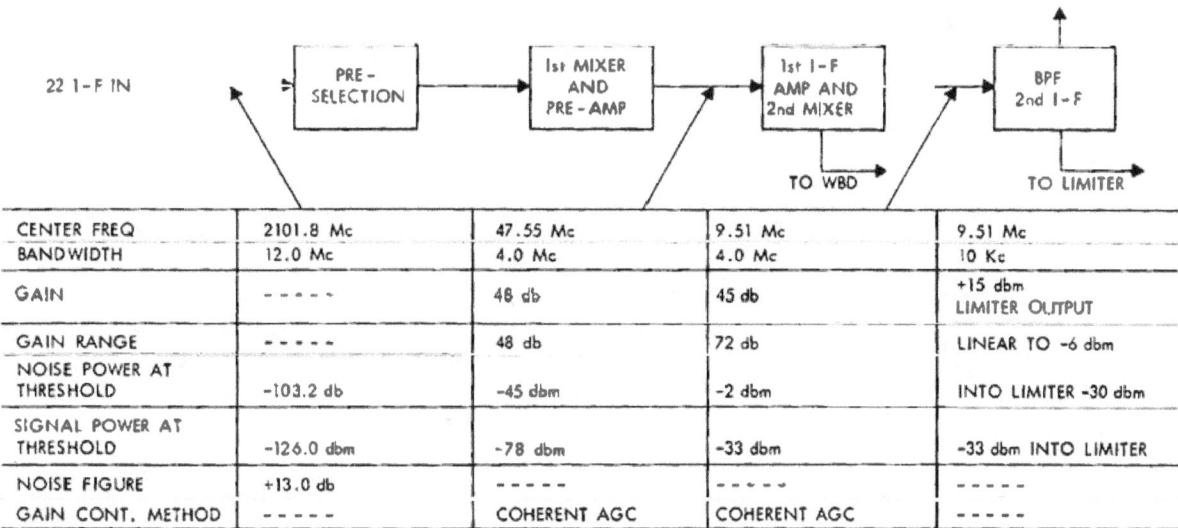

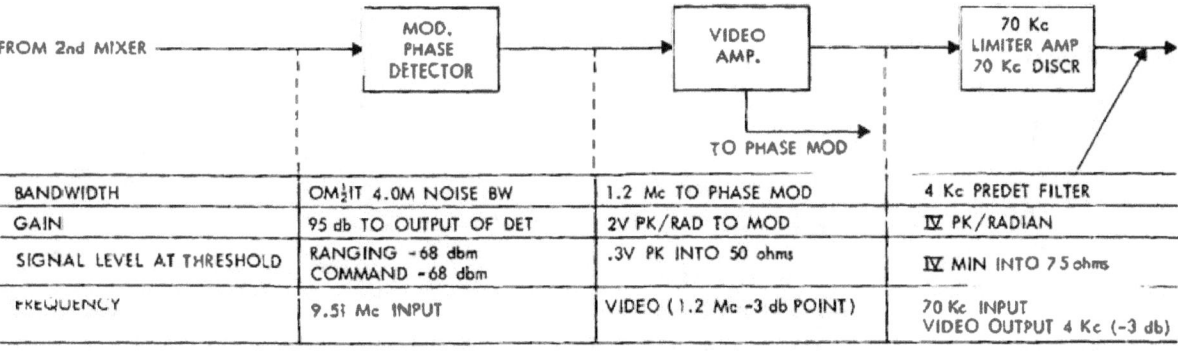

Figure 4—Goin, bandwidth power distribution.

Table 2

CCS Electrical Characteristics.

Frequency Received	2101.8 cps
Frequency Transmitted	2282.5 cps
Offset Ratio	240/221
Noise Figure (including preselector)	13db
Threshold Noise Bandwidth (B_{Lo})	400 cps
Strong Signal Noise Bandwidth	2330 cps (2BL)
Carrier Threshold (S/N = +6db)	-126db
Dynamic Range	100db (120db capability)
Transmitter Power	500 mw
VCO Gain (K_v)	400 cps/volt
Phase Detector Gain (k_c)	0.35 volt/degree
Phase Multiplication (M)	110.5
Tracking Range	±185 kc
Input Power	28 ± 4 VDC @ 32.5 watts
Weight	20.5 lbs.
Size	5.4 x 9.8 x 14.9 inches

Table 3

CCS Environmental Characteristics.

Temperature (Operating and Storage)	-20°C to +85°C
Vibration:	
Random Noise	20-59 cps at 0.04 g^2/cps/S
(5 minutes in each plane)	59-126 cps at 9db/octave
	126-700 cps at 0.4 g^2/cps/S
	700-900 cps at 18db/octave
	900-2000 cps at 0.09 g^2/cps/S
Shock:	50 g for 11 milliseconds with 8 shocks in three perpendicular planes
Acceleration:	100 g for one minute in three perpendicular planes
Vacuum and Pressurization:	Less than 1.0 psi leakage per 24 hour period when pressurized to 15 psig and subjected to a vacuum of 1.5 X 10^{-6} millimeters of mercury.
RF Interference:	MIL-I-6181D

ANTENNA SYSTEM

The antenna system for CCS is being designed to meet three requirements:

1. To provide adequate coverage for pre-launch and launch conditions.

2. To provide the wide angular coverage required at low altitude during parking orbit, where roll maneuvers will be performed

3. To provide adequate coverage of all stations which are optically visible during the period from injection to spacecraft separation.

To meet these requirements, a system consisting of an omnidirectional antenna pair for the receiver and an omnidirectional pair and a directional antenna for the transmitter are being designed. The omnidirectional pair for the receiver will be permanently connected and will maintain essentially omnidirectional angular coverage for the receiver from launch throughout the mission. The transmitter will operate into an omnidirectional pair from launch through injection to a point approximately 10,000 kilometers from the earth. At this distance, the path loss becomes great enough to require additional gain from the vehicle antenna, and the transmitter is then switched to a directional antenna having a gain of about 6 db initially. The beamwidth of this antenna will be adjustable, so that it may be reduced as the range becomes greater, the angle subtended by the earth becomes smaller, and the required gain increases. The orientation of the directive antenna pattern remains fixed with respect to the vehicle, and directing of the pattern toward the ground stations is accomplished by attitude control of the vehicle.

The omnidirectional antenna elements are rectangular half-loops, located diametrically opposite on the vehicle body and driven with equal power. The directional antenna is a 3×3 array of Archimedian spirals, connected through a switchable system of phasing and power-dividing transmission lines. The antenna system is shown in Figure 5. As can be seen, the CCS transponder and the UHF TM transmitter may be switched to share an omniantenna or to their own separate directional antenna as described above.

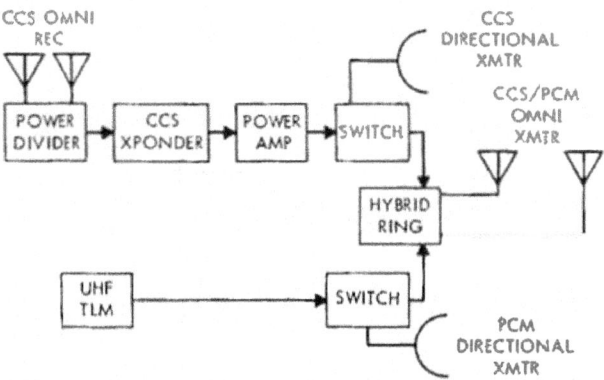

Figure 6—Antenna system.

PHASE LOCK LOOP ANALYSIS

The criteria that determined the CCS RF carrier-tracking loop design was the required 20,000 kilometer communication link range and the received frequency dynamics caused by a combination of vehicle velocity, acceleration, and ground station acquisition procedures.

The following standard phase-lock equations were used to calculate the loop errors and characteristics under the combination of signal strength and frequency dynamics listed with the equations:

1. Threshold loop bandwidth, B_{L0}

$$B_{L0} = \left(\frac{405}{400} \dot{f}\right)^{1/2}; \quad B_{L0} = 400 \text{ cps}.$$

2. Loop bandwidth, B_L

$$B_L = B_{L0}\left(\frac{1}{3} + \frac{2\alpha}{3\alpha_0}\right).$$

3. Loop noise bandwidth, B_N

$$B_N = 2B_L$$

4. Signal suppression factor, α

$$\alpha = \frac{1}{\left[1 + \frac{4}{\pi}\left(\frac{N}{S}\right)\right]^{1/2}} \quad \frac{S}{N} < 10\text{db}$$

$\alpha = \alpha_0$ at threshold where

n = noise into the limiter

s = signal into the limiter.

5. Phase error due to acceleration, θ_A

$$\theta_A = \left(\frac{405}{4B_L^2}\right)(\dot{f})\left(\frac{\alpha_0}{\alpha}\right)$$

where

$\dot{f}$ = doppler frequency offset due to acceleration.

6. Phase error due to velocity, θ_V

$$\theta_V = \frac{(\text{maximum doppler frequency})}{(K_d)(K_{vco})(m)(\alpha)} \times 360°$$

where phase detector gain = .35 volt/cycle, VCO gain = 400 cycles/volt

M = multiplication ratio = 110.5 .

7. Available tracking loop gain, G_A

$$G_A = (360)(K_d)(K_{vco})(m)(\alpha).$$

8. Required tracking loop gain, G_R

$$G_R = \frac{\text{maximum doppler frequency offset}}{\text{threshold phase error at detector}}$$

$G_A > G_R$ if the loop is to be functional.

The two doppler dynamic conditions considered are:

$$\theta_A + \theta_V \leq \pm 24 \text{ degrees}$$

$$B_{LO} = 400 \text{ cps}$$

$$\text{Case I} \begin{cases} \dot{f}_1 \leq 63 \text{ KCS}^2 \\ fd_1 \leq \pm 181.5 \text{ KCS} \end{cases}$$

$$\text{Case II} \begin{cases} \dot{f}_2 \leq 35 \text{ KCS}^2 \\ f_{d2} \leq \pm 90 \text{ KCS} \end{cases}$$

Table 4 is the tabulated velocity and acceleration errors calculated to be produced by the transponder under case I doppler dynamics. Table 5 is the calculated errors in the transponder under case II doppler dynamics. Figure 6 shows the phase errors tabulated for the two doppler

Table 4.

Tabulated Velocity and Acceleration Errors Produced by Transponder Under Case I Doppler Dynamics. $f_d + 181.5$ KCS; $\dot{f} = 63$ KCS^2

α	θ_A	θ_V	θ_L	INPUT SIGNAL LEVEL (dbm)	PRE. DET. S/N (db)	AVAILABLE LOOP GAIN	REQUIRED LOOP GAIN
.242	39.8°	46.5°	400	-132.0	-11.0	1.339 x 10⁶	2.573 x 10⁶
.260	37.0°	43.8°	420	-131.4	-10.4	1.441 x 10⁶	2.573 x 10⁶
.280	34.4°	42.0°	442	-130.7	- 9.7	1.550 x 10⁶	2.573 x 10⁶
.300	32.1°	39.0°	464	-130.0	- 9.0	1.662 x 10⁶	2.573 x 10⁶
.320	30.1°	37.0°	486	-129.4	- 8.4	1.770 x 10⁶	2.573 x 10⁶
.340	28.3°	34.5°	508	-128.4	- 7.4	1.880 x 10⁶	2.573 x 10⁶
.360	26.8°	32.5°	530	-128.2	- 7.2	1.996 x 10⁶	2.573 x 10⁶
.380	25.3°	30.5°	552	-127.7	- 6.7	2.110 x 10⁶	2.573 x 10⁶
.400	24.1°	29.1°	574	-127.1	- 6.1	2.220 x 10⁶	2.573 x 10⁶
.450	21.3°	26.1°	629	-125.5	- 4.5	2.448 x 10⁶	2.573 x 10⁶
.500	19.3°	23.4°	684	-124.7	- 3.7	2.775 x 10⁶	2.573 x 10⁶
.550	17.6°	21.2°	744	-123.3	- 2.3	3.050 x 10⁶	2.573 x 10⁶
.600	16.0°	19.5°	798	-122.4	- 1.4	3.330 x 10⁶	2.573 x 10⁶
.650	14.8°	18.0°	853	-121.3	- 0.3	3.600 x 10⁶	2.573 x 10⁶
.700	13.8°	16.8°	908	-120.1	+ 0.9	3.870 x 10⁶	2.573 x 10⁶
.750	12.9°	15.6°	963	-118.9	+ 2.1	4.156 x 10⁶	2.573 x 10⁶
.800	12.0°	14.6°	1018	-117.4	+ 3.6	4.440 x 10⁶	2.573 x 10⁶
.850	11.3°	13.8°	1073	-115.7	+ 5.3	4.710 x 10⁶	2.573 x 10⁶
.900	10.7°	13.0°	1128	-113.6	+ 7.4	4.990 x 10⁶	2.573 x 10⁶
.950	10.1°	12.3°	1183	-109.8	+11.2	5.260 x 10⁶	2.573 x 10⁶
.980	9.8°	11.3°	1138	-106.0	+15.0	5.540 x 10⁶	2.573 x 10⁶

Table 5

Tabulated Velocity and Acceleration Errors Produced by Transponder Under Case II Doppler Dynamics.
$f_d = \pm 90$ KCS: $\dot{f} = 35$ KCS^2

α	θ_A	θ_L	B_L	INPUT SIGNAL LEVEL (dbm)	PRE. DET. S/N RATIO (db)	AVAILABLE LOOP GAIN
.242	22.1°	24.4°	400	-132.0	-11.0	1.339 x 10^6
.260	20.5°	22.5°	420	-131.4	-10.4	1.441 x 10^6
.280	19.1°	21.0°	442	-130.7	- 9.7	1.550 x 10^6
.300	17.8°	19.5°	464	-130.0	- 9.0	1.662 x 10^6
.320	16.7°	18.1°	486	-129.4	- 8.4	1.770 x 10^6
.340	15.7°	17.1°	508	-128.4	- 7.4	1.880 x 10^6
.360	14.9°	16.1°	530	-128.2	- 7.2	1.996 x 10^6
.380	14.0°	15.4°	552	-127.7	- 6.7	2.110 x 10^6
.400	13.4°	14.6°	574	-127.1	- 6.1	2.220 x 10^6
.450	12.0°	13.1°	629	-125.5	- 4.5	2.448 x 10^6
.500	10.7°	11.7°	684	-124.7	- 3.7	2.775 x 10^6
.550	9.8°	10.7°	744	-123.3	- 2.3	3.050 x 10^6
.600	8.9°	9.75°	798	-122.4	- 1.4	3.330 x 10^6
.650	8.2°	9.0°	853	-121.3	- 0.3	3.600 x 10^6
.700	7.6°	8.4°	908	-120.1	+ 0.9	3.870 x 10^6
.750	7.1°	7.7°	963	-118.9	+ 2.1	4.156 x 10^6
.800	6.7°	7.3°	1018	-117.4	+ 3.6	4.440 x 10^6
.850	6.3°	6.9°	1073	-115.7	+ 5.3	4.710 x 10^6
.900	5.9°	6.5°	1128	-113.6	+ 7.4	4.990 x 10^6
.950	5.6°	6.15°	1183	-109.8	+11.2	5.260 x 10^6
.980	5.4°	5.83°	1238	-106.0	+15.0	5.540 x 10^6

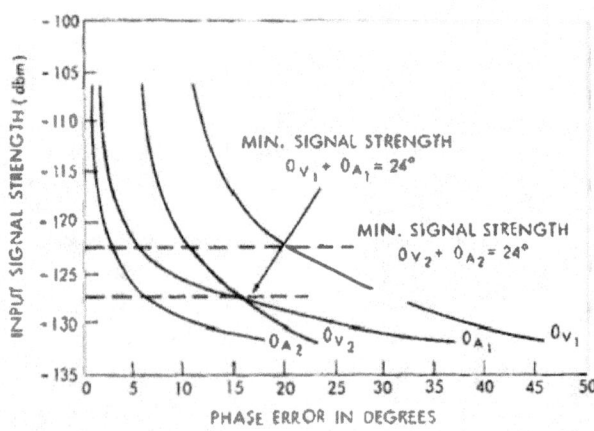

Figure 6—Phase errors for doppler dynamic conditions versus input signal level where O_{V_1} and O_{A_1} represent a frequency deviation (fd) of ±181.5 kilocycles and a center frequency (f) of 63 kilocycles^2, and where O_{V_2} and O_{A_2} represent an fd of ±90 kilocycles and an f of 35 kilocycles^2.

dynamic conditions plotted versus transponder input signal level. The errors calculated in Tables 4 and 5 used the approximate formula for the suppression factor, alpha. The approximate formula was used to make the error calculations a simple computer run. The approximate formula actually produces a greater phase error as shown by Figure 7. Figure 8 is the exact and actual suppression factor plotted versus signal level. In all cases the approximate suppression factor is equal to or less than the exact suppression factor. Since error due to velocity is inversely related to the suppression factor, the approximate solution will yield a greater error than the exact solution under the same doppler dynamics.

SYSTEM MARGINS

The system margin calculations were performed for the up-link command, range code and carrier and for the down link telemetry, range code and carrier.

Standard communications equations were used to arrive at the circuit margins.

1. Assumptions:

   a. These calculations were based upon the sub-carrier frequencies being 1.024 megacycles for telemetry and 70 kilocycles for the command. The modulation indices were taken to be 1.22 for both sub-carriers and 0.6 for the range code

   b. The sub-carrier demodulators were assumed to be standard Foster-Seely type discriminators.

   c. The circuit losses that were estimated are so marked.

   d. Maximum range is 27,000 nautical miles or 48,600 kilometers.

2. Equations:

   a. Carrier power

   $$PC = \cos^2 M_1 J_0^2 M_2$$

   $M_1 = .6$ rad.

   $M_2 = 1.22$ rad.

   b. Ranging power

   $$PR = \sin^2 M_1 J_0^2 M_2$$

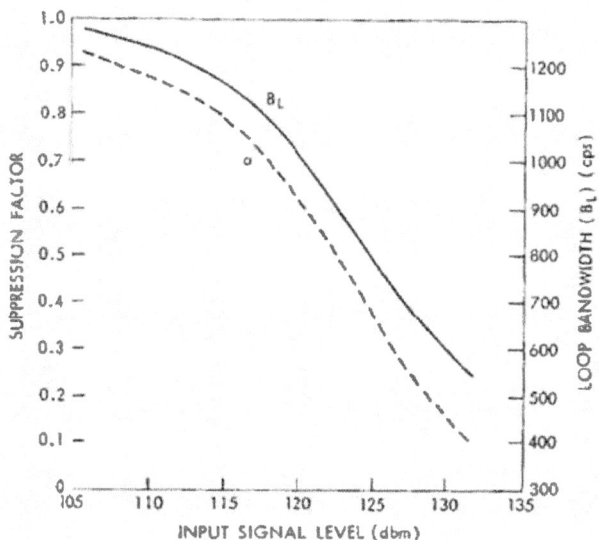

Figure 7—Loop bandwidth and suppression factor versus input signal level, where $\alpha = (1 + 4/\eta \; N/S)^{-1/2}$, $B_L = B_{LO}(1/3 + 2\sqrt{3}\alpha_0)$, $\alpha_0 = 0.242$, and $B_{LO} = 400$ cps.

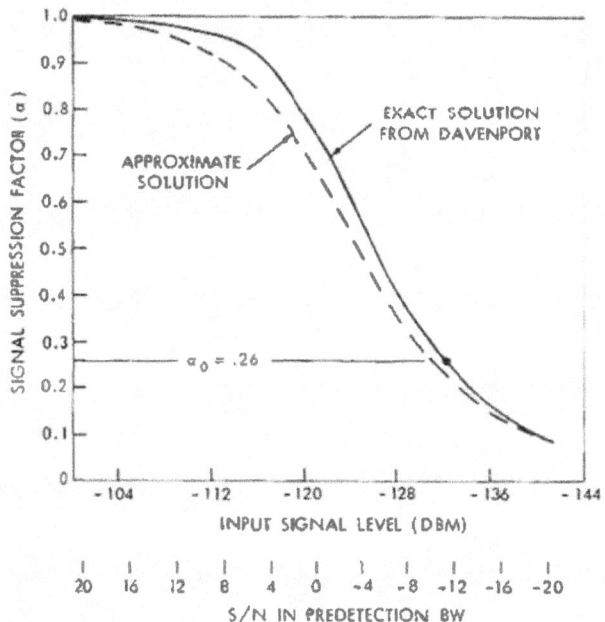

Figure 8—Signal suppression versus input signal level and S/N in predetection BW, when $\alpha = \sqrt{1/(1 + 4/\eta \; N/S)}$ (approximate) where N equals KTBF (noise power in predetection BW), S equals signal power, B_{LO} equals 400 cps, B_{IF} equals 10 kilocycles, and F_n equals 13db.

c. Sub-carrier power

$$PSC = 2(\cos^2 M_1 \, J_1^2 M_2)$$

d. Modulation loss carrier

$$MLC = 10 \log_{10} \frac{PT}{PC}$$

e. Modulation loss, ranging

$$MLR = 10 \log_{10} \frac{PT}{PR}$$

f. Modulation loss, sub-carrier

$$MLSC = 10 \log_{10} \frac{PT}{PSC}$$

g. Required S/N predetection

$$S/N \text{ Req.} = \left(\frac{3}{2}\right)\left(\frac{BW_{sc}}{fm_{sc}}\right)\left(B_{sc}^2\right)\left(S/N \text{ Input}\right)$$

h. Noise spectral density

$$N_B = KT_e B; \quad T_e = 290°K \text{ per cycle across 50 ohms}$$

TYPICAL ACQUISITION PROCEDURE

by
R. H. Newman
Goddard Space Flight Center

ABSTRACT

The problem of acquisition is considered from the ground operational aspect during earth orbit for each of the following tasks: (1) Spacecraft illumination, (2) Two-way RF acquisition, (3) Data acquisition, (4) Range acquisition, (5) Angle acquisition, and (6) Station-to-station hand-over.

The time required to perform each of these tasks, total acquisition summary chart, and ground acquisition aids are discussed.

INTRODUCTION

The typical Apollo Unified S-Band acquisition procedure presented consists of the following tasks:

1. Spacecraft illumination - to position the ground antenna properly for spacecraft antenna reception of the ground transmitted up-link.

2. Two-way RF acquisition - to obtain a two-way RF lock between the ground transmitter, spacecraft transponder and ground receiver.

3. Data acquisition - to lock the ground demodulators and the pulse-code modulation (PCM) decommutator after RF lock.

4. Range acquisition - to provide the network with a range reading.

5. Angle acquisition - to provide the network with accurate angle information.

6. Station-to-station handover - to handover RF lock from one station to another without losing down-data or voice from the spacecraft.

SPACECRAFT ILLUMINATION

Spacecraft illumination depends on the position of the spacecraft antenna and the pointing accuracy of the ground antenna. Here we will assume that the spacecraft antenna is properly oriented and only the task of positioning the ground antenna will be considered.

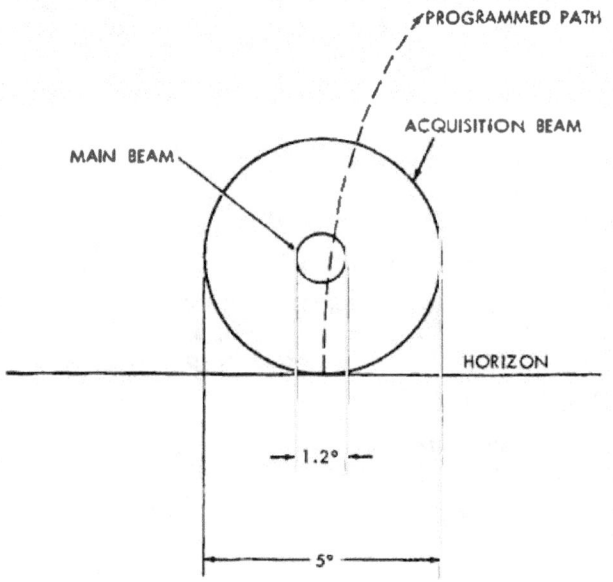

Figure 1—Ground transmit and receive pattern in space.

Figure 1 is an attempt to depict the ground transmit and receive pattern in space. The acquisition antenna actually has a 10-degree receive beam but is not shown because the spacecraft would always be within the acquisition beam shown for a normal mission. If other than a normal condition exists and a signal is not observed in the 10-degree acquisition beam, the antenna control will be changed to one of its search modes.

During the initial acquisition phase, the antenna program track mode will be utilized. In this mode, the antenna will be positioned in real-time in accordance with a predicted spacecraft flight profile computed prior to launch during the launch phase. The main beam will be positioned at the horizon initially and will follow the programmed path. The antenna programmer can be up-dated as required prior to launch by insertion of time and angle offsets. After the launch phase, up-dated predicts will be sent to all stations if required. If necessary, the antenna may be deviated from its predicted path by manually adding X, Y and time offsets while in this mode.

It should be noted that if the spacecraft is within the acquisition beam shown, the spacecraft and both ground receiving systems will be illuminated at the RF horizon.

TWO-WAY RF ACQUISITION

The unified S-band two-way RF acquisition can be achieved faster and more reliably by locking the up-link first, because the ground receiver cannot maintain a down-link lock when the spacecraft receiver acquires the up-link. A method of achieving up-link lock first is given below:

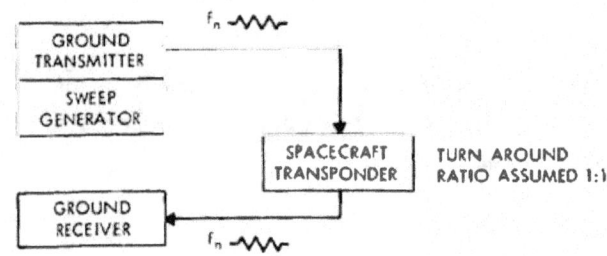

Figure 2—Diagram illustrating procedure for two-way RF acquisition.

Sweep the ground transmitter over a frequency range sufficient to lock the spacecraft receiver. After the up-link is locked, the spacecraft transmitter frequency follows the ground transmitter sweep. The sweep continues until the ground receiver locks to the spacecraft transmitter. Two-way lock is indicated by observing that the ground receiver voltage controlled oscillator (VCO) is following the ground transmitter sweep. Figure 2 illustrates this procedure. The sweep about the nominal frequency (F_n) should have a range sufficient to cover the frequency uncertainty associated with both the spacecraft and ground receiver. This procedure would require modification to accommodate the transponder in earth orbit due to doppler shift.

Figure 3 shows the worse case (overhead pass) of doppler shift anticipated in earth orbit. The doppler, as seen by the spacecraft, varies from plus 60 kilocycles at the acquisition horizon to minus 60 kilocycles at the loss-of-lock horizon. The ground receiver, if the spacecraft is locked to the ground transmitter, will see two-way doppler shift or twice that observed by the spacecraft receiver.

Therefore, a sweep range covering the total doppler and receiver frequency uncertainty range or the proper selection of frequency offsets for the ground receiver and transmitter plus a sweep will allow the ground system to acquire the spacecraft at any time during a pass. Figure 4 shows an example of how acquisition can be accomplished at the horizon for an overhead pass using receiver and transmitter bias (frequency offsets).

By biasing the ground transmitter f_n minus 60 kilocycles to counteract the f_n plus 60 kilocycles doppler, the spacecraft appears to receive a signal at its nominal frequency.

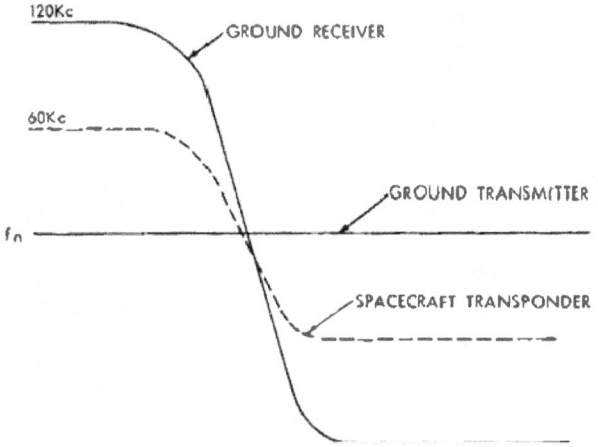

Figure 3—Worse case (overhead pass) doppler anticipated in earth orbit.

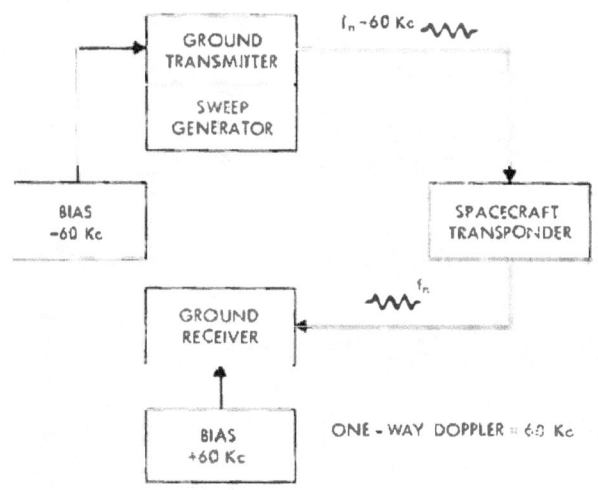

Figure 4—Procedure for acquisition at horizon for an overhead pass using receiver and transmitter bias.

If the sweep is adequate to cover all frequency uncertainties of both receivers, the spacecraft should acquire the up-link. The spacecraft output will sweep in step with the ground transmitter after lock about the nominal frequency. The ground receiver has an f_n plus 60 kilocycles bias to offset the f_n plus 60 kilocycles doppler. Therefore, this combination (frequency offsets plus sweep) will allow the ground system to acquire lock at anytime during the pass.

In order to accommodate this feature and also to provide the capability for reliable acquisition procedures, the ground system flexibility shown in Table 1 is being provided.

Automatic bias decay is necessary so that the ground receiving and transmitting system can operate at the nominal frequency. This minimizes the stress in all receiver loops and

Table 1

Ground Acquisition System Capabilities.

Variable sweep range
Variable sweep rate
Thirteen selectable bias levels
Variable bias
One operation selects proper bias for ground transmitter and both receivers
Automatic bias decay

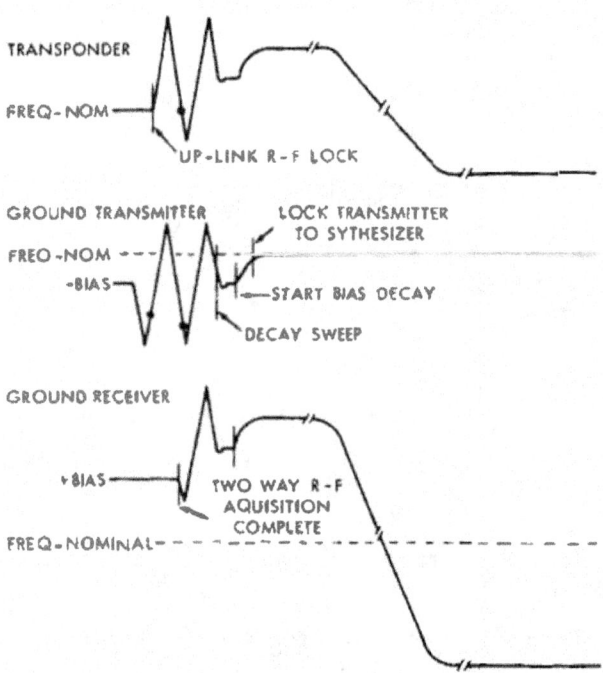

Figure 5—Typical RF acquisition sweep procedure.

allows the ground transmitter to return to a stable frequency necessary for accurate ranging information (Figure 5).

The acquisition system also has the capability to sweep the ground receivers for down-link locks only, if required.

Figure 5 represents a typical sweep and bias system that could be utilized for an overhead pass during horizon acquisition in earth orbit. It is drawn to scale and would require three selectable bias levels to accommodate reacquisition during all phases of an overhead pass. Biasing is utilized in order to minimize transponder stress during acquisition. The sweep characteristics are:

1. Ground transmitter bias f_n - 30 kilocycles
2. Ground receiver bias f_n + 60 kilocycles
3. Sweep range f_n + or - 60 kilocycles
4. Sweep rate -35 kilocycles per second/second

The following should be noted when observing the sweep diagram:

1. At the point where two-way acquisition is indicated, down-data is being received and up-link modulation should be initiated.
2. Two-way lock is recognized on the ground by noting the ground receiver static phase error following the transmitter sweep. The operator can also note the spacecraft telemetered AGC and demodulator in-lock indicators.
3. The sweep decay function is normally initiated immediately, although the diagram shows the sweep continuing.
4. A time lag is shown between sweep decay and bias decay. During normal operation the sweep decay is manual but the bias decay would automatically start as soon as the sweep decay reaches zero.
5. The point shown, lock transmitter to synthesizer, is a manual operation that is initiated after bias decay (push-button operation).

6. After the synthesizer and exciter VCO are locked, range rate is available and range acquisition can be initiated.

DATA ACQUISITION

Data acquisition is the task required to lock the data demodulators and PCM decommutator. This operation is completely automatic. The time required for each function is given in Table 2. The PCM decommutator lock time is the time required for bit-error rates of 1×10^{-4} or better.

Table 2
Demodulator Lock Times.

Voice demodulator	1.0 sec
Telemetry demodulator	1.0 sec
Pcm decom (51.2 KBS)	0.16 sec
(1.6 KBS)	3.7 sec
On-site data processor	0.05 sec

RANGE ACQUISITION

Range acquisition should be attempted only after the exciter has been locked to the synthesizer (time standard) as indicated in Table 3. The exciter control operator should initiate range modulation on the up-link. Immediately after the range code modulation is initiated, the range receiver will automatically lock. Next, the range subsystem operator initiates code acquisition, which is automatic. The range reading is then sent to the network equipment automatically.

Table 3
Range Acquisition Procedure.

Initiate range modulation (manual).
Lock range receiver (automatic).
Initiate code acquisition (manual).
Acquire code (automatic).

ANGLE ACQUISITION

Accurate angle information is not considered available until RF lock has been achieved and the antenna controlled in the auto-track mode.

If it is assumed the spacecraft is illuminated at the horizon with the acquisition and main beam, two-way lock should occur in both receivers (acquisition and main) within one sweep period. As soon as lock is observed in both receivers, signal levels should be compared in both receivers to ascertain that the main receiver channel is not locked on a side lobe. Table 4 gives the time required to accomplish this function.

The antenna servo control operator must decide when to switch from the program-track mode to the auto-track mode. The effects of multipath at low elevation angles will determine when the switch can be made.

It is anticipated that the auto-track mode will be feasible when the antenna is approximately two to five degrees above the horizon. Multipath will be the only constraining factor after RF lock before this mode can be initiated.

STATION-TO-STATION HAND-OVER

The task required for station-to-station hand-over is that of transferring up-link lock from one station to another without down-link loss of lock and with a minimum loss of up-data. This can be accomplished during mission phases that have sufficient overlapping RF antenna coverage to allow that:

1. The up-range station (station A) has a two-way lock with the spacecraft transponder.

2. The down-range station (station B) acquires the down-link in both receiver channels (acquisition and main receiver).

3. Initiation of up-link transmission from station B at a predetermined time.

4. Transfer lock of spacecraft receiver to station B up-link transmission and terminate transmission from station A.

Table 4

Typical Worst Case Time After Spacecraft Illumination Before Angle Data Available.*

One transmitter sweep period	—
Switch to auto-track (decision time included)	3.0 sec
Antenna slew time	1.0 sec
Antenna settling time	1.5 sec
Total sweep period	+5.5 sec

*Multipath effects not considered.

The task required to achieve a down-link lock with stations A and B simultaneously is relatively simple. The main problem is that of making the spacecraft transponder change from one up-link to another without switching to the auxillary oscillator mode which is controlled by AGC. If the transponder did switch to the auxillary oscillator mode, both station A and station B would lose down-link lock because of the sudden discrete frequency change caused by the switchover. Therefore, station B must place its up-link in the transponder receiver pass band before station A terminates transmission.

The step-by-step procedures listed below will assume station A has a two-way RF lock and station B must acquire without loss of down-data and with a minimum loss of up-data during hand-over. Sufficient overlapping RF coverage between stations to accommodate hand-over will also be assumed.

Step 1. (a) Antenna in the program track mode.

(b) Exciter VCO manually set to the spacecraft nominal frequency minus the sum of stations A and B predicted one-way doppler.

(c) Receiver (acquisition and main), VCO set to spacecraft nominal frequency minus station A's predicted one-way doppler plus station B's predicted one-way doppler.

(d) Transmission through the main antenna.

(e) Sweep characteristics set for a normal initial acquisition procedure.

Step 2. (a) Lock the acquisition and main receiver. Lock should occur automatically without sweep at the RF horizon.

(b) If a down-link RF lock is not established within a predetermined time, initiate a normal acquisition procedure (push button operation).

Step 3. (a) Observe the signal power in both receivers and ascertain that the main receiver channel is not locked on a sidelobe.

(b) Ascertain all demodulators locked.

(c) Switch antenna to the auto-track mode as soon as possible (when multipath permits).

Step 4. (a) At a predetermined time initiate the up-link without modulation and with the transmitter set for the maximum power out.

(b) An audible beat should be observed at each station as soon as the up-link is initiated. Station B changes the exciter VCO frequency to obtain a zero beat. Station A ceases transmission and station B initiates up-link modulation as soon as a zero beat is observed. (Beat is caused by the presence of two carriers in the transponder pass band that modulates the down-link.)

(c) Initiate an initial acquisition procedure if an audible beat is not observed as the up-link is energized.

Step 5. (a) Offset exciter VCO to spacecraft nominal frequency. Switch to synthesizer control when the exciter VCO static phase error is approximately zero.

(b) Lock range clock receiver.

Step 6. (a) Initiate range code acquisition.

(b) Continue to monitor spacecraft and ground receiver static phase error and AGC throughout pass.

The above step-by-step procedure for hand-over was presented in detail to demonstrate the total task required for a ground station. It should be noted that this procedure is applicable to a normal horizon acquisition except for the bias level requirements and the fact that the exciter VCO would be utilized in a sweep mode.

The manual mode presented is recommended for hand-over because:

1. If the down-link is locked utilizing the ground receiver sweep mode, the possibility of locking one of the turned around up-link subcarriers (30 and 70 kilocycles) is better than that of acquiring the carrier. This can be recognized but a time-consuming procedure would be required to ascertain carrier lock. The manual mode appears feasible because the spacecraft VCO is being controlled by station A's time standard. Therefore, the down-link frequency and transponder receiver pass band can be determined very accurately during this controlled period and the maximum error will be that contained in prior information given to each station which should be within the spacecraft and ground receiver pass bands.

2. The transfer of up-link is presented as a manual operation because it appears to be the fastest way to reliably hand-over the up-links. Utilizing a sweep mode at station B will

cause transfer of lock when the signal at station B is stronger than that of station A as the sweep passes through the transponder receiver pass band. However, the question is, when is the signal stronger?

ESTIMATED ACQUISITION TIME CHART

Table 5 gives an estimation of the time required to accomplish the different tasks discussed. The time shown for each function represents the time required from the RF horizon considering worst case conditions.

Table 5

Estimated Acquisition Time Chart.*

Function	Comments	Time after S/C illumination
(1) S/C illumination	At the RF horizon with transmission through main antenna	0.0 sec
(2) Two-way RF lock	Sweep exciter VCO (acquisition and main receiver locked)	8.0 sec
(3) All demods locked	Automatic	9.0 sec
(4) Initiate up-link modulation	2 sec recognition time allowed after function (2)	10.0 sec
(5) Decay sweep	Manual operation	14.0 sec
(6) Offset exciter VCO to nominal frequency	Automatic function after sweep decay	16.0 sec
(7) Lock exciter VCO to synthesizer	Manual operation	17.0 sec
(8) Range rate	Available only when the exciter VCO is controlled by the synthesizer	17.0 sec
(9) Range code acquisition		20.0 sec
(10) Angle acquisition	Auto track mode and antenna settled. Available as soon as multipath problems cease	—

*Worst case without considering the effects of multipath.

IMPACT OF APOLLO UNIFIED S-BAND SYSTEM ON NASA COMMUNICATIONS NETWORK

by
W. Dickinson
Goddard Space Flight Center

ABSTRACT

The engineering aspects of the data communications buildup required of the NASA Communications Network in order to support the Apollo Unified S-Band (USB) System are presented. USB system needs for a worldwide high-speed data transmission capability in terms of sites, circuit facilities, data rates, and transmission system reliability, flexibility, and restorability are outlined. The specific hardware being used to build the system is described, including HF radio and wireline data modems, data error-detection and correction equipment, data error-detection terminals, data quality monitors, audio (circuit) and digital (data) switching systems and data technical control and test systems. Performance design goals of the final network and the capability for expansion of the persent network are discussed.

INTRODUCTION

The NASA Communications Network (NASCOM) provides operational communication lines and facilities carrying mission-related information for the conduct of NASA programs and projects. These lines interconnect such facilities as NASA's foreign and domestic tracking, telemetry, and command control sites; launch areas; test sites; and mission control centers. The present NASCOM network consists of approximately 600,000 route-miles of facilities including voice, teletype, and high-speed data circuits.

The Unified S-Band (USB) System which NASCOM will support requires transmission of three basic data streams: telemetry and tracking information in an inward or site-to-mission control center direction and command information in an outward or control center-to-site direction. The quantity of data handled via the interconnecting communications network is such that existing teletype facilities cannot handle the volume of data required as has been done during support of the Mercury and Gemini projects. Therefore, it has been necessary to design and engineer a world-wide network of high-speed data transmission facilities interconnecting nearly 30 overseas locations and five locations in continental North America, including tracking sites, communications switching centers, prime communications carrier terminal locations, and mission control centers.

These facilities consist of voice bandwidth channels derived from combinations of landline, microwave radio, submarine cable, and high frequency radio communications systems which in

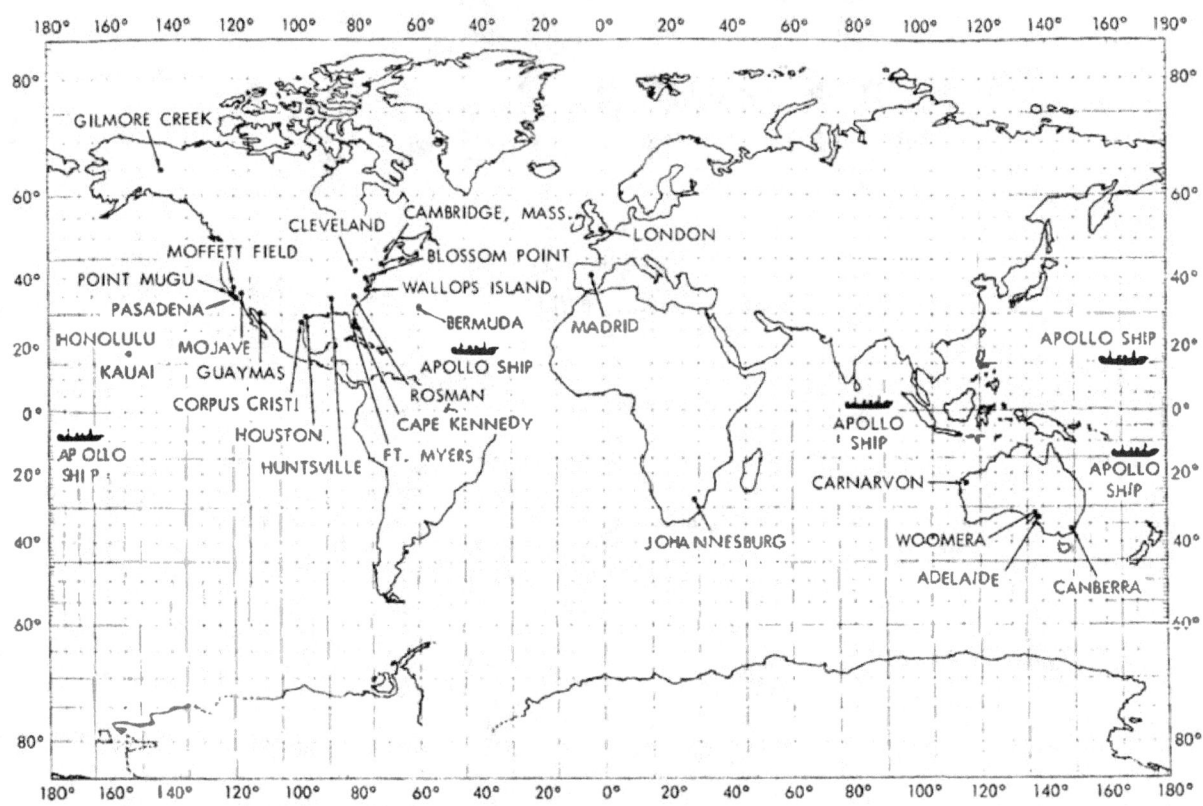

Figure 1—High-speed and wideband networks, present and planned.

most cases are leased, but in some cases are government-owned. Most of these circuits have required considerable special treatment in order to condition them properly for efficient digital data service. In this connection, since a large portion of the overseas facilities are leased, it has been necessary to maintain close coordination and free exchange of information with communications carriers.

The resulting data transmission network provides a minimum of two full-duplex data circuits to every USB site. Operation at 600 or 1200 bits per second is possible to all sites and operation at 2400 bits per second can be achieved to some. By means of diverse circuit routing, full-period channel performance monitoring, end-to-end error detection, and a concept of center-to-center and center-to-subcenter circuit trunking with subcenter-to-site tributaries, a highly reliable, rapidly restorable, and extremely flexible data communications network has been designed. This is absolutely essential in order to support a concept of fully remoted site operations to sites where no on-site flight controllers are employed.

IMPLEMENTATION

Communication Subcenters

The first step in the expansion of the NASCOM network was to consider communications subcenters in areas of the world which would provide convenient and efficient concentration of a number of data circuits. Four such subcenters were established, located in Canberra, Australia; Honolulu, Hawaii; London, England; and Madrid, Spain. These exercise control over facilities in the Australian/Indian Ocean, the Pacific, the English/African, and the European/Atlantic areas respectively.

In Honolulu and London it was only necessary to expand existing subcenters; while in Canberra and Madrid completely new facilities were designed and are being installed. All are operated by the foreign national licensed communications carriers having jurisdiction in these respective areas. The prime communications center located at the Goddard Space Flight Center controls all trunks to the London and Honolulu subcenters; trunks to other centers located at Cape Kennedy, Florida and at Houston, Texas; and tributaries to sites in the continental North American area. The normal configuration for such circuits is to route through the Goddard Space Flight Center for facilities control purposes and then to terminate at the appropriate mission control center (MCC), such as the MCC at Houston for USB system support.

In addition to providing more effective control, the communication subcenters also provide a location for reconstruction of data as necessary by means of regenerative repeating equipment on circuits which pass through the subcenter. This permitted a solution to a technical and operational problem encountered during the design phase of this system — that of maintaining adequate control of the end-to-end equalization characteristics of the circuits in order that reliable operation at rates up to 2400 bits per second might be achieved.

In the initial investigation of this problem, it was discovered that because of the variety of facilities employed and the multiplicity of controlling agencies involved, no single licensed carrier, domestic or foreign, would assume responsibility for the end-to-end characteristics of a circuit — from Carnarvon, Australia to the Goddard Space Flight Center, for example. Also, if each carrier is responsible for only a segment of the total circuit, the equalization specifications for each segment would be so severely stringent that it would be doubtful they could be met initially and nearly impossible to maintain them under operational conditions. Consequently, the approach of data regeneration at all subcenters was adopted, with the result that the longest circuits or circuit legs which must meet end-to-end equalization specifications are between sites and subcenters, between subcenters, or between subcenters and the Goddard Space Flight Center.

In nearly all cases, this segmentation places the entire trunk or tributary circuit within the jurisdiction of a single leasing agency and maintenance of high-quality service is more readily assured. Another very important advantage is that any combination of trunks, tributaries, or alternate routes can be interconnected to provide service between any two points in the network, the only limiting factor being the additive accumulation of bit errors from each leg of the composite circuit. This flexibility obviously would not be available if regeneration were not employed during an attempt to maintain a quite stringent end-to-end circuit equalization characteristic.

Equipment Employed in Implementation

There are three principal classes of equipment employed in the implementation of this system: data modems (modulator/demodulators), which serve to condition the data suitably for transmission via the communications channel; error control equipment, which in some cases only detects, but in other cases also corrects, bit errors; and technical facilities control equipment which provides appropriate test gear and patching access to the circuits to permit proper maintenance.

Data Modems

There are two types of data modems used for USB System support — a wireline modem designated the 205A built by the Western Electric Company and a high-frequency radio modem built to NASA Communications Division specifications by Stelma, Inc. These were chosen specifically to solve the problems of implementing the USB support effort and are becoming standards for the rest of NASCOM data network obligations.

The 205A is a synchronous, phase-modulated, single-channel or single-carrier modem basically capable of 1200- or 2400-bit per second operation and, with a manufacturer's modification, also capable of 600-bit per second operation. It is designed to perform with a bit-error rate of 1 in 10^5 or better at 2400 bits per second on channels which have a signal-to-noise ratio as low as 12db and which meet American Telephone and Telegraph (AT&T) schedule 4B equalization specifications. Because of the phase-modulation scheme employed, the modem is relatively immune to typical levels of impulse-type noise. The high-stability timing system (± 0.0005 percent) used by the modem permits back-to-back regenerative operation without excessive accumulation of phase jitter and will allow a complete line dropout of from one to five minutes or more before system synchronization is lost. The modem is fully transistorized, takes about 24 vertical inches of 19-inch wide cabinet space and is capable of being fully remote controlled. In the continental United States and Hawaii, this equipment is leased from and maintained by Bell System-affiliated companies. In all other locations, the modems are procured and maintained by NASA.

The Stelma HF radio data modem is also synchronous and uses a phase-modulation technique but employs 12 tone channels spaced throughout the audio spectrum, each operating at 200 bits per second. This modem is capable of 600-, 1200-, or 2400-bit per second operation with a bit-error rate performance of 1 in 10^5 at 2400 bits per second with a signal-to-noise ratio of 17db. A high-stability clocking system is also used to permit back-to-back regenerative operation and allow maintenance of system synchronization through long fades or dropouts. The modem is fully transistorized, takes about 38 vertical inches of 19-inch wide cabinet space, and is capable of being fully remote controlled. Other features include internal or external system self-check, full-diversity receiver operation, and Doppler shift or spectrum translation correction resulting from a moving source (an aircraft for example), multipath ionospheric propagation, or a mistuned radio receiver. The HF and 205A modems are fully compatible on the DC side and consequently may be placed back-to-back for interconnection of wireline and HF radio services. The HF modem will be used primarily to support the USB tracking ships in the Atlantic, Pacific, and Indian Ocean areas.

Error Control Equipment

There are three types of error-control equipment being implemented. One of these, a forward-acting error detection and correction system, will be used only on the HF radio circuits and will improve the quality of digital data being sent over this media by detecting and correcting bit errors which originate in the data transmission path. This is accomplished using algebraic, parity-like, digital encoding and decoding techniques. Characteristics of this system are that one-half of the bits being transmitted through the transmission media are error-control bits and that the decoder has two to three seconds of serial data storage in order to permit calculation, location, and correction of bit errors which have occurred. The resulting bit-error rate, however, which may have been as bad as 1 in 10^2 or 1 in 10^3 on the radio path, will be between two and three orders of magnitude better after correction, or nearly equivalent to the performance expected of a wireline circuit. Particularly significant is the fact that long bursts of bit errors caused by typical HF path-fade durations of one second or less can be completely corrected. This feature is expected to materially improve the reliability of the HF radio portions of the communications system.

In addition to the error control scheme just described, a powerful error detection arrangement will be used to determine data quality and overall circuit performance on an end-to-end basis. The number of additional bits required for this function is in the order of 10 percent, much lower than for the HF radio error control scheme. However, the reliability of operation is such that the undetected bit error rate is never worse than 1 in 10^9 and is usually in the neighborhood of 1 in 10^{12} to 1 in 10^{15}. Present plans call for implementation of this feature on all data channels handling tracking data.

Another device to assist in the real time assessment of circuit performance is the high speed data quality monitor (DQM). This is a relatively inexpensive piece of equipment that can be used for continuous monitoring of operational data traffic at many points throughout the network. The DQM is programmed to recognize the format of the data being transmitted and to measure bit errors that occur in known sequences of bits within the data block. The measured bit-error rate is then displayed on the front panel and updated at appropriate intervals. The DQM can also generate its own test pattern which can then be sent through a channel and measured by other DQM's along the length of the circuit. This equipment is completely transistorized and occupies seven vertical inches in a 19-inch wide rack.

An example of the use of this equipment can be provided by considering a data circuit which originates at Carnarvon on the west coast of Australia, routes through Canberra, Honolulu, and the Goddard Space Flight Center for regeneration and circuit control purposes, finally terminating at the mission control center in Houston. DQM's would be placed on-line at all locations, including the originating station, in order to determine whether or not the data was of good quality at that point. If the error rate was low at Carnarvon and Canberra but high at Honolulu, the faulty portion of the circuit would then be isolated to a trunk between Canberra and Honolulu, and be replaced by a spare. Continuous monitoring will, in most cases, permit replacement of a poor quality segment of the circuit before that particular segment has completely failed.

Technical Facilities Control System

The technical facilities control system is designed to permit complete circuit and equipment monitoring and/or testing. This function is fully implemented at the Goddard Space Flight Center and at all of the subcenters and partially implemented at each of the sites. Test equipment in this facility include an audio signal generator, vacuum tube voltmeter, oscilloscope, digital pattern generator, equalization-measuring equipment, and a digital counter. The DQM's and appropriate patching arrangements for the required monitoring functions are located in this facility also.

PERFORMANCE

In terms of system performance the following goals have been set:

1. Wireline circuit bit-error rate of 1 in 10^5 or better.

2. Basic HF circuit bit-error rate of better than 1 in 10^3, 50 percent of the time, and better than 1 in 10^2, 90 percent of the time.

3. HF circuit bit-error rate with error correction of 1 in 10^4 or better 90 percent of the time.

4. End-to-end undetected bit-error rate of 1 in 10^9 or better.

CONCLUSION

From the foregoing, it can be seen that the impact of USB system support requirements on the NASCOM network is indeed quite large. Many new problems have been encountered and have been solved either by new approaches to data communications network design or by application of new techniques in specific pieces of equipment. Because of the concept of a segmented network and adherence to standard equipment and techniques at all locations that are supported by the NASCOM network, a very flexible and easily expandable data communications network has been created.

ROLE OF APOLLO SHIPS

by
M. D. Greene
Goddard Space Flight Center

ABSTRACT

This presentation describes the Apollo ships program planned to provide support for the insertion/injection and reentry phases of the Apollo space flights to the moon. Discussion covers characteristics of the ships and an outline of their instrumentation systems, their coverage areas, and time schedules for completion, operational readiness, and integration into the Manned Space Flight Network (MSFN). Organizational responsibility for the Apollo ships program is also treated.

INTRODUCTION

Under the Apollo ships program, five ships are now undergoing conversion to provide support for the insertion, injection, and reentry phases of the Apollo moon flights. These ships will be operated by the U.S. Navy and will fill critical gaps in support requirements which cannot be met by land stations. The first ship is scheduled to be operational by mid-1966 and all are scheduled to provide full mission support by early 1967. The three ships which will support the insertion/injection phase are converted T-2 tankers (Figures 1 and 2). The two reentry support ships were Victory ships used for tracking at the Pacific Missile Range and are being modified for the Apollo program (Figure 3).

Figure 1—T-2 tanker being converted for Apollo program coverage.

Figure 2—Drawing of completed Apollo program ship providing insertion/injection coverage.

Figure 3—Drawing of completed Apollo program reentry ship.

NEED FOR SHIPS COVERAGE

The requirement for ships to provide coverage during the insertion portion of the Apollo flight stems from the need to track the spacecraft continuously until after the C-5 booster cuts off and the spacecraft is inserted into an earth orbit. As will be seen from Figure 4, land stations do not provide coverage for the final portion of the insertion phase. This requirement is also shown in Figure 5, which depicts coverage on the vertical plane during launch and various recovery modes. The need for ships coverage of injection into lunar trajectories will be readily apparent from Figure 6. The basic coverage at the start of injection is indicated by the circles within the dark lines. The lines define the limits of the area from which injection can be made into a lunar trajectory. It may be noted that a large portion of this is over the open sea. Ship coverage is indicated by the darker circles. Figure 7 illustrates coverage seven minutes after the start of injection.

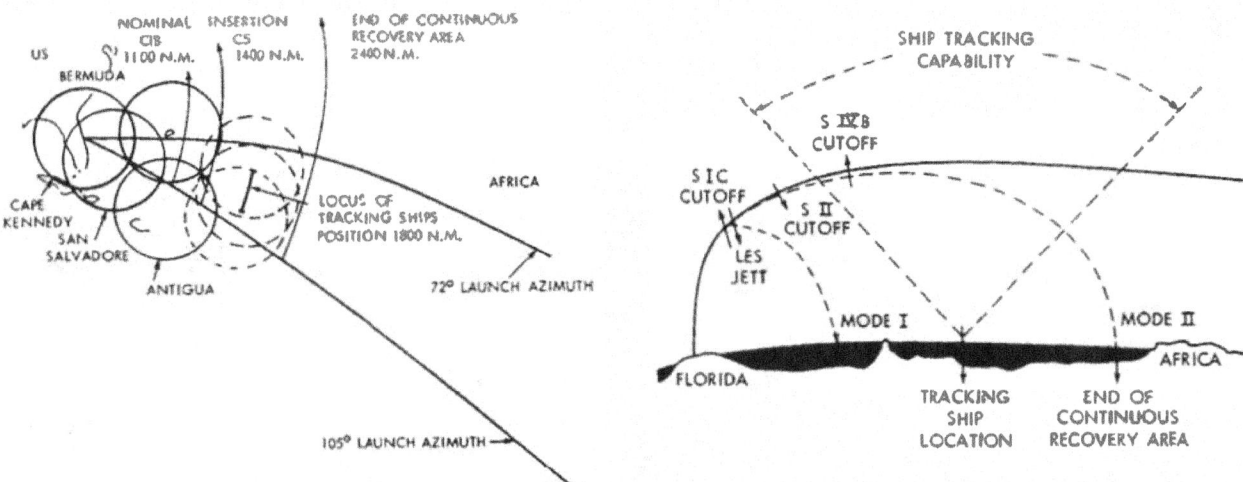

Figure 4—Station coverage during insertion phase.

Figure 5—Coverage on vertical plane during launch and various recovery modes.

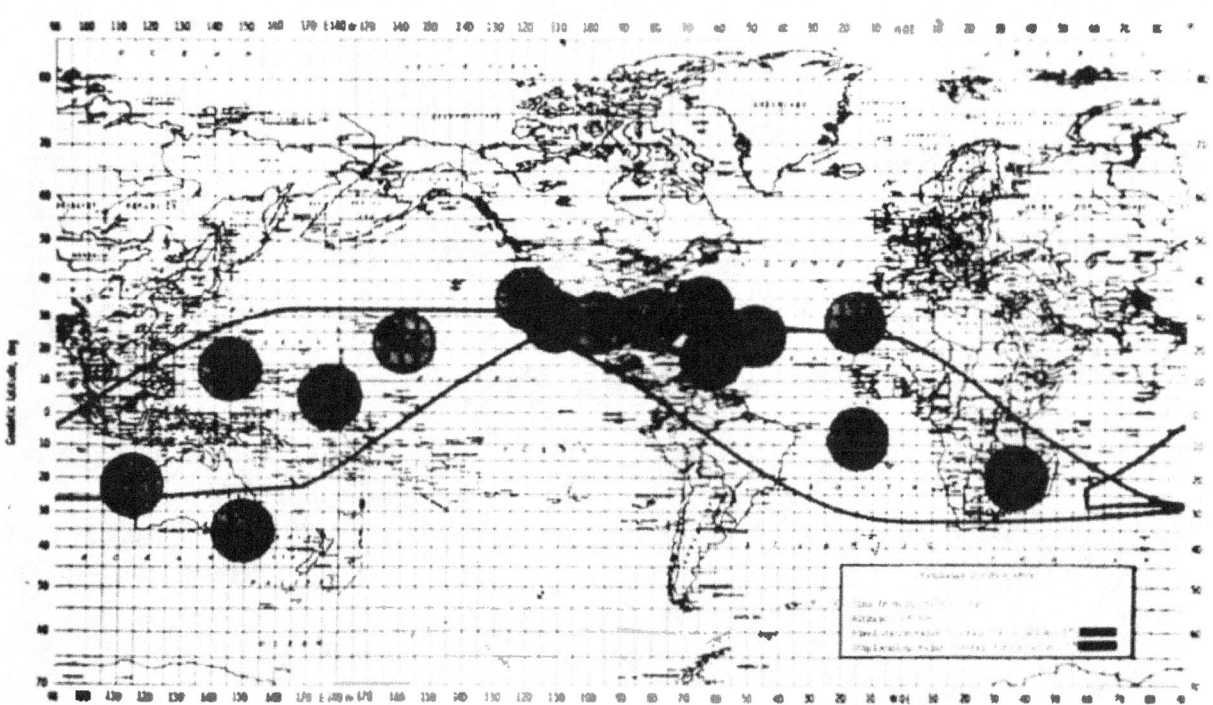

Figure 6—0 minute injection coverage.

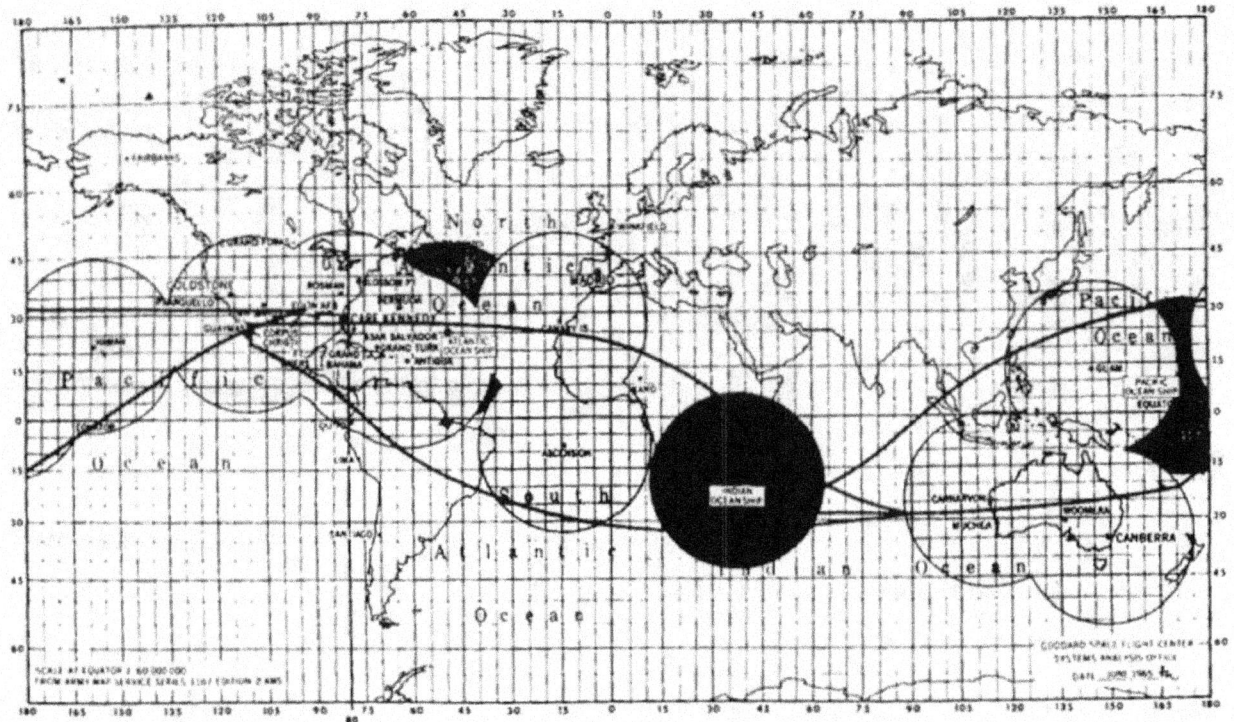

Figure 7—7 minute injection coverage.

IMPLEMENTATION OF APOLLO SHIPS PROGRAM

Organization Responsibility

The organization responsible for the conversion of the Apollo ships is the Instrumentation Ships Project Office of the Office of Naval Material. This organization, which was established by a joint Department of Defense-NASA agreement, is headed by a Navy captain, who is assisted by two deputies, one from the U.S. Air Force and the other from NASA. The prime ships contractor for the three insertion/injection ships, which are being converted at Quincy, Massachusetts, is the General Dynamics Corporation Electronic Division. The subcontractor is General Dynamics Electric Boat Division. Ling-Temco-Vought Company has the contract for the two reentry ships, which are being converted at Avondale, Louisiana.

Schedule for Implementation

Figure 8 gives the conversion and operational schedule for the Apollo ships. It is hoped to have the insertion ship on station by about July 1966, but at that time they will have only a rather limited support capability. It is expected that the ships will be capable of fully supporting the Apollo mission by early 1967.

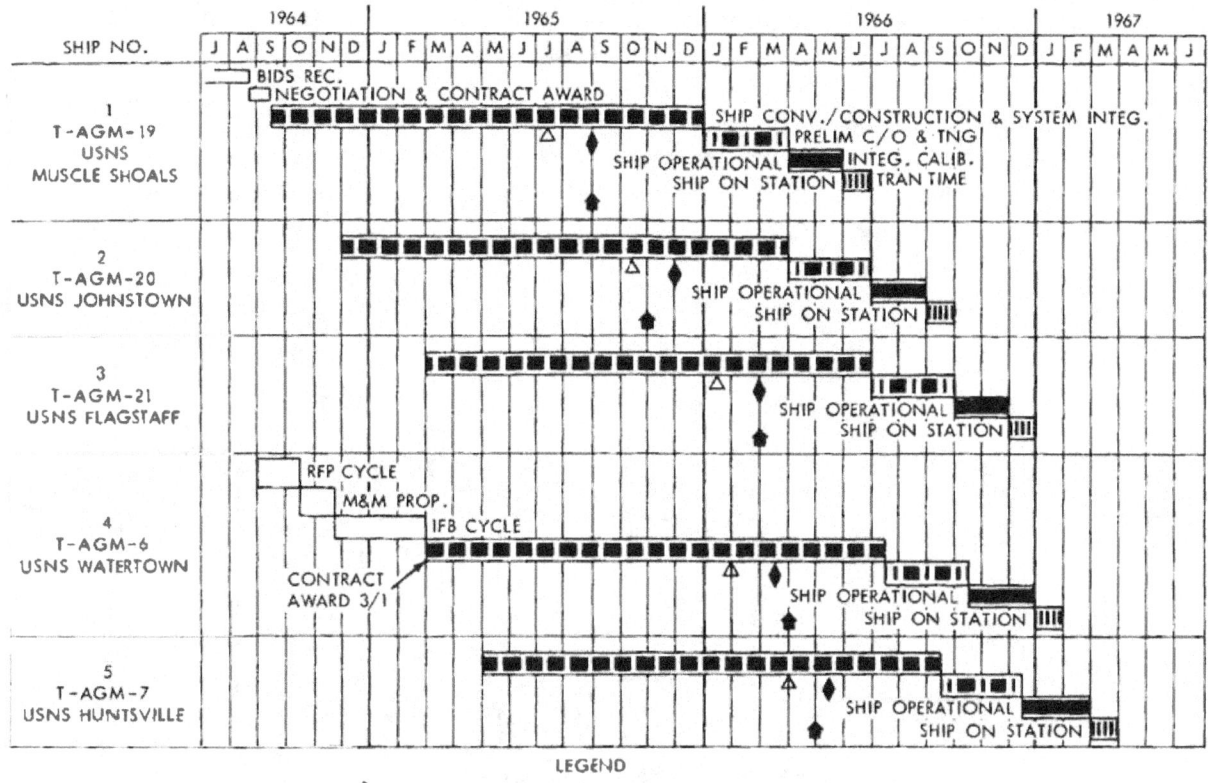

Figure 8—Apollo ships conversion and operational schedule.

SHIPS INSTRUMENTATION SYSTEMS

Instrumentation systems for the Apollo ships include:

1. Communications
2. C-Band Radar Tracking
3. Unified S-Band
4. Navigation and Stable Reference
5. Meteorological
6. Frequency Monitoring
7. Acquisition
8. Command
9. Telemetry
10. Timing
11. Data Handling
12. Display and Control Center

Figure 9 is a block diagram of the various systems and Figure 10 indicates their location on an insertion/injection ship.

The Unified S-Band System is the key system on all five ships. The prime telemetry antenna, shown in Figure 10, is a wideband antenna ranging from 200 to 2300 megacycles. There is also a backup antenna for the telemetry system. The S-band equipment is located below-decks on the third deck, which is almost completely taken up by instrumentation systems. The data processing room is located a deck lower, on the first platform. Also located below-decks are the command control system, C-band radar, and the navigation and stable reference system, which includes the ships inertial navigation system (SINS), star tracker, and flexure monitor system.

The two reentry ships do not have a command system and their telemetry systems are considerably reduced from those on the insertion/injection ships. There is a timing system on both types of ships. Datahandling and the display and control center on the reentry ships are also considerably reduced.

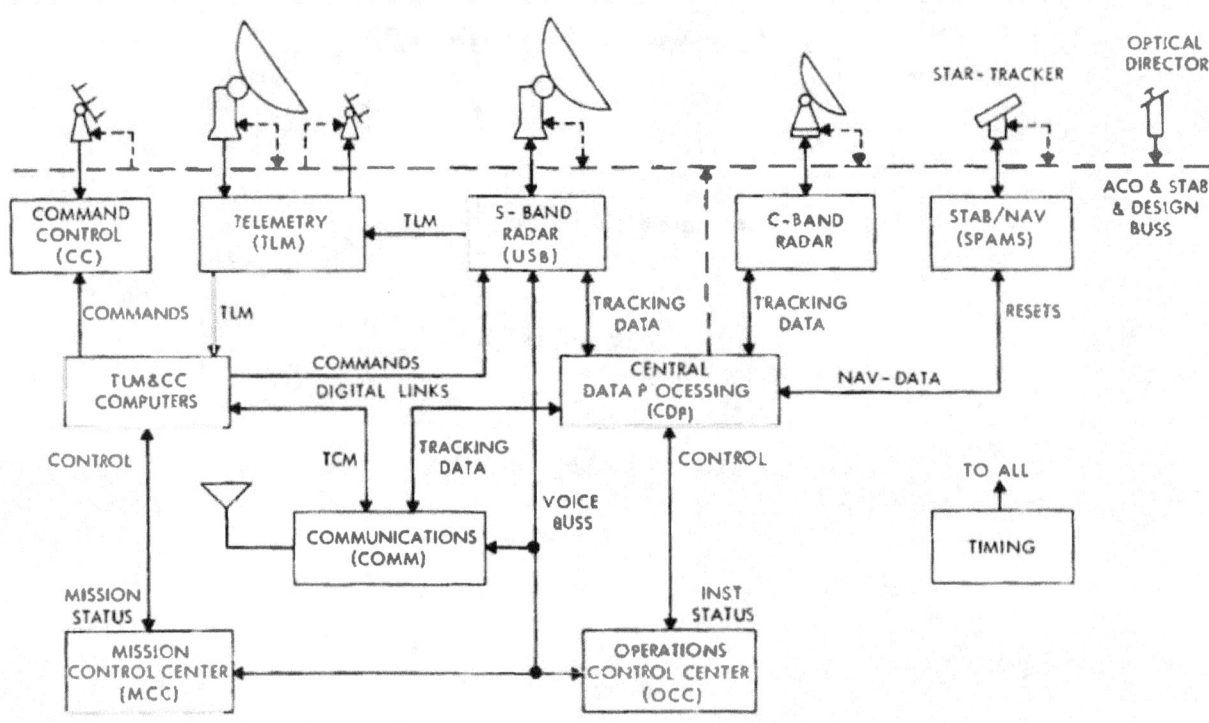

Figure 9—Block diagram of Apollo ships instrumentation systems.

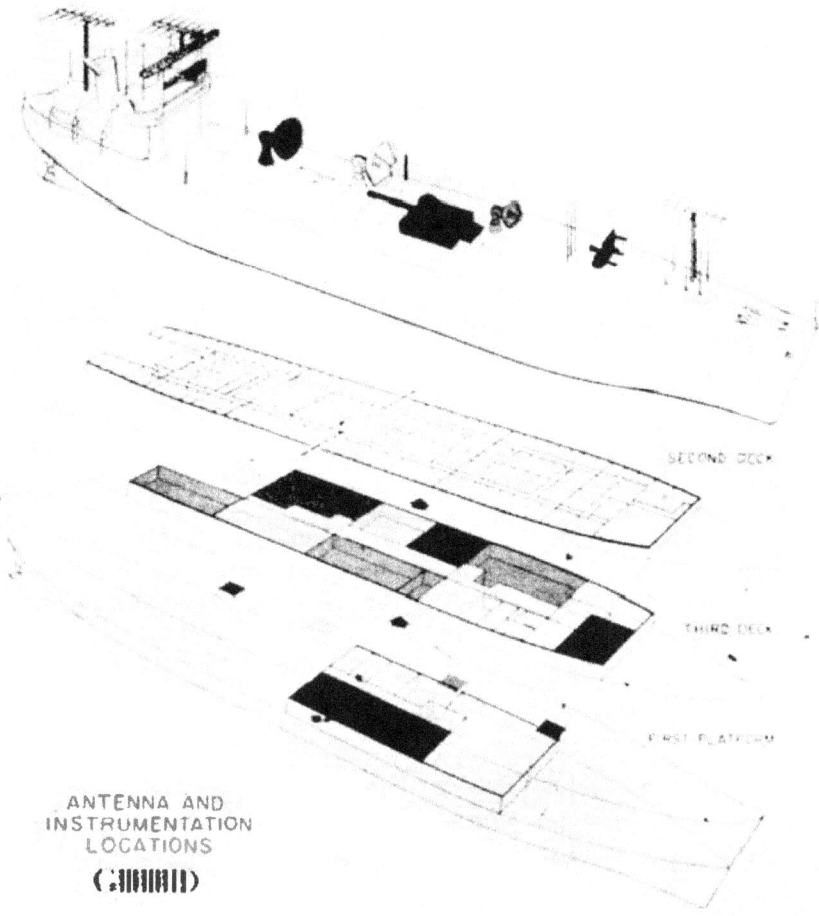

Figure 10—Location of instrumentation systems on ship providing insertion/injection coverage.

APOLLO/RANGE INSTRUMENTED AIRCRAFT

by
L. C. Shelton
Goddard Space Flight Center

ABSTRACT

The functions of aircraft in providing communications coverage in support of the Apollo mission, the reasons for using aircraft to supplement land and ship station coverage and factors determining the number of aircraft required are discussed. A sample operational aircraft deployment plan for lunar mission is presented. A functional description of the onboard instrumentation by major subsystems and operator positions is given. Technical problem areas including antenna-size limitations, multipath interference and radio frequency interference and proposed solutions to these problems are treated.

INTRODUCTION

The objective of the Apollo instrumented aircraft project is to provide a fleet of eight aircraft specially instrumented to maintain inflight voice and telemetry communications with the Apollo spacecraft during the injection and reentry phases of the Apollo mission.

The U.S. Air Force Electronic Systems Division at Hanscom Field, Massachusetts has procurement responsibility for the aircraft. The Air Force National Range Division at Patric Air Force Base will operate the aircraft. Two contractors, Collins Radio Company and Dougl Aircraft Company, are now conducting competing program studies.

These aircraft are intended to supplement, not replace, ships and land stations. During certain phases of the Apollo mission they will serve as links in the communications chain between the spacecraft and the Mission Control Center at Houston, Texas. They will be specif ically used during the following mission phases:

1. One minute before injection burn, during the injection burn period, and continuing for three minutes after the end of injection burn.

2. Those portions of the reentry during which S-band communications are possible.

3. Rapid deployment to any critical area which may require voice and telemetry coverage

Without aircraft, 20 to 30 additional surface stations, ship and shore combined, would be required to maintain communications coverage of all possible injection areas for a particular mission. The number of these areas is considerable because of the variable launch azimuth and parking orbit injection options for the Apollo lunar mission.

SUPPORT REQUIREMENTS FOR INJECTION

The support requirements for injection consist only of voice relay and telemetry recording. The voice relay is required for real-time mission control and the telemetry data is required for postflight data analysis. There are no plans for obtaining tracking data or real-time telemetry data processing on board the aircraft.

The aircraft is required to provide two-way voice communications with the spacecraft by the Unified S-Band (USB) System and a VHF/AM voice link. Communications with the ground are by conventional HF radio. The aircraft acts as an automatic voice relay station. When a transmission is made from the Mission Control Center to the spacecraft, the aircraft receives the transmission from the ground by HF radio and automatically retransmits the voice to the spacecraft via the USB system or VHF. Vice versa, the aircraft automatically relays voice to a ground station and to the Mission Control Center. Facilities are provided on board the aircraft to permit an aircraft operator to communicate directly with the spacecraft or ground, should the normal automatic relay not function.

The aircraft will be required to record telemetry transmitted from the Apollo spacecraft and the S-IV-B booster. The telemetry to be recorded consists of the USB and VHF from the command module and S-band (PCM/FM) and VHF from the S-IV-B booster.

The current planning for an Apollo lunar mission is such that injection can occur over any part of a large area of the earth's surface. The exact point at which the injection will occur may not be decided until a short time prior to the injection, thereby requiring either a great quantity of stations to cover a large area, or a small fleet of highly mobile stations. The area of injection is narrowed considerably as a specific launch day and launch time is chosen. Still further narrowing occurs after launch when a particular orbit is chosen from which injection takes place. The following discussion will emphasize the need for the highly mobile airborne stations in order to give the needed mission flexibility.

Figure 1 shows the entire range of parking orbits and launch azimuths for the earth orbit phase prior to injection. The labels on the orbits indicate the launch azimuth and the orbit number. The launch azimuths shown (72° to 108°) are the approximate Cape Kennedy range safety limitations. A variable launch azimuth is required in order to offset the earth's rotation and eliminate the need for plane changes. For a particular launch, the variable azimuth is actually reduced to a maximum of 26°, which is the limit of coverage provided by the insertion ship. The 26° corresponds to approximately three hours of hold time.

The mission rules further state the injection can occur on the second or third orbit for an Atlantic planned injection; or the first, second, or third orbit for a Pacific injection. Also, the decision to delay injection from one orbit to another can be made just prior to the scheduled burn time.

Figure 2 shows a typical Atlantic injection area, bounded on the east by orbit 2, 72° launch azimuth, and on the west by orbit 3, 98° launch azimuth. The north-south limits of this area correspond to one minute prior to injection burn and three minutes after injection burn. The Pacific injection area is somewhat larger since there is also a possibility of injecting from the first orbit.

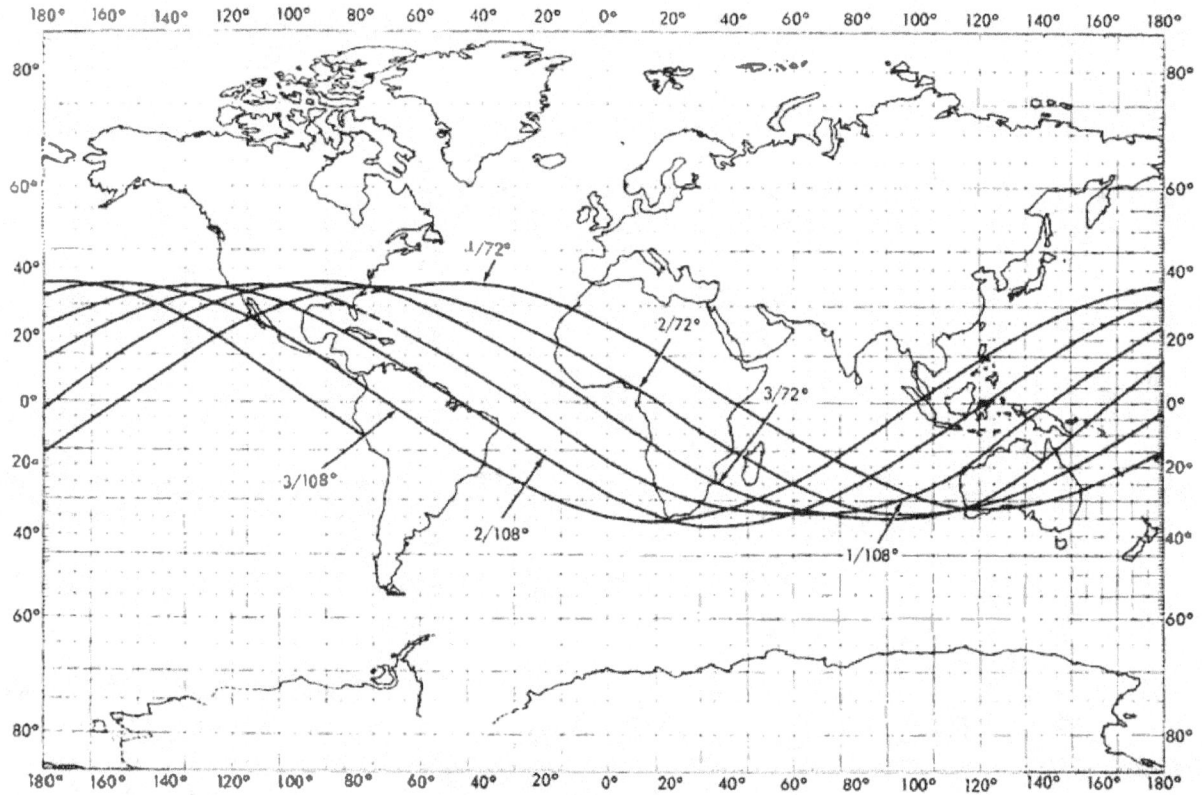

Figure 1—Range of parking orbits and launch azimuths for earth orbit phase.

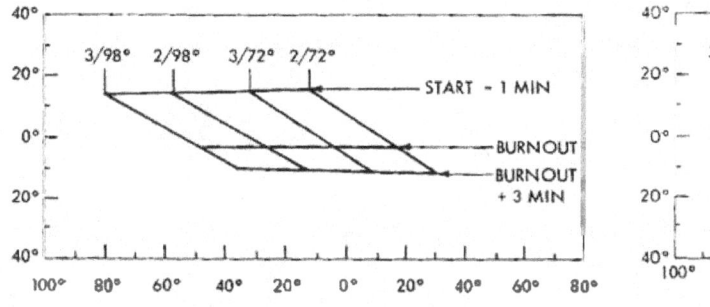

Figure 2—Typical Atlantic injection area.

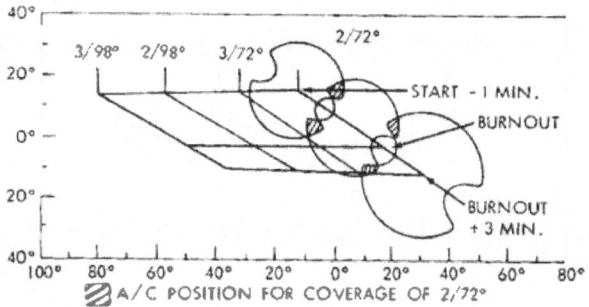

Figure 3—Aircraft line-of-sight coverage.

The original planning for aircraft support for injection coverage included the assumption that the aircraft would be able to establish communications with the spacecraft at or very near line-of-sight distances. Figure 3 illustrates the method used to locate the aircraft for coverage of the ten-minute injection burn data interval. The radii of the three circles are equal to the line-of-sight distances between the aircraft and spacecraft at the beginning, middle, and end of the data interval. Spacecraft altitudes of 100 nautical miles prior to injection and 200 nautical miles three minutes after injection were used. Using an aircraft altitude of 35,000 feet, the line-of-sight coverage is approximately 900 nautical miles one minute prior to injection burn and 1400 nautical miles three minutes after burn. The area of overlap of these circles defines the location of the aircraft. Two aircraft located in the shaded overlap areas can therefore provide the required line-of-sight coverage. The "notches" in the circles represent approximations for the expected antenna nulls at the front of the spacecraft and rear of the S-IV-B booster.

FACTORS DETERMINING THE NUMBER OF AIRCRAFT REQUIRED

In order to provide coverage of all possible launch azimuths and orbits within the data area, the aircraft must perform a complicated deployment pattern. The aircraft flight patterns have been devised so that maximum use is made of the aircraft's mobility, thus decreasing the number of aircraft required.

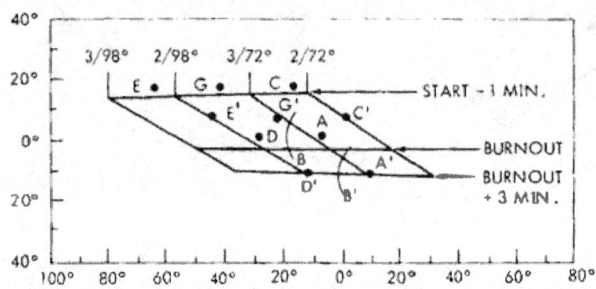

Figure 4—Illustration of key aircraft locations for coverage of atlantic injection data area.

The following discussion will illustrate how the aircraft are deployed in a typical situation. Figure 4 shows some of the key aircraft locations for four possible injection ground tracks.

Points A and A' correspond to the shaded portion of Figure 3 and are the aircraft locations for a 72° orbit 2 injection. Let us assume that the injection is actually planned for orbit 2 and the launch is planned for 72°. The aircraft would loiter along the curved lines B and B' until the launch takes place. The reason for initially locating the aircraft to the west of the planned injection will be shown later in this illustration. If the launch did occur on schedule and at the 72° launch azimuth, the aircraft would fly eastward to points A and A', taking their onstation positions just before the spacecraft pass. The distance to be flown is approximately 700 nautical miles, and there is adequate time to cover this distance, since the spacecraft will have traveled over one orbit prior to the pass. If a decision were made at the last moment to postpone the injection to orbit 3, the aircraft would have to assume positions D and D' or C and C' on Figure 4. Actually, only the aircraft at A could reach C' or D' in the one and one-half hours required for another orbit, and a third aircraft would be required at either C or D.

Now, consider the case of a hold occurring near the scheduled launch time, and continuing for the full three hours. In this case, the injection ground track would appear to sweep from east to west across the data area at about 900 knots at the equator. This speed is greater than that of the C-135's, which is the reason for initially positioning the aircraft to the west of the planned injection. In order to reach positions for coverage of 98°, which in this example corresponds to maximum hold time, the aircraft must fly to points G and G'. Although this distance is almost 2000 nautical miles, the aircraft can make it, taking advantage of the three hour maximum hold, plus time for one orbit. Should a decision be made to delay injection until orbit 3, the aircraft at G would fly to E', and an additional aircraft would be required at point E.

From the above discussion it is seen that three operational aircraft are required to cover the data area in the Atlantic. One backup is required for a total of four aircraft. In the Pacific, six operational aircraft plus two backups are required. The additional aircraft in the Pacific are required because of the necessity to cover orbit 1 as well as orbits 2 and 3. A total of eight aircraft are thus required to cover the "worst case" Apollo injection.

AIRCRAFT COMMUNICATIONS SYSTEM AND EQUIPMENT

The primary objective in determining performance requirements was that the aircraft must be able to communicate on all links at the maximum line-of-sight limitation. As previously indicated, this is 900 nautical miles at the beginning of the data interval and 1400 nautical miles at the end of the data interval. Normal signal margin calculations yielded the following aircraft antenna gain requirements: 30db for S-band and 12db for VHF. These antenna gains would theoretically yield good quality communications at 900 nautical miles and somewhat marginal communications at 1400 nautical mile range, providing the S-band transponder in the spacecraft is operating in the high power mode. Higher antenna gains are not specified because of the limited available antenna aperture on the C-135 aircraft.

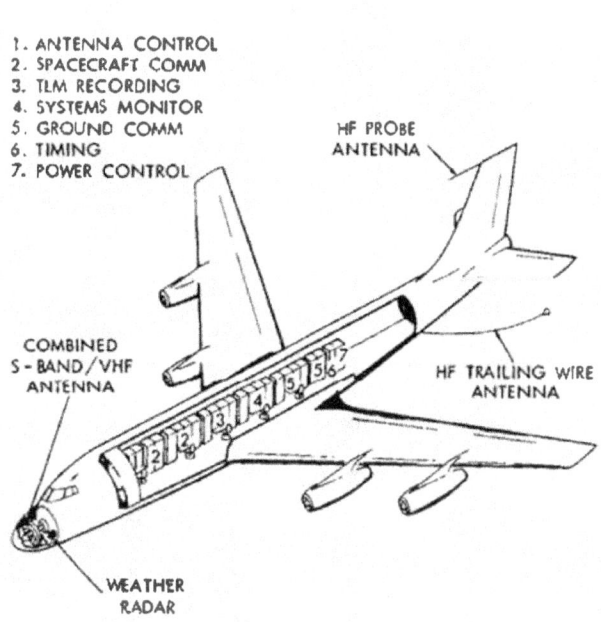

Figure 5—Aircraft equipment arrangement.

Figure 5 shows the approximate location of the equipment inside the plane. There are about 20 racks of equipment, with five primary operator positions. The spacecraft communications antenna will be located in the nose of the airplane as shown, and it will be an S-band and a VHF antenna mounted on a common axis. Polarization diversity will be included on both frequency bands. The fact that both frequencies are mounted on one axis will enhance acquisition procedures. During injection, it is planned to initially acquire the spacecraft on VHF, probably using the S-IV-B booster telemetry. This will automatically point the narrow beam S-band antenna on target, permitting acquisition of the spacecraft S-band transponder. A complicating factor in installing the spacecraft communications antenna is that it is necessary to include the existing C-135 weather radar. The antenna will have the capability of tracking in azimuth plus or minus 80 degrees, and in elevation, up 80 degrees and down 30 degrees.

Figure 5 also illustrates the five operator positions presently envisioned. The function of the Systems Monitor is to direct the activities of the aircraft instrumentation operators, and to directly converse with the spacecraft or Mission Control Center as necessary.

Figures 6 through 9 illustrate the aircraft communications system. Figures 6 and 7 illustrate the polarization diversity technique which will be used in receiving S-band and VHF

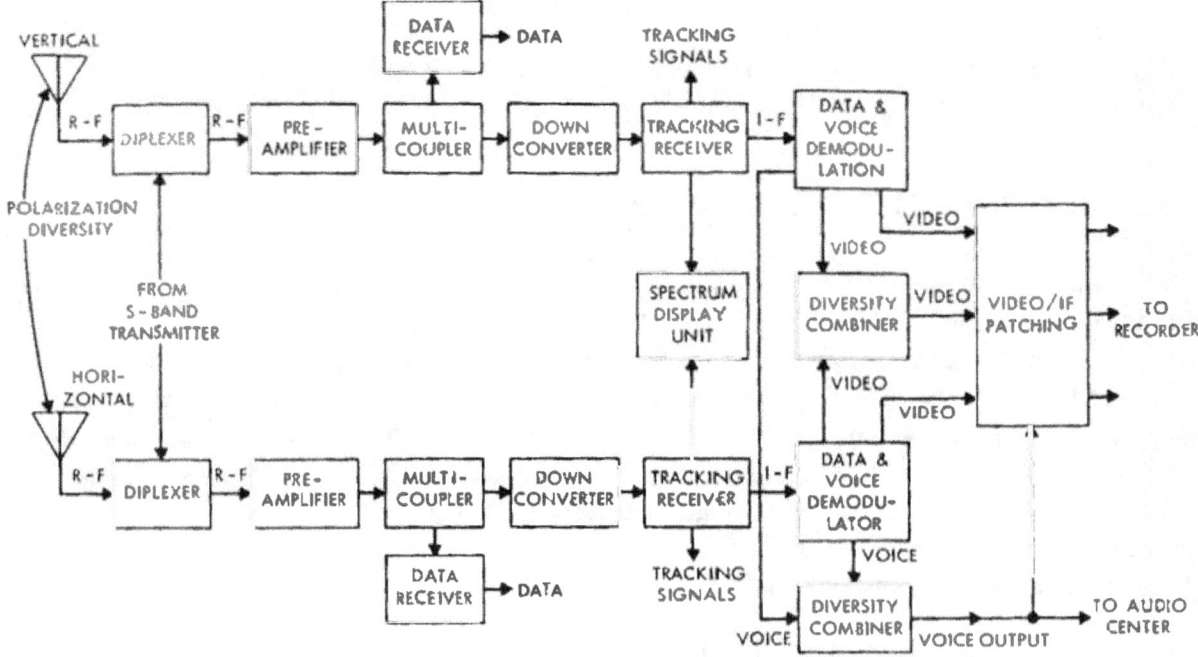

Figure 6—S-band receiving system block diagram.

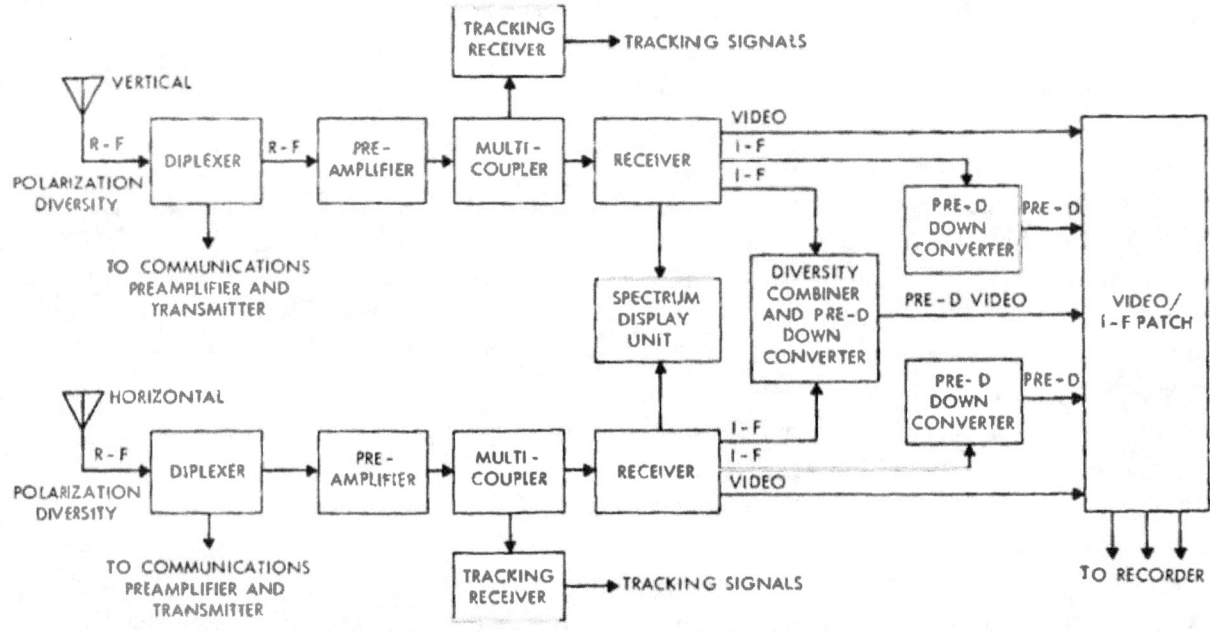

Figure 7—VHF telemetry receiving system block diagram.

data. It is expected that the use of polarization diversity will somewhat reduce the signal multipath problem the aircraft will encounter. The capability will be provided to select either vertical/horizontal linear on left/right circular polarization prior to the mission.

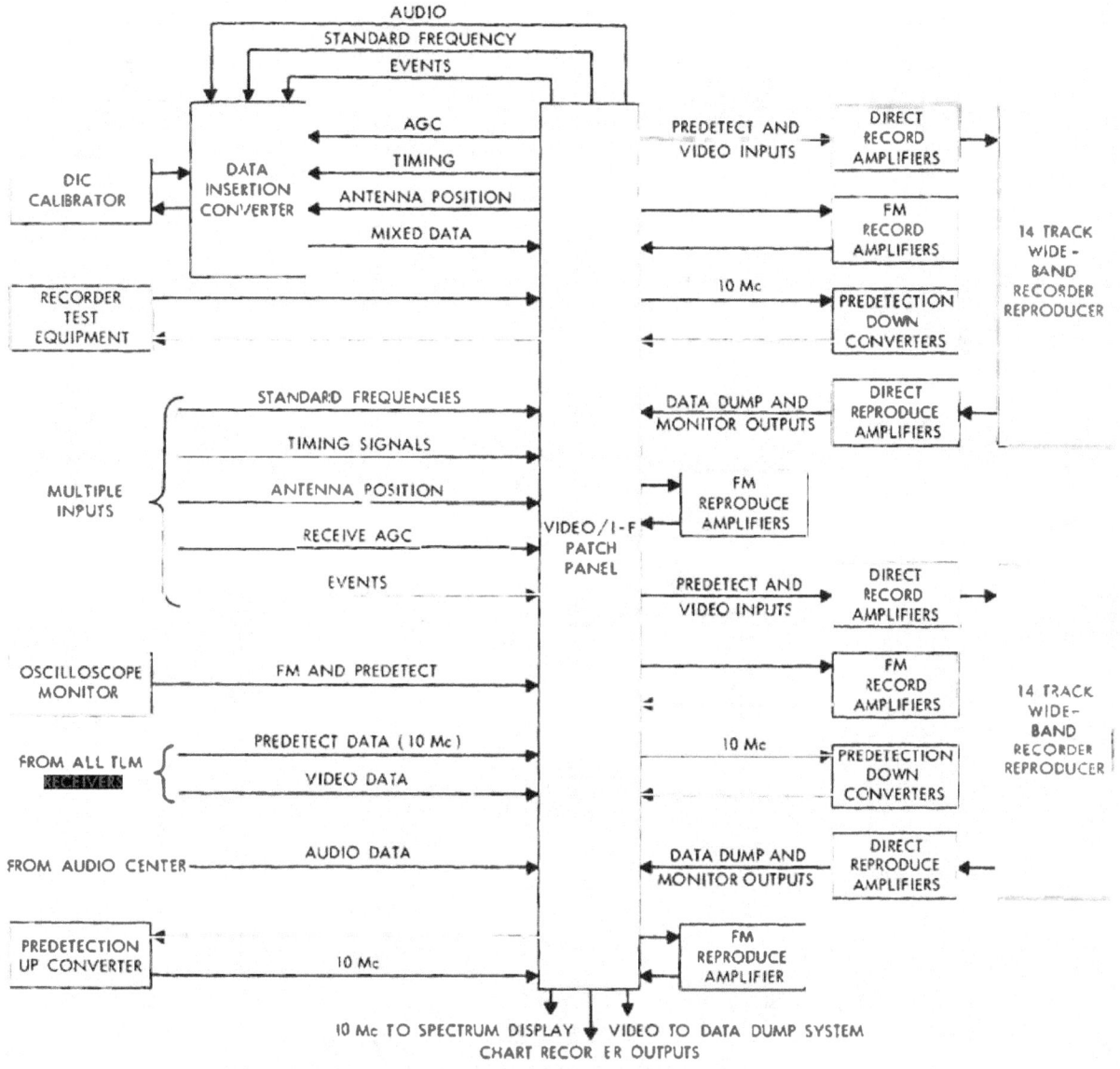

Figure 8—Video/IF distribution block diagram.

The aircraft S-band system is similar to the ground unified S-band system, with the exception that no ranging data is transmitted to or received from the spacecraft. As shown in Figure 6, the unified S-band telemetry data is demodulated and routed to the recording system and the demodulated voice is routed to the HF transmitters for automatic relay to ground. The voice received from ground is transmitted to the spacecraft via a 100-watt S-band transmitter with the same S-band antenna used for receiving.

The VHF telemetry receiving system, shown in Figure 7, accounts for a large portion of the aircraft instrumentation, since up to ten VHF data links may be received from the S-IV-B and Apollo spacecraft. Two receivers will be used for each data link, as shown.

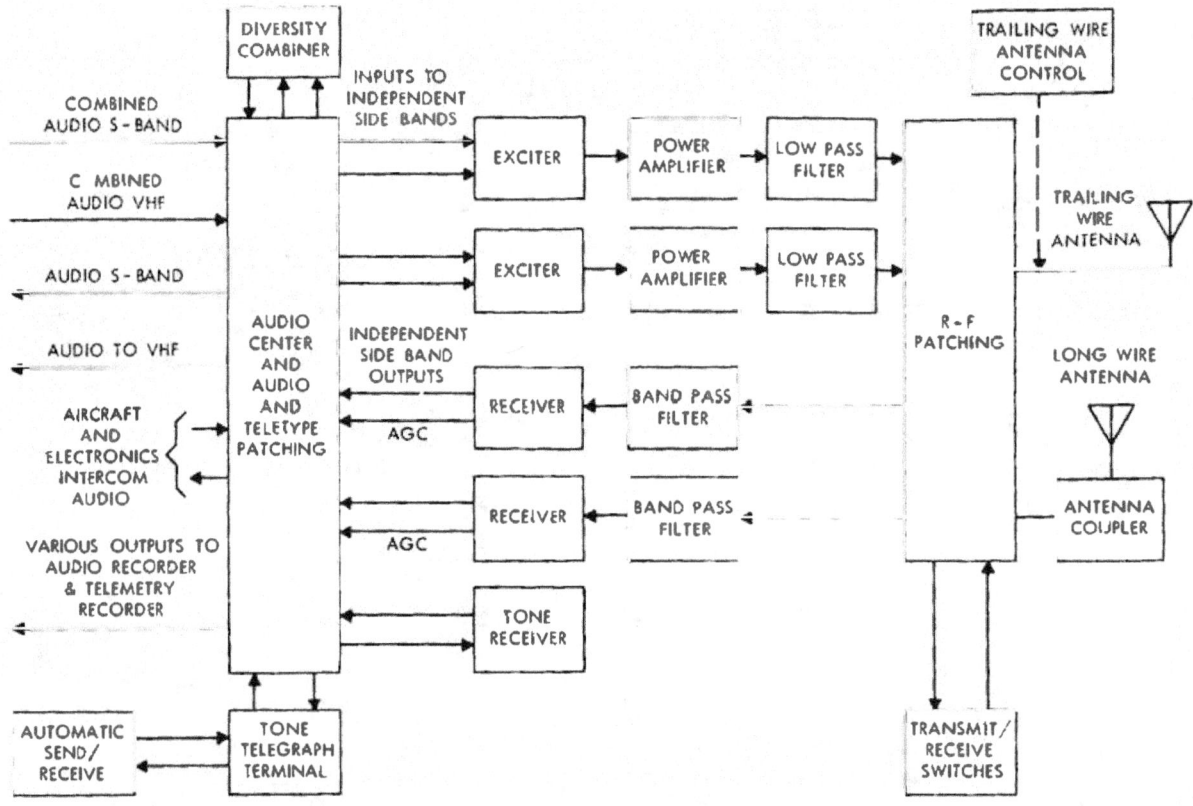

Figure 9—Aircraft/ground communications block diagram.

Figure 8 illustrates the types of data which will be recorded on board the aircraft. It is tentatively planned to use three 14-track wideband tape recorders. Data outputs of individual receivers and diversity combiners will be recorded in order to enhance reliability.

The HF ground communications system is shown in Figure 9. The capability to use both frequency diversity and sideband diversity will be provided. The HF teletype equipment will be used to provide the aircraft with updated deployment and antenna pointing data.

A trailing wire antenna will be used to increase HF transmission efficiency. A combination of wing tip probes and fuselage-long wire antennas will also be used.

PROBLEMS IN USE OF AIRCRAFT AS COMMUNICATION STATIONS

There are several rather serious technical problem areas inherent in the use of aircraft as communication stations. Mainly, these problems are: antenna size limitation, multipath interference at low antenna elevation angles, and radio frequency interference.

The antenna size limitation is mainly a function of how much aircraft performance degradation can be permitted. The seven-foot diameter aperture in the nose of the C-135 aircraft causes two to four percent decrease in the aircraft's maximum range. When this is added to the drag caused by the HF antennas (probes and trailing wires) the range of the aircraft may be reduced enough to prevent deployment for Apollo injection coverage. Other aerodynamic factors such as stability, weight and balance, and pilot vision are also involved.

The seriousness of the multipath interference problem in aircraft telemetry reception has been observed by NASA and DOD in previous aircraft telemetry receiving missions. There are three primary reasons why airplanes have a serious multipath problem: (1) the high reflection coefficient of sea water, (2) the increased height of the aircraft, which causes greater frequency of multipath nulls, and (3) the broader beam width of relatively low-gain aircraft antennas. Some measures which are being considered for reducing the multipath problem include the use of high-capture ratio data receivers and antenna beam tilting. Multipath interference at S-band is expected to occur at frequencies from 10 to 250 cycles per second. These nulls would make phase-lock capture of the spacecraft transponder by the aircraft very difficult.

A well-recognized problem in aircraft communications is radio frequency interference. RFI problems are most prevalent when attempting to transmit high power HF while simultaneously receiving weak signals. This is just the case in the Apollo aircraft when relaying voice from the spacecraft to ground. It is felt that the RFI problem, through careful design, can be minimize if not eliminated.

Other operational problem areas which will require work include: increasing the reliability of ground-to-aircraft communications, optimizing the aircraft deployment flight patterns, and overall maintenance of the aircraft and electronic systems on a near world-wide basis.

It is expected that the first Apollo aircraft will be fully operational in early 1967. A sufficient amount of operational experience should be gained with the aircraft on later 200 series missions and early 500 series missions, before actually providing support for the lunar mission.

STATUS OF THE APOLLO PROGRAM

by

Ozro M. Covington
Deputy Assistant Director
Office of Tracking and Data Systems
Goddard Space Flight Center

From the beginning of this program we have attempted to move from tried and true systems and facilities only when we could prove new developments and integrate them into the existing system. We started out by expanding as much as possible the existing facilities and equipment capabilities used for Mercury, to meet the requirements of the Gemini support program. Similarly, we are going to start the Apollo program with primary support from the Gemini systems, and move to the Unified S-Band (USB) commitment only after we have had a chance to check the system thoroughly, integrate it, and qualify it for manned flight support.

I would like to point out the very obvious by saying that the Mercury program was, by comparison with Gemini, and certainly with Apollo, a relatively smaller support task than we now have with Apollo. With Mercury we had a single vehicle, and our specific planning was for a three orbit mission. We were able to accomplish the support test, to a great extent, by using existing low-speed data and analogue techniques. We registered our first major departure from this course in Gemini, when we moved to a capability to support at least two vehicles simultaneously.

Added to this was a complete order of magnitude requirement increase, in terms of data to be transmitted, processed, and displayed in real-time. The result is, to my mind, that except for the additional facilities which we have added, the main change made in Gemini was in the move toward digital techniques. To that end we installed PCM telemetry, digital up-data links, and on-site digital data processors.

We have moved from Mercury to Gemini, and as we look forward to Apollo, we have established a goal for ourselves of transmitting essentially all the data from the station back to the control center, so that truly centralized flight control can be accomplished. You have heard some talk about the data communications network here today. I think that the words which strike hardest are "HF communications and data error correction," when we think about the job of communications and the goal we have set of sending large quantities of data back to the control center. Quite frankly, and I hope I am not saying anything I shouldn't at this point, we don't see any answer to really centralized control until we can depend on satellite communications for all the stations which are not now tied to main trunk circuits on cables and hardwire lines.

Looking at Mercury and Gemini and moving on to Apollo, I think the major aspects of the change affecting our planning and our work to date have been the requirements for an addition in the number of network stations.

Table 1 lists those stations which were identified as Mercury support stations and those we are including in the Gemini network. We could talk to some extent about the changes here, but I think it is quite clear that the Apollo requirement does call for many new facilities.

Table 1

Network evolution.

Station	Mercury	Gemini	Apollo	Station	Mercury	Gemini	Apollo
Bermuda	X	X	X	CNV (MCC)	X	X	X
Canary Island	X	X	X	SAL	X	R	
Kano	X	L*	L	GBI	X	X	
Zanzibar	X			GTK	X	X	L
Tananarive		L	L	ANT	X	X	X
Guaymas	X	X	X	ASC			X
Merritt I. (CNV)			X	Canton	X	L	L
Guam			X	Hawaii	X	X	X
Goldstone			X	Pt. Arguello	X	R	R
Goldstone - JPL			BU*	White Sands	R	R	R
Madrid			X	Corpus Christi	X	X	X
Madrid - JPL			BU	Eglin	R	R	R
Canberra			X	Rose Knot	X	X	L
Canberra - JPL			BU	Coastal Sentry	X	X	L
Muchea	X			Apollo Ship 1			X
Woomera	X	R*	R	Apollo Ship 2			X
Carnarvon		X	X	Apollo Ship 3			X
				Apollo Ship 4			X
				Apollo Ship 5			X

*L - Limited   R - Radar Only   BU - Backup

The change is also in support of three Apollo modules, whereas Gemini has only two which have called for additional equipment at each site, and in many cases the capability of handling simultaneously two equally complex problems, as well as additional capability within elements of station equipment. For example, we must move to a considerably larger on-site computing complex than was used to support Gemini. In addition, there is the requirement for a capability to provide tracking and data acquisition at lunar distance. I believe the USB discussions pretty well cover what we plan to do.

With that, then, as a little of the background and our early thoughts on the subject, let us quickly get to the status of implementation of the network. We have based all of our schedule planning on three or four different time parameters. In the beginning, we decided that the most difficult job we asked the contractors of the support team to do was that which went to Collins. We felt that the most complex task and the one that would take the longest, in terms of design and manufacture, was the manufacture of the USB equipment, and we geared our equipment availability to that contract schedule.

In each case, we add one month to the schedule completion date for shipment, if the station is a continental U. S. station, and based on past experience, we have added two months to overseas sites. We have allowed seven months for the installation and check-out of the electronic equipment, after the facilities are available, and we are still able to get away with insisting on six months network simulation before we are ready to commit these stations as prime support capability for a manned flight mission.

Another factor which turns out to be a key determining factor in scheduling availability of operating stations is the construction of facilities program. Our job in scheduling, then, has been to consider all of these factors together and to keep optimizing the possible station sequence availability against the mission requirements, as best we can.

In Table 2, we show two on-site dates. The first is the time when we expect all these sites to have the equipment installation and check-out complete. The final completion data is the point at which we say that the stations will be ready for any and all prime support missions.

We are not insisting that all of this work be finished before we are willing to commit these stations to test and qualification runs. We would like, however, to insist that the block of stations which is required for any mission have six months for network simulation and mission training of the people before we commit it as the source of prime data on which the success of the mission will depend.

The status, as you have heard, is that a 30-foot USB station is operating at Collins now. It is being used to check-out the integration and interface problems which we will have to face at the land sites. A second USB station in combination with ships' antennas is being set up right now. This will be dedicated primarily to finding the interface problems that we will encounter on board ship. Two of the 30-foot antenna structures are actually in shipment, one to Guam and one to Carnarvon. The first USB electronic equipment to go to the Guam site will be delivered this month. We expect that the USB will be checked out for missions support during 1966. There is some indication we will have an Apollo mission flight earlier than that. However, we will be able to do very little, if anything, on that mission. We hope to complete the check-out by early 1967 at all the prime sites, and begin to commit these as prime facilities from early to mid-1967.

Table 2

Apollo MSFN schedule.

Station	Equipment Installation & On-Site Check-Out Complete	Network Simulation Complete
Guam	1 May '66	1 Nov '66
Carnarvon	1 Jun '66	1 Dec '66
Bermuda	15 Jun '66	15 Dec '66
Hawaii	1 Aug '66	1 Feb '67
Mila	1 Aug '66	1 Feb '67
Texas	1 Oct '66	1 Apr '67
Guaymas	15 Oct '66	15 Apr '67
Ascension	15 Dec '66	15 Jun '67
Grand Bahama	1 Jan '67	1 Jul '67
Antigua	1 Apr '67	1 Oct '67
Grand Canary	1 Apr '67	1 Oct '67
Goldstone	15 Oct '66	15 Apr '67
Canberra	15 Jan '67	15 Jul '67
Madrid	15 Mar '67	15 Sep '67
Ship #1	1 Apr '66	1 Jul '66
Ship #2	1 Jul '66	1 Oct '66
Ship #3	1 Oct '66	1 Jan '67
Ship #4	15 Oct '66	1 Feb '67
Ship #5	15 Dec '66	1 Apr '67
1 Aircraft	15 Dec '66	15 Mar '67
8 Aircraft	1 Oct '67	1 Jun '68

Appendix A

Glossary of Abbreviations

AMC	— aeromedical monitor console	Modem	— modulator/demodulator
APP	— antenna position programmer	MPAD	— mission planning and analysis division
CAM	— computer address matrix	MSCC	— manned space control center
CCC	— command computer console	MSFN	— manned space flight network (Mercury or Apollo)
CCS	— command control system	PAM	— pulse-amplitude modulation
CDP	— command data processor	PCM	— pulse-code modulation
CRT	— cathode ray tube	PDM	— pulse-duration modulation
CSM	— command and service module	PMP	— premodulation processor
D/A	— digital/analog	PPM	— pulse-position modulation
DCS	— digital command system	PRN	— pseudo-random noise
DQM	— data quality monitor	SCO	— subcarrier oscillator subsystem
DRK	— data request keyboard	SDC	— shipboard Doppler counter
DTU	— data transmission unit	SINS	— ships inertial navigation system
I/O	— input/output	SPE	— static phase error
IR	— infrared	SPS	— service propulsion system
LEM	— lunar excursion module	SSC	— spacecraft system console
LES	— launch escape system	TDP	— tracking data processor
LOR	— lunar orbit rendezvous	USB	— unified S-band
MCG	— memory character generator	VCO	— voltage controlled oscillator
MCVG	— memory character vector generator	VSWR	— voltage standing wave ratio
MM	— maintenance monitor	VTVM	— vacuum tube volt meter

Appendix B

List of Attendees

Abernethy, F. G.
Adis, George
Albanes, N. J.
Albin, Frank
Allen, K. J.
Allen, Joseph J.
Allinder, Dick
Allison, James E.
Allred, Val W.
Alvey, John D.
Anderson, D. W.
Arndt, E. A.
Arslanian, John G.
Ashcroft, R. T.
Augenstein, R. J.
Bacque, Cramer
Bahan, C. W.
Barr, Thomas A.
Barnes, Robert T.
Bartel, Donald R.
Bannerman, James
Basson, David
Baumiller, William
Bauer, Harry
Beall, George
Beagan, Charles V.
Begenwald, A. J.
Behuncik, John
Bell, Holland
Behonick, John
Bellcomm, H. Stanly
Benjamin, Vernon E.
Benzel, Cliff R.
Best, Al
Bickford, Fred H., Jr.
Bienko, Mitchell
Billig, Lewis S.
Bilyk, Zenon
Birks, William E. Lt. Col.
Bislip, Robert P.
Blanchard, Roger L.
Blenis, Ronald D.
Block, Arthur
Boden, W.
Bond, A. C., Jr.
Bonnell, Dean W.
Bonney, Lyle D.
Bonton, Richard
Bosmajian, C. P.
Boulander, J. L.
Boykin, F. M.

Brooks, David R.
Brown, B. Porter
Brown, Carl W.
Brown, Dwight
Brown, Robert H.
Brown, Robert M.
Brown, W.
Brown, Wilfred III
Broughton, Thomas G.
Brumber, P.
Brunker, Lester A.
Bryant, Fred B.
Bryant, William C., Jr.
Buckley, Edward
Buehler, Richard K.
Bugg, William M.
Bunce, Robert C.
Bunda, Frank E.
Burke, M. L.
Burke, P. G.
Burkepile, James M.
Burnham, R. E.
Burton, Walter G., Jr.
Burnham, R. E. Lt. Col.
Burns, Richard W.
Burrows, Lloyd
Busche, H. C.
Butler, David
Byer, David L.
Byrd, S. S.
Byrne, Frank
Campbell, Willis S.
Call, D. W.
Campenni, Robert D.
Camgel, Joe
Canqel, J.
Carlson, Arthur William, Jr.
Carlton, A. George
Carey, Charles W.
Carney, Wayne K.
Carr, Ronald
Carson, Thomas M.
Chalk, Charles F.
Chandler, George P.
Charpentier, George Henry
Chase, William
Chenowith, H. D.
Chi, Andrew
Chicoine, E. L.
Clark, Howard E.
Clark, Melvin C.

Clark, Orval
Coates, Robert J.
Cochran, Earl
Conner, James
Connolly, G. E.
Connors, William E.
Corwin, Stanley
Courtright, Morris, Jr.
Covington, Ozro M.
Cox, H. R.
Cox, R. T.
Cox, R. T.
Crump, Arnold W.
Curkendall, D. W.
Dantzig, Henry P.
Dauphin, V. M.
D'Ausilio, Robert F.
Davis, Charles
Debris, R., Dr.
Dentel, Wm. A.
DiLosa, Vincent J.
Dinwoodie, Jack
Donahoo, M. L.
Dorman, Donald K.
Downs, James E.
Drechsler, William G.
Dudney, Richard K.
Dungan, Larry J.
Dunn, Johnny
Ealick, Perry L.
Eastman, S. R.
Eaves, Robert
Eggers, Arthur L.
Emmons, Paul
Engels, P.
Estridge, Philip D.
Eurmrous, Paul
Fadden, B. J.
Falotico, Anthony
Fariss, George W.
Farkas, Leonard
Federico, Paul R.
Feiden, R. L.
Fellerman, K. D.
Ferrick, Eugene, Jr.
Findell, Max (Lt. Col.)
Fink, Charles
Fisher, J. Prestley, Jr.
FitzPatrick, Joseph F.
Flaherty, John B.
Flowers, John

Folsom, W. L.
Fosque, Hugh S.
Fratkin, Sidney J.
Friedman, R. F.
Friel, Fred J.
Gantt, Alphonso M.
Garabedian, A.
Gaston, George
Gardner, Virgil F.
Garrison, Charles
Gatto, Philip R.
Georgeadis, George N.
Gibbons, Thomas J.
Godfrey, Robert D.
Golden, David E.
Goldberg, Vernon
Goodwin, Paul S.
Goolsby, Lee
Gordon, Millard F.
Granata, R. L.
Grandinetti, Anthony F.
Greene, Edward
Greene, E. P.
Gregg, James C.
Griffin, Thomas M.
Griffin, Tom
Grisham, William
Grossman, Bernard
Guthrie, Jon
Guthrie, John
Guy, Thomas
Guy, J.
Habib, Edmund J.
Haggerty, W. H.
Halcomb, J. K.
Hancock, A.
Hass, Donald
Hafer, Lt.-Col. Frederick L.
Hahn, William E.
Halpeny, Owen S., 1st Lt, USAF
Hancock, A. F., Capt.
Harner, Walter S.
Haugen, Glenn
Hayes, Robert D.
Heller, N. R.
Hepler, David S.
Heyroth, J.
Hightower, Lloyd E.
Healy, J. J.
Helm, Ted
Herrburger, E. C.
Hibbert, John J.
Hildebrand, Carl E.
Heitzman, D. E.
Hill, J. D.
Hogg, David N.

Hood, Ben H.
Horton, Jack A.
Hughes, Aeft
Hocking, William M.
Holt, Richard L.
Hoover, Alvie
Horton, William P.
Hulce, Richard L.
Humprey, E.
Hunter, Dan
Hunt, Valerio R.
Jackson, James C.
Johnson, Joseph E.
Jondahl, Lee
Joyce, F. J.
Kadar, Ivan
Jacobs, James
Johnson, Philip S.
Jones, O. L.
Jumper, Fred
Kaerner, Murray
Kalil, Ford
Keesey, Walter T.
Keith, Ernest L.
Kerstetter, Don
Kessinger, H. E.
Kay, Robert
Kehl, Robert P.
Kirshur, Carl
Kessinger, H.
Kiebler, J.
Kindgr, William J.
Kirby, Lt.-Col. Julian
Knox, Carl B.
Koerner, Murray A.
Kramer, Robert
King, Leonard
Kirwan, Emil R.
Koeritz, Kenneth W.
Koos, Raymond K., Jr.
Kramer, O.
Krasnecan, J.
Krumpelman, J. L.
Kruger, Bodo
Kuykendall, W. E.
Kyle, Howard C.
Lane, J. H.
LaFleur, Walter
Laios, S.
Lamb, Wilson G.
Lampl, S.
Lanzkron, Dr. R. W.
Lantz, Paul A.
La Rosa, Roy
Laudermilch, Ray
La Vance, Cecil

Laudermilch, R. H.
Laulman, W.
Lauman, B.
Lawhead, N. L.
Leidy, Kenneth
Leigh, Robert T.
Levine, A.
Levy, Harold H.
Liebermann, Ralph W.
Lindley, P. L.
Logan, F.
Looney, C. H.
Mackey, Robert J., Jr.
Maehr, E.
Maehr, T. E.
Mahoney, Michael
Mallory, Gene H.
Malnati, Robert J.
Maness, Clyde T.
Mansur, Dr. G. F.
Markham, Allan
Markham, A. S.
Martin, John B.
Martin, William P., Lt. Col.
Martin, Wm.
Matlick, Thomas L.
Mayfield, Samuel O.
McCaul, P.
McClannahan, Jack T.
McCaffery, R. J.
McCaffery, Robert
McCombe, David H.
McDaniel, Wes
McDonald, T.
McDonald, T.
McElroy, James B.
McIntyre, Paul, Capt.
McKenzie, Joe
McKiernan, John W.
McMillian, M. W.
McNealus, A.
McNealus, A., Mr.
McNealus, Arthur L.
McNealus, L. A.
Meszaros, John P.
Miller, Ralph
Miller, Robert D.
Miller, M. C. Col. Lt.
Miller, C. M. Lt.
Miller, R. D.
Miller, W.
Miller, Wm. E., Jr.
Miller, Warner H.
Mitnick, Herbert
Moore, Carl Mr.

Morian, John A.
Morrison, C. R.
Morrison, Louis H.
Morton, F.
Muller, V.
Nall, R. D.
Naylor, D. A.
Nelson, Donald A.
Nebb, Warren
Nelson, Robert M.
Newman, R. H., Jr.
Nizko, Henry S.
Niner, Edgar P.
Nuttall, Robert R.
Okada, Koyo
Olden, Wm.
Ottinger, George W.
O'Keefe, John J.
O'Shaughnessy, S. J.
Owen, D. H.
Packham, L. E.
Packham, Leonard E.
Painter, Parker
Paris, Paris B.
Parker, Charles D., Capt.
Paddack, S. J.
Page, Robert K.
Paradiso, A. M.
Parker, James E.
Parry, William R.
Parsons, W.
Pashby, Paul J.
Pauley, Willis J.
Payne, Mary H.
Peake, Gerald M.
Pendley, David B.
Perry, Robert A.
Peavey, Ross D.
Peters, K.
Pfeiffer, W. A.
Piccoli, P. E.
Pixley, P. T.
Politzer, Lt. - Col.
Pope, Robert H.
Potter, Philip D.
Pinckernell, H.
Poland, William B., Jr.
Polking, Urban H.
Potter, Lee
Pu, Robert T.
Quinn, M. J.
Raleigh, James T.
Rand, Robert L.
Randolph, R. Z.
Reed, B. R.
Reich, Bruno W.
Rand, Robert

Robert, Randolph C.
Reed, R. B.
Regusters, Herman
Reisch, F. E.
Reisch, F. E.
Rende, J.
Roberts, Carl
Roper, J. C.
Rosenthal, Henry
Reising, Paul E.
Richard, Herbert L.
Roberts, T.
Rosenberg, Harold R.
Ross, Edward
Roy, Mrs. Melba L.
Rush, Martin
Russell, Walter C., Jr.
Ryan, James
Runge, George
Russell, H. K.
Rutledge, C. K.
Rycroft, Michael J. Dr.
Safman, L. W.
Salmon, R. F.
Salzberg, I. M.
Sanderson, Kenneth C.
Sapper, Larry W.
Scanlon, Robert T.
Schachne, H. S.
Sanborn, S.
Santos, Severino Z.
Saunders, Mark C.
Scharf, Allen R.
Schauer, Karl
Schempp, Len
Schiesser, Emil R.
Schroeder, C.
Schwartz, John J.
Scott, Richard M.
Schruder, K.
Schmid, Paul E.
Schroth, R.
Schwartz, Robert F.
Schwartz, Victor A.
Sebastia, Miss
Shaw, Kurt
Shawhan, John J.
Sheboya, B.
Simas, Victor R.
Selden, R. L.
Shaw, Rollo H.
Shaughnessy, James P.
Siller, Joe
Slick, E. P.
Smith, Jahn A.
Smor, Paul

Soar, W. S.
Sos, John Y.
Spintman, Daniel A.
Staggs, Fred
Smith, Philip T.
Snow, R.
Sohner, Harold
Spafford, M.
Satterfield, Jim
Staniloff, A. S.
Stevens, James M.
Stilmar, Robert L.
Stock, Richard L.
Stodola, E. King
Stanley, Robert R.
Stevenson, R. H.
Stiver, Willard Carl, Jr.
Stockwell, E. J.
Strum, Louie W., Jr.
Stuart, Thomas A.
Sulester, James
Sullivan, Cornelius J.
Suomala, John B
Susman, Dr. S. M.
Swan, Col. Harry C.
Swee, Conley
Taggler, R.
Talbott, John W.
Tanner, Walter E.
Taylor, Robert C.
Temkin, Aaron
Thompson, Henry F.
Toler, James C.
Travis, A. D.
Tell, J. D.
Thomas, Edgar F. (Capt.)
Thompson, J. T.
Toukdarian, Richard
Trinter, V. E.
Tucker, Allen G.
Turnbull, T. D.
Underwood, C. H.
Uvaas, Charles M.
Vavra, Paul H.
Vesper, Ralph D.
Wachsman, R. H.
Walker, C. B.
Walch, Rudy
Wallston, Don
Waranch, M.
Weaver, J.
White, P. David
Wigand, R. C.
Williams, A. D.
Ward, Robert J.
White, J.

Whitson, Roy
Wilkes, Morgan E.
Willis, Don
Wilson, C. P.
Witson, R.

Wittgartner, Dan
Woolston, D.
Wright, Eugene L.
Yageldwich, J.

York, Will
Zimmerman, Elliott
Zink, Eugene
Zugler, T.

www.ingramcontent.com/pod-product-compliance
Lightning Source LLC
Chambersburg PA
CBHW081719170526
45167CB00009B/3629